MX

Cell Imaging

METHODS EXPRESS

The **METHODS EXPRESS** series

Series editor: B. David Hames

Faculty of Biological Sciences, University of Leeds, Leeds LS2 9JT, UK

METHODS EXPRESS

Bioinformatics

Biosensors

Cell Imaging

DNA Microarrays

Immunohistochemistry

PCR

Protein Microarrays

Proteomics

Whole Genome Amplification

MX

MX

Cell Imaging

METHODS EXPRESS

edited by D. Stephens

Department of Biochemistry, University of Bristol,

Bristol, UK

Scion

First published 2006

A CIP catalogue record for this book is available from the British Library.

ISBN 1 904842 046 (paperback)
ISBN 1 904842 267 (hardback)

Scion Publishing Limited
Bloxham Mill, Barford Road, Bloxham, Oxfordshire OX15 4FF
www.scionpublishing.com

Important Note from the Publisher

Typeset by Phoenix Photosetting, Chatham, Kent, UK
Printed by Biddles Ltd, King's Lynn, UK, www.biddles.co.uk

Cover image:
A HeLa cell fixed and labeled by immunofluorescence with specific antibodies directed against EB1 (green, labeling microtubule plus-ends) and Sec24Cp (red, labeling COPII-coated secretory cargo export sites on the endoplasmic reticulum). The nucleus was counterstained with DAPI (blue). An image stack was acquired at 0.1 μm axial (*z*) spacing and 0.134 mm pixel size (*x*–*y*). The stack was deconvolved using an iterative deconvolution algorithm with a calculated point spread function.

Contents

Contributors

Branco, Miguel R. MRC Clinical Sciences Centre Faculty of Medicine, Imperial College London, Hammersmith Hospital Campus, Du Cane Road, London W12 0NN, UK. E-mail: miguel.branco@csc.mrc.ac.uk

Burgoyne, Robert The Physiological Laboratory, The University of Liverpool, Crown Street, Liverpool L69 3BX, UK. E-mail: burgoyne@liverpool.ac.uk

Conchello, José-Angel Molecular, Cell, and Developmental Biology, Oklahoma Medical Research Foundation, 825 NE 13th Street, Oklahoma City, OK 73104, USA. E-mail: jose-conchello@omrf.ouhsc.edu

Hasdemir, Burcu Department of Surgery and Physiology, University of California, San Francisco, CA 94143-0660, USA.

Jepson, Mark A. Cell Imaging Facility, Department of Biochemistry, University of Bristol, School of Medical Sciences, University Walk, Bristol BS8 1TD, UK. E-mail: m.a.jepson@bris.ac.uk

Lane, Jon D. Department of Biochemistry, School of Medical and Veterinary Sciences, University of Bristol, University Walk, Bristol BS81TD, UK. E-mail: jon.lane@bristol.ac.uk

Martin, Sonya MRC Clinical Sciences Centre Faculty of Medicine, Imperial College London, Hammersmith Hospital Campus, Du Cane Road, London W12 0NN, UK. E-mail: sonya.martin@csc.mrc.ac.uk

Mayor, Satyajit National Centre for Biological Sciences (TIFR), UAS-GKVK campus, Bellary Road, Bangalore 560 065, India. E-mail: mayor@ncbs.res.in

Miura, Kota Cell Biology and Biophysics, European Molecular Biology Laboratory, Meyerhofstrasse 1, 69117 Heidelberg, Germany. E-mail: miura@embl.de

Pombo, Ana MRC Clinical Sciences Centre Faculty of Medicine, Imperial College London, Hammersmith Hospital Campus, Du Cane Road, London W12 0NN, UK. E-mail: ana.pombo@csc.mrc.ac.uk

Presley, John F. Department of Anatomy and Cell Biology, McGill University, 3640 University Street, Montreal, Quebec, Canada H3Z 2L9. E-mail: john.presley@mcgill.ca

Rietdorf, Jens Advanced Light Microscopy Facility, European Molecular Biology Laboratory, Meyerhofstrasse 1, 69117 Heidelberg, Germany. E-mail: rietdorf@embl.de

Roquemore, Elizabeth P. Discovery Systems Development, GE Healthcare Biosciences, Cardiff CF14 7YT, UK. E-mail: liz.roquemore@ge.com

Stebbings, Howard Peninsula Medical School, Washington Singer Laboratories, Perry Road, Exeter EX4 4QG, UK. E-mail: H.Stebbings@ex.ac.uk

Stephens, David Department of Biochemistry, University of Bristol, School of Medical Sciences, University Walk, Bristol BS8 1TD, UK. E-mail: david.stephens@bristol.ac.uk

Suhling, Klaus Department of Physics, King's College London, Strand, London WC2R 2LS, UK. E-mail: klaus.suhling@kcl.ac.uk

Tepikin, Alexei The Physiological Laboratory, The University of Liverpool, Crown Street, Liverpool L69 3BX, UK. E-mail: a.tepikin@liverpool.ac.uk

Varma, Rajat Program in Molecular Pathogenesis, Skirball Institute for Biomolecular Medicine, New York University School of Medicine, 540 First Avenue, New York 10016, USA. E-mail: varma@saturn.med.nyu.edu

Xie, Sheila Q. MRC Clinical Sciences Centre Faculty of Medicine, Imperial College London, Hammersmith Hospital Campus, Du Cane Road, London W12 0NN, UK. E-mail: sheila.xie@csc.mrc.ac.uk

Zenisek, David Department of Cellular and Molecular Physiology, Yale University School of Medicine, 333 Cedar Street, New Haven, CT 06520, USA. E-mail: david.zenisek@cmp.yale.edu

Zimmermann, Timo Advanced Light Microscopy Facility, European Molecular Biology Laboratory, Meyerhofstrasse 1, D-69117 Heidelberg, Germany. E-mail: tzimmerm@embl.de

Preface

The diversity of microscopes and imaging modes means that it is more important than ever to keep up with and understand the way in which imaging techniques can be applied to biological problems. It is with this fundamental aim in mind that we have compiled *Cell Imaging: Methods Express*. Specifically, we have aimed to include a diverse range of techniques; the equipment for the majority of these approaches is widely available, notably in organizations that benefit from a centralized microscopy facility. Each chapter includes basic explanations as well as detailed protocols which will help lead less experienced readers through these techniques. Each chapter is written by an expert in their field and includes protocols that are typically used in their own work. In addition, benefits and limitations of each approach are discussed within the chapters. One of the main aims of the book is to drive readers towards thinking of ways of investigating their own research areas, by highlighting both classical and cutting-edge cell imaging techniques.

David Stephens
October 2005

Abbreviations

ADC	analogue-to-digital converter
AM	acetoxymethyl ester
AOBS	acousto-optic beam splitter
AOTF	acousto-optic tunable filter
Arf1	ADP-ribosylation factor
BAPTA	bis-(aminophenoxy)ethane-tetraacetic acid
BFP	blue fluorescent protein
BODIPY	4,4-difluoro-5,7-dimethyl-4-bora-3a,4a-diaza-*s*-indacene-3-propionic acid
BrdU	bromodeoxyuridine
BrRNA	BrUTP-labeled RNA
BrUTP	bromouridine triphosphate
BSA	bovine serum albumin
CaBPs	calcium-binding proteins
CCCP	carbonyl cyanide *m*-chlorophenylhydrazone
CCD	charge-coupled device
CFD	constant fraction discriminator
CFP	cyan fluorescent protein
CLSM	confocal laser scanning microscopy
COPI	coat protein I
Cy	cyanine
DABCO	1,4-diazabicyclo(2,2,2) octane
DAPI	6-diamidino-2-phenylindole
DIC	differential interference contrast
DiO	3'-dioctadecyloxacarbocyanine perchlorate
DMSO	dimethyl sulfoxide
dSNR	detector signal-to-noise ratio
EC_{50}	50% effective concentration
EDTA	ethylene diamine tetraacetic acid
EGFP	enhanced GFP
EM	electron microscopy
EM–ML	expectation-maximization maximum likelihood
ER	endoplasmic reticulum
EYFP	enhanced YFP

FAD	flavin adenine dinucleotide
FBS	fetal bovine serum
FCCP	*p*-trifluoromethoxyphenylhydrazone
F1AsH	fluorescein arsenical helical binder
FISH	fluorescence *in situ* hybridization
FITC	fluorescein isothiocyanate
FLIM	fluorescence lifetime imaging
FLIP	fluorescence loss in photobleaching
FPR	fluorescence photobleaching and recovery
FRAP	fluorescence recovery after photobleaching
FRET	fluorescence resonance energy transfer
FTS	Fourier transform spectroscopy
GFP	green fluorescent protein
GPCR	G protein-coupled receptor
GPI	glycosylphosphatidylinositol
GST	glutathione *S*-transferase
IC_{50}	50% inhibitory concentration
IPTG	isopropyl-1-thio-β-D-galactopyranoside
IRF	instrumental response function
JvC	Jansson–van Cittert
KChIP1	K^+ channel-interacting protein 1
LOPAC	library of pharmacologically active compounds
LM	light microscopy
LUT	look-up table
MAP	maximum *a posteriori* probability
MCA	multi-channel analyzer
MEM	minimal essential medium
ML	maximum likelihood
MP	multi-photon
MPL	maximum penalized likelihood
MPPI	Moore–Penrose pseudo-inverse
mRFP	monomeric red fluorescent protein
MSD	mean square displacement
NA	numerical aperture
NADH	nicotinamide adenine dinucleotide
NDB	nitrobenzoxadiazole
NP-EGTA	nitrophenyl ethylene glycol tetraacetic acid
OTF	optical transfer function
PA-GFP	photoactivatable GFP
PBS	phosphate-buffered saline
PFA	paraformaldehyde
PI3K	phosphatidylinositol 3-kinase
PKC	protein kinase C
PMT	photomultiplier tube
PSF	point spread function
ReAsH	resorufin arsenical helical binder

RI	refractive index
ROI	region of interest
siRNA	small interfering RNA
SNR	signal-to-noise ratio
SPT	single-particle tracking
SSC	saline sodium citrate
SM	sphingomyelin
SVD	singular value decomposition
TAC	time-to-amplitude converter
TCSPC	time-correlated single-photon counting
TIRF	total internal reflection fluorescence
TIRFM	total internal reflection fluorescence microscopy
TR-FAIM	time-resolved fluorescence anisotropy imaging
TRITC	tetramethylrhodamine isothiocyanate
VAREL	variable relief contrast
VE-DIC	video-enhanced DIC
VSV-G	vesicular stomatitis virus G protein
WFM	wide-field microscopy
WHF	Wiener-Helstrom filter
YFP	yellow fluorescent protein

Color section

Chapter 1. An introduction to cell imaging

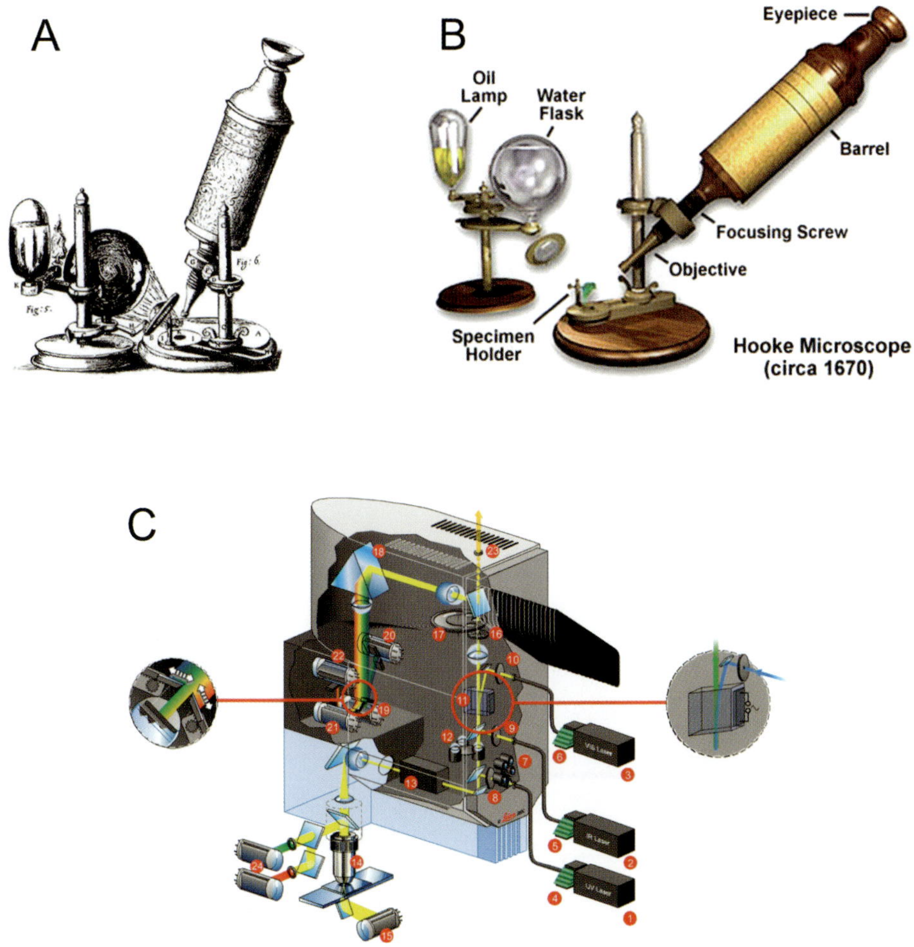

Figure 2. Microscopy from the 17th to the 21st centuries (see page 10).
(*A*) Robert Hooke's microscope as detailed in *Micrographia*, published in 1665 (23). Image reproduced with permission of the Charles Deering McCormick Library of Special Collections at Northwestern University, USA. (*B*) A schematic of Hooke's microscope showing the relevant components. Image reproduced with permission from the Molecular Expressions Microscopy Primer (www.microscopy.fsu.edu/primer/museum/hooke.html). (*C*) Leica Microsystems' TCS AOBS confocal scan head, commercially available from 2002 (www.leica-microsystems.com). The numbered components are as follows: *1*, UV laser; *2*, IR laser; *3*, visible range laser; *4*, UV acousto-optical tunable filter; *5*, IR electro-optical modulator; *6*, visible range acousto-optical tunable filter; *7*, UV adaptation optics; *8*, UV excitation pinhole; *9*, IR excitation pinhole; *10*, VIS excitation pinhole; *11*, primary beam splitter; *12*, adjustable pupil illumination; *13*, 'K'-scanner with rotator; *14*, microscope and objective; *15*, transmitted light detector; *16*, confocal detection pinhole; *17*, analyzer wheel; *18*, spectrophotometer prism; *19*, photomultiplier channel 1; *20*, photomultiplier channel 2; *21*, photomultiplier channel 3; *22*, photomultiplier channel 4; *23*, external optical port; *24*, nondescanned reflected light detectors.

Chapter 3. Image quantification and analysis

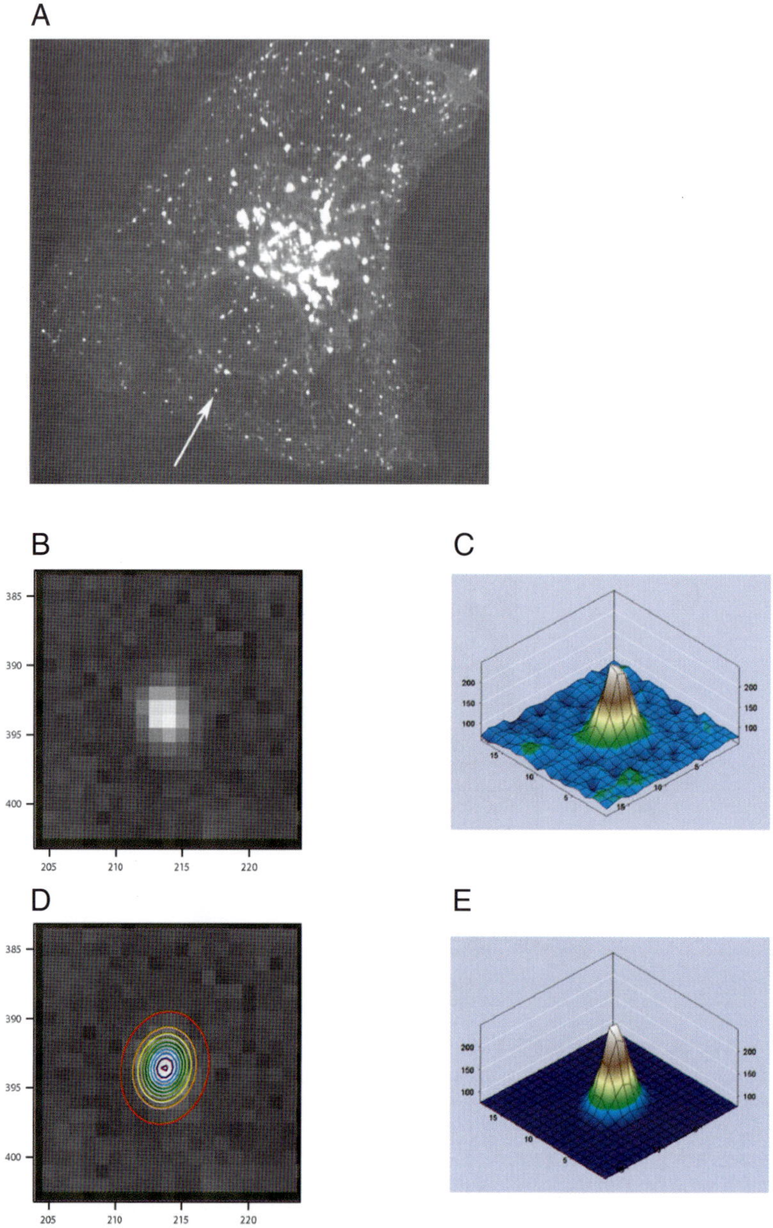

Figure 1. Gaussian fitting method (see page 56).
In (*A*), a Vero cell expressing vesicular stomatitis virus G protein fused to yellow fluorescent protein is shown, a model protein frequently used for studying intracellular vesicle transport. The arrow indicates a vesicle, which is shown enlarged (*B*) and by 3D plotting (*C*). In (*C*), signal intensity was converted to the height in the *z*-axis. Fitting of this vesicle image to a 2D Gaussian equation in IGOR PRO resulted in images (*D*) and (*E*), showing the contour and 3D plot of the fitted curve, respectively. Note how well the fitting works, as one can see by the 3D shape of the peak. From the fitted parameters of the 2D Gaussian curve, the position coordinate of the vesicle was determined as (213.7, 393.7).

Chapter 5. Spectral imaging techniques for fluorescence microscopy

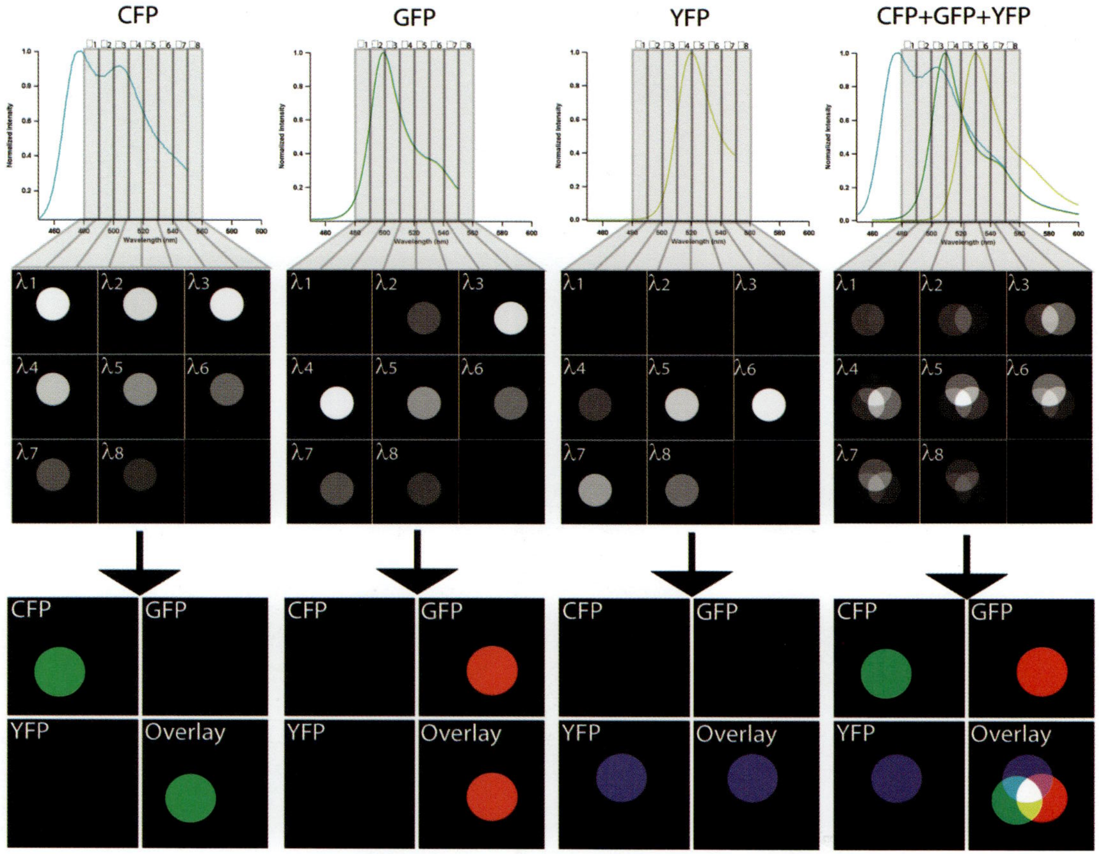

Figure 2. Spectral imaging and linear unmixing of fluorescence signals (see page 101).
Contributions of the fluorescent proteins cyan fluorescent protein (CFP), green fluorescent protein (GFP) and yellow fluorescent protein (YFP) to eight successive spectral channels are shown. The distribution of the emission signal to the channels is a direct representation of the fluorophore emission spectrum and thus constitutes its spectral signature. Using these spectral signatures as reference, even combined and mixed signals can be clearly separated by linear unmixing into the fluorophores that contribute to the total signal.

Chapter 5. Spectral imaging techniques for fluorescence microscopy

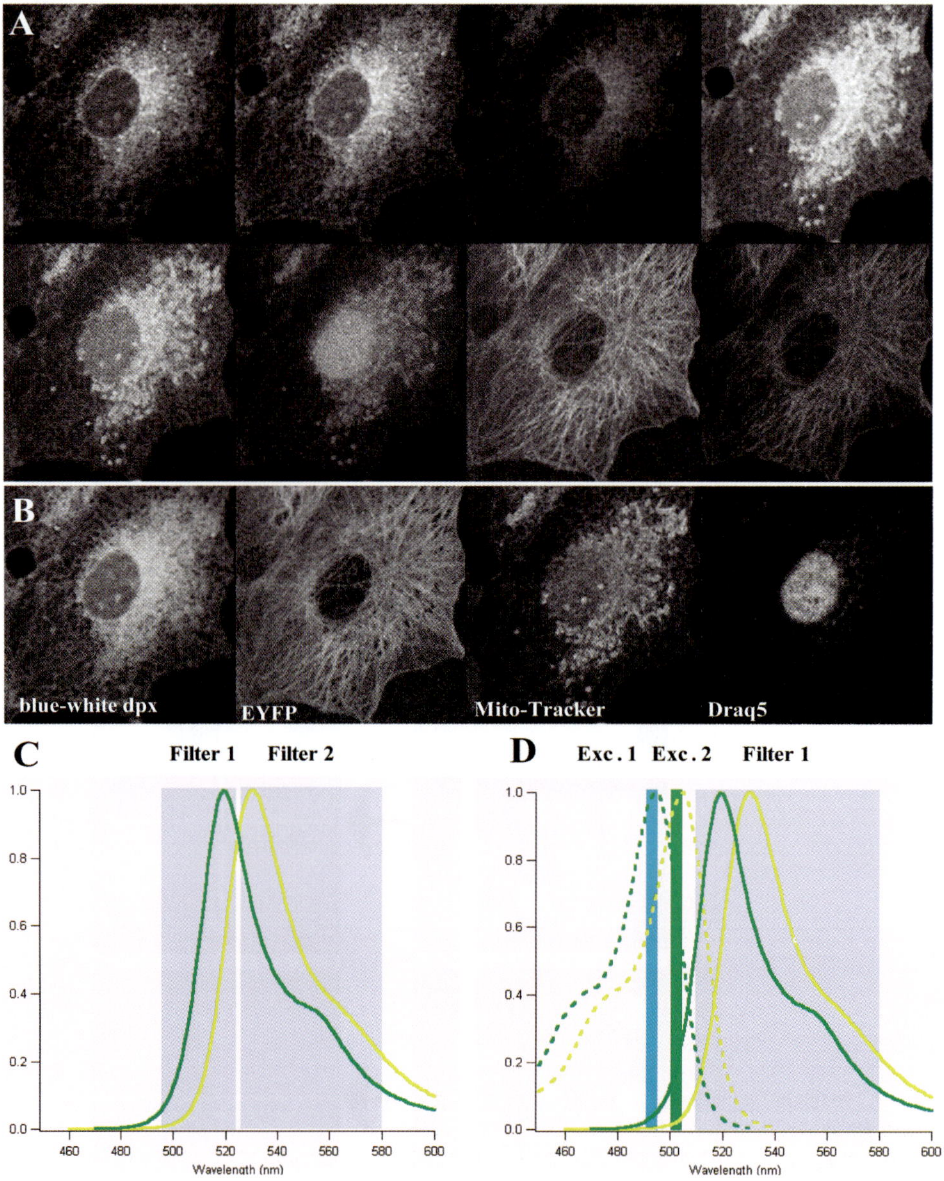

Figure 3. Application of spectral imaging and linear unmixing to living samples stained with multiple fluorophores (see pages 100–105).

The data were acquired on a Zeiss LSM 510 Meta confocal microscope using eight user-defined Meta-channels. Acquisition was performed in two sequential steps (channels 1–6 and channels 7–8). (*A*) Live sample labeled with Blue-White DPX (endoplasmic reticulum), enhanced YFP–tubulin (microtubules), MitoTracker (mitochondria) and DRAQ5 (nucleus). No photobleaching was observable during 50 subsequent frames (frame size 60 × 60 μm, 512 × 512 pixels). (*B*) After application of linear unmixing algorithms, overlapping signals from different fluorophores are completely separated. (*C, D*) Schematic of setups for linear unmixing on the emission (*C*) and excitation (*D*) side.

Chapter 5. Spectral imaging techniques for fluorescence microscopy

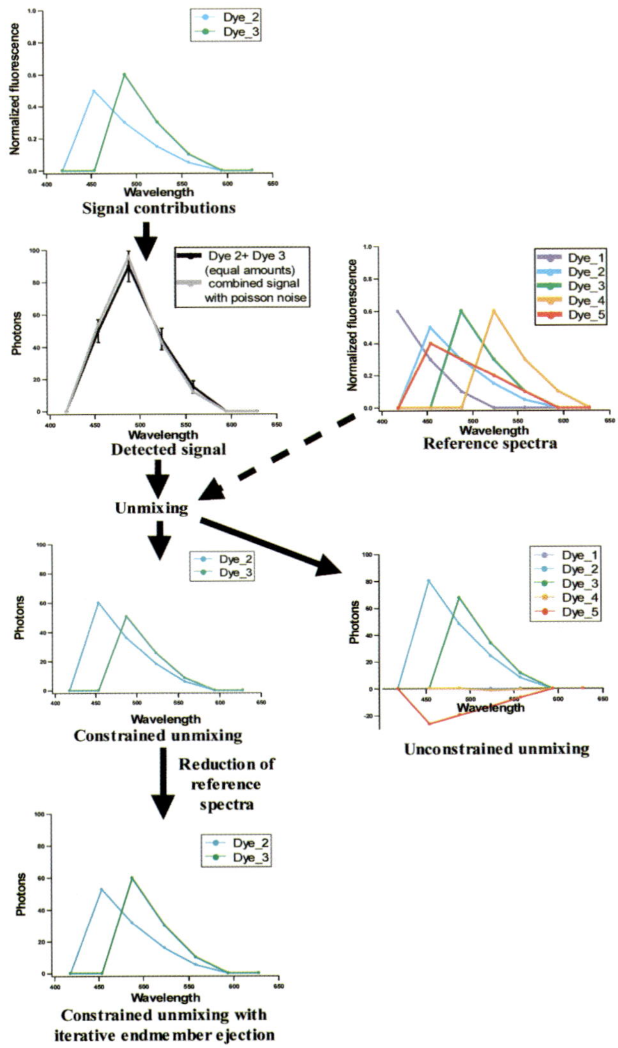

Figure 4. Possible improvements for linear unmixing calculations (see page 115).
Spectra are shown as they would be detected in seven spectral detectors. (*Row 1*) The two contributing fluorophore signals are shown separately. (*Row 2, left*) The combined signal of the two dyes with standard deviations according to the photon shot noise in the channels (black) and the combined signal with simulated photon shot noise (gray). (*Row 2, right*) The reference spectra of five fluorophores that could be present in the sample. (*Row 3, left*) The result after constrained unmixing excluding negative contributions. (*Row 3, right*) The result after unconstrained unmixing allowing negative contributions. Note how the two positive signals in unconstrained unmixing are higher to compensate for the calculated negative contributions. The constrained result distributes all of the available signal only on the two positive contributions. However, due to the photon shot noise in the signal, the contributions of the two dyes are not identical to the input signal (compare with *row 1*). (*Row 4*) The result after constrained unmixing with iterative endmember ejection. The contributions are now similar to the input signal (*row 1*).

Chapter 6. Measurement of protein motion by photobleaching

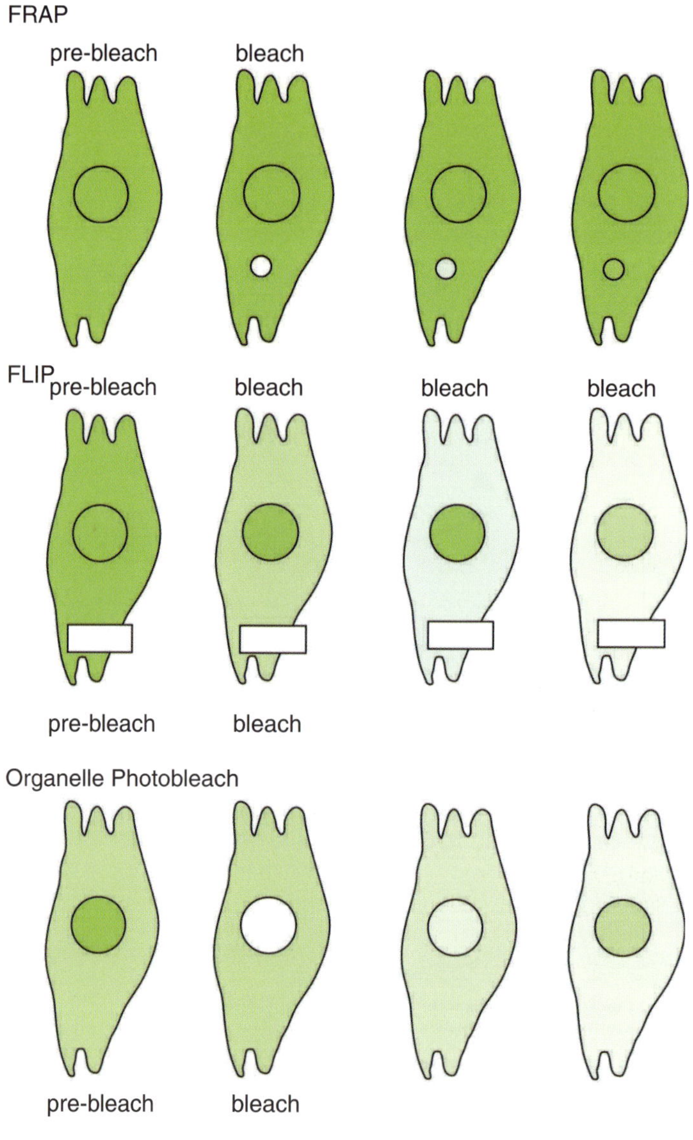

Figure 1. Major categories of photobleaching experiments described in the text (see page 122).

Chapter 6. Measurement of protein motion by photobleaching

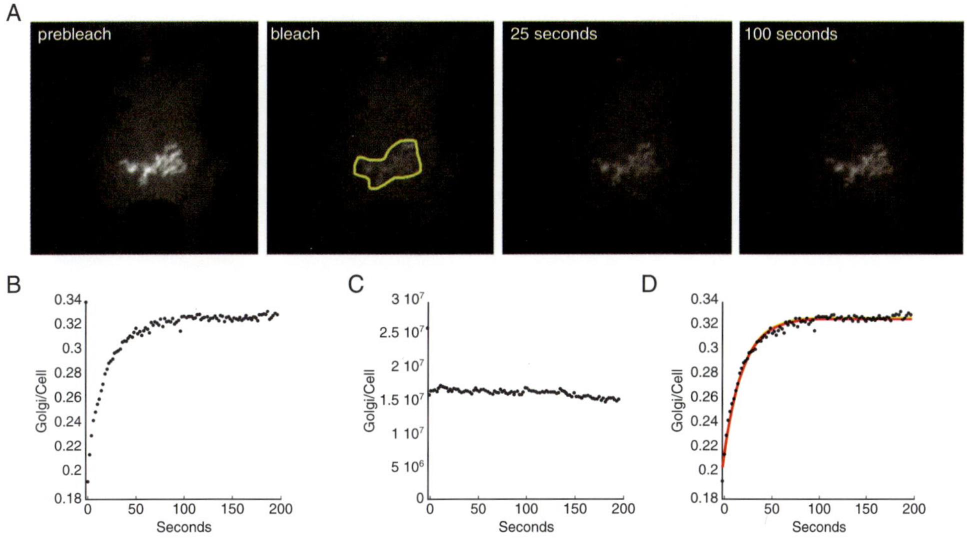

Figure 2. Example of an organelle photobleaching experiment (see page 126).
The experiment was performed using a Zeiss LSM 510 confocal microscope with a 25× 0.8 NA oil-immersion objective and a fully open pinhole. (*A*) Photobleaching of Arf1–GFP localized on the Golgi apparatus of a normal rat kidney cell. Note that a substantial pool of ADP-ribosylation factor 1 (Arf1)–GFP can be detected on the Golgi apparatus even after the bleaching as this protein cycles between the cytosol and the Golgi on a timescale that is comparable to the total bleaching time. (*B*) Recovery curve for the Golgi area shown as a fraction of total cell fluorescence. (*C*) Total cell GFP fluorescence as a function of time. The drop in fluorescence between the first and second frame is due to photobleaching of the Golgi-associated pool. Total cell fluorescence remains roughly constant for the remainder of the experiment, even though redistribution between the Golgi and the cytosol occurs. This is an important test of whether all fluorescence in the cell is accounted for. (*D*) Curve fitted to date from (*B*) (as described in *Protocol 5*).

Chapter 7. Imaging calcium and calcium-binding proteins

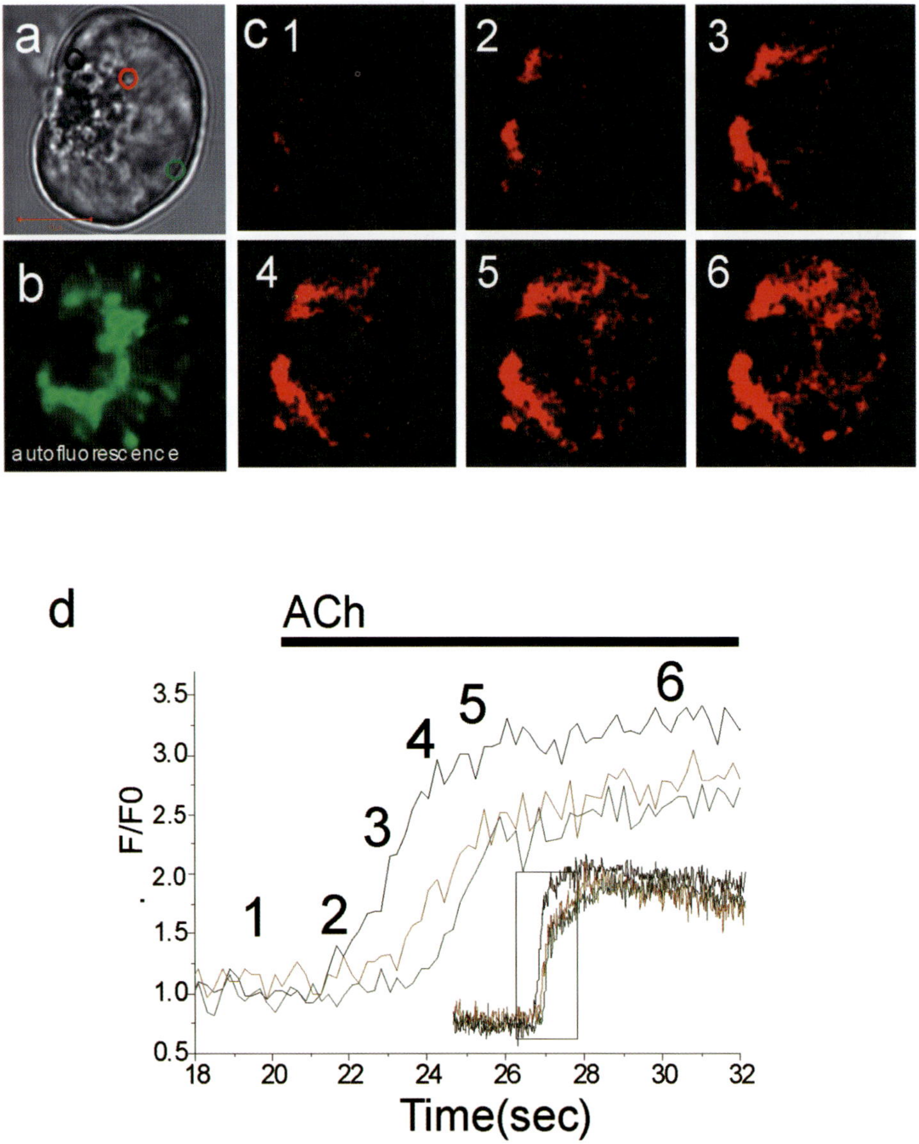

Figure 2. Mitochondrial Ca^{2+} uptake in a pancreatic acinar cell, measured with rhod-2 (32) (see page 148).

Changes in rhod-2 fluorescence were imaged before and during maximal acetylcholine (ACh, 10 μM)-elicited Ca^{2+} release from the ER. (*A*) Transmitted light picture showing the cell under investigation, with the three areas of interest identified by colored circles. Bar, 10 μm. (*B*) Autofluorescence of NADH showing mitochondrial localization. (*C*) Six images showing mitochondrial Ca^{2+} concentration. Image 1 shows the situation just before the start of stimulation. Image 2 shows that immediately after acetylcholine application, there was Ca^{2+} uptake into mitochondria very close to the apical membrane. A little later the whole perigranular mitochondrial belt (the structure specific for pancreatic acinar cells) was revealed (images 3 and 4), and finally (images 5 and 6) all mitochondria in the cell took up Ca^{2+}. (*D*) Time course of the mitochondrial Ca^{2+} uptake in the three regions identified in (*A*) using the appropriate color coding.

Chapter 10. Correlative microscopy using Tokuyasu cryosections

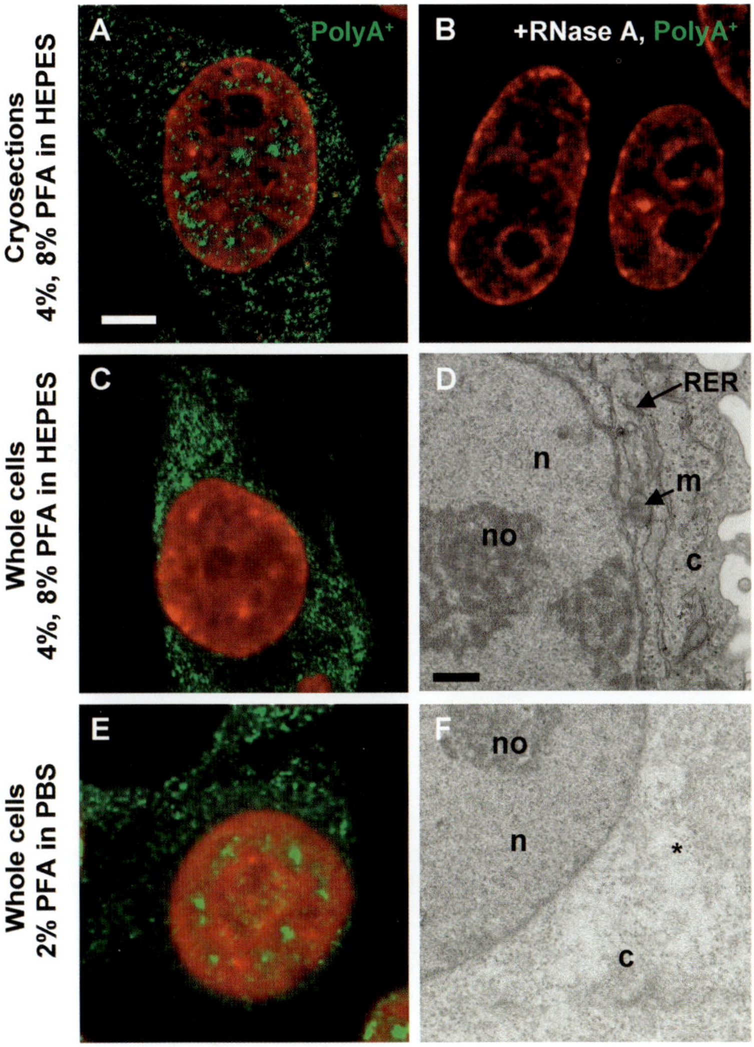

Figure 3. RNA FISH in HeLa cells (see page 207).
(*A*, *B*) HeLa cells were fixed in 4 and 8% PFA in 250 mM HEPES, embedded in sucrose, cryosectioned (100–120 nm thick; see *Protocol 1*), and permeabilized with 0.1% Triton X-100 for 10 min before RNA FISH using an FITC–poly(dT) oligonucleotide probe, as described in *Protocol 4*. Both nuclear and cytoplasmic pools of poly(A)$^+$ RNA are efficiently detected (*A*), and treatment of sections with RNase A shows the specificity of detection (*B*). (*C*)–(*F*) HeLa cells were grown on coverslips, fixed as for cryosections, and permeabilized with 0.1% Triton X-100 (*C*, *D*) or fixed in 2% PFA in PBS for 15 min (28) and permeabilized with 0.5% Triton X-100 for 5 min (*E*, *F*). After fixation, cells were processed for RNA FISH according to *Protocol 4* (*C*, *E*) or for ultrastructural analysis by EM (*D*, *F*) (7). Cells fixed in a stronger fixative showed good preservation of cellular morphology (*D*), but resulted in reduced probe accessibility, such that nuclear RNAs were not detected (*C*). Cells fixed in a weaker fixative provided improved probe accessibility (*E*), but exhibited poor preservation of ultrastructure (*F*), such as disrupted rough endoplasmic reticulum (RER) and mitochondria (*). n, Nucleus; no, nucleolus; c, cytoplasm; m, mitochondria. Bars, 4 μm (LM) and 1 μm (EM).

Chapter 11. Fluorescence lifetime imaging

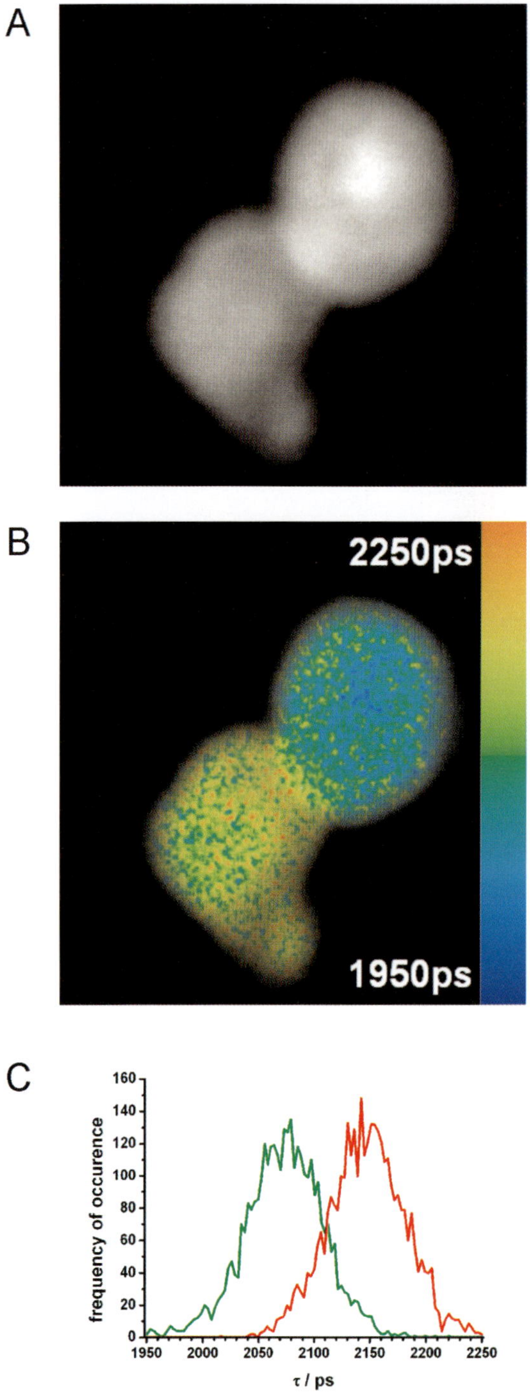

Figure 5. Preliminary results from FLIM experiments with fibroblast growth factor receptor tagged with enhanced GFP and fibroblast receptor substrate 2 tagged with mRFP (see page 224).
(*A*) Intensity image of GFP-tagged proteins (donors) in the presence of RFP-tagged proteins (acceptors) in cells. (*B*) Corresponding FLIM image (single exponential decays). The average GFP fluorescence lifetime is encoded in a continuous rainbow color scale from 1950 ps (blue) to 2250 ps (red). Contrast according to the fluorescence lifetime can clearly be discerned. (*C*) Each cell has a different average GFP fluorescence lifetime: a fluorescence lifetime histogram for the cell on the left (red histogram) shows an average GFP fluorescence lifetime of 2160 ps, and for the cell on the right (green histogram) shows an average GFP fluorescence lifetime of 2070 ps. This may be due to FRET with varying efficiency. The images were obtained by TCSPC-based confocal scanning FLIM on a Leica TCS SP2 inverted microscope with 470 nm diode laser excitation, a 525 ± 25 nm emission filter, and the pinhole fully open. (Samples from Mark Williams and Zamal Ahmed of the Department of Biochemistry & Molecular Biology, University College London.)

Chapter 12. Homo-FRET measurements to investigate molecular-scale organization of proteins in living cells

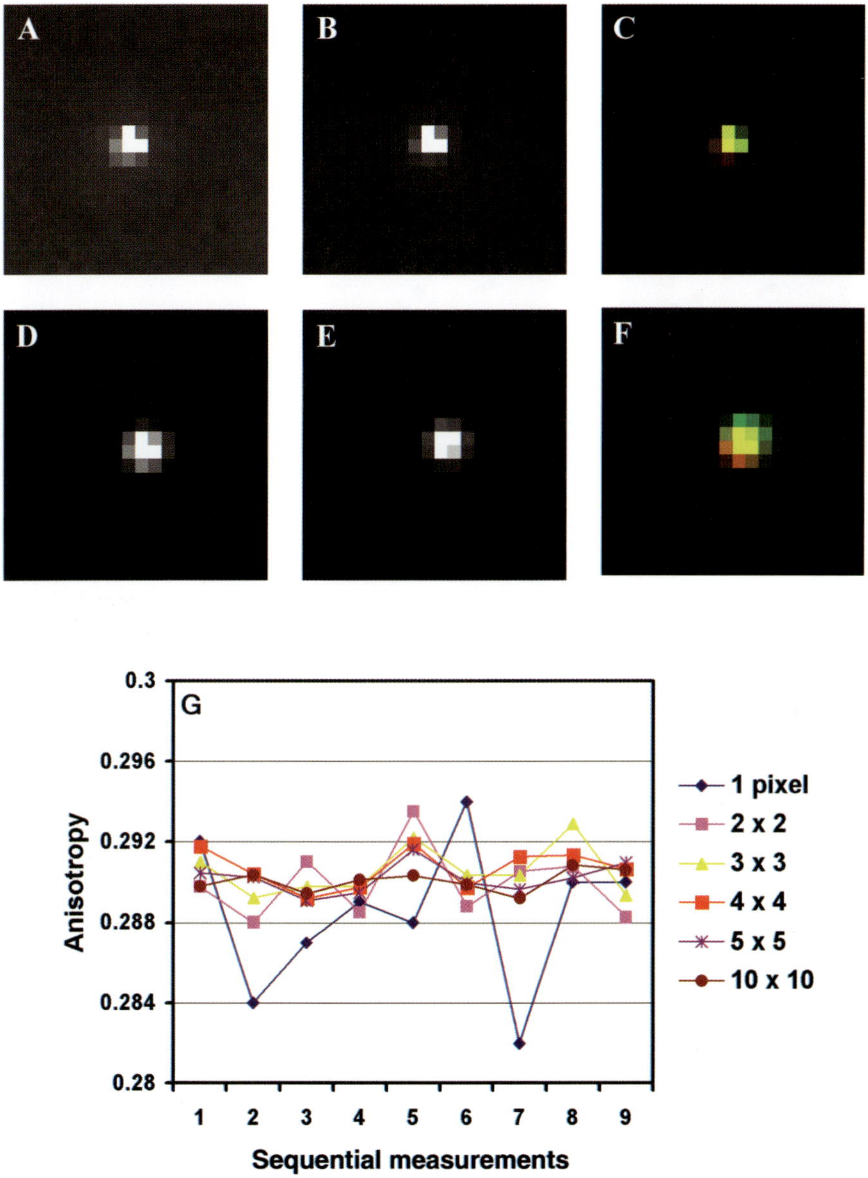

Figure 4. Spatial resolution in anisotropy measurements (see page 260).
(*A*)–(*C*) Images of subresolution beads under 3 × 3 binning showing the overlap of the PSFs collected in the parallel (*A*) and perpendicular (*B*) directions. (*D*)–(*F*) Images of subresolution beads taken at no binning showing the nonoverlap of the PSFs for parallel (*D*) and perpendicular (*E*) directions. (*C*) and (*F*) are fluorescence overlays of parallel and perpendicular directions by pseudo-coloring the parallel and perpendicular images red and green. (*G*) Effect on the measurement of anisotropy of averaging over a larger number of pixels. Fluctuations in anisotropy of rhodamine in glycerol for areas on the images ranging from 1 pixel to 100 pixels are shown. Anisotropy can be determined accurately for an area of 4 × 4 or larger.

Chapter 12. Homo-FRET measurements to investigate molecular-scale organization of proteins in living cells

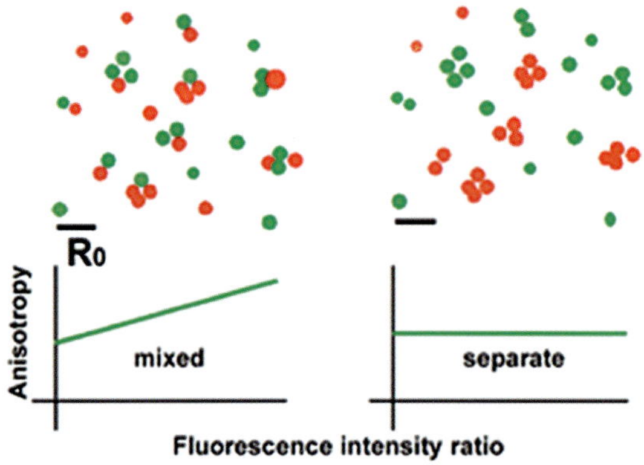

Figure 6. Co-localization at the nanometer scale using homo-FRET (see page 263). In the schematic, if two different proteins (red and green circles) occupy the same cluster (left panel), increasing expression of one protein (red circles) will lead to a decreasing number of homo-FRET events for the second protein (green circles). As a result, homo-FRET between the green species will decrease. Consequently, there will be an increase in emission anisotropy of the green protein species being monitored. Alternatively, if the different protein species are present in separate clusters (right panel), there will be no change in the anisotropy of the fluorescent species being monitored with increased expression of either of the proteins.

Chapter 13. High-content and high-throughput screening

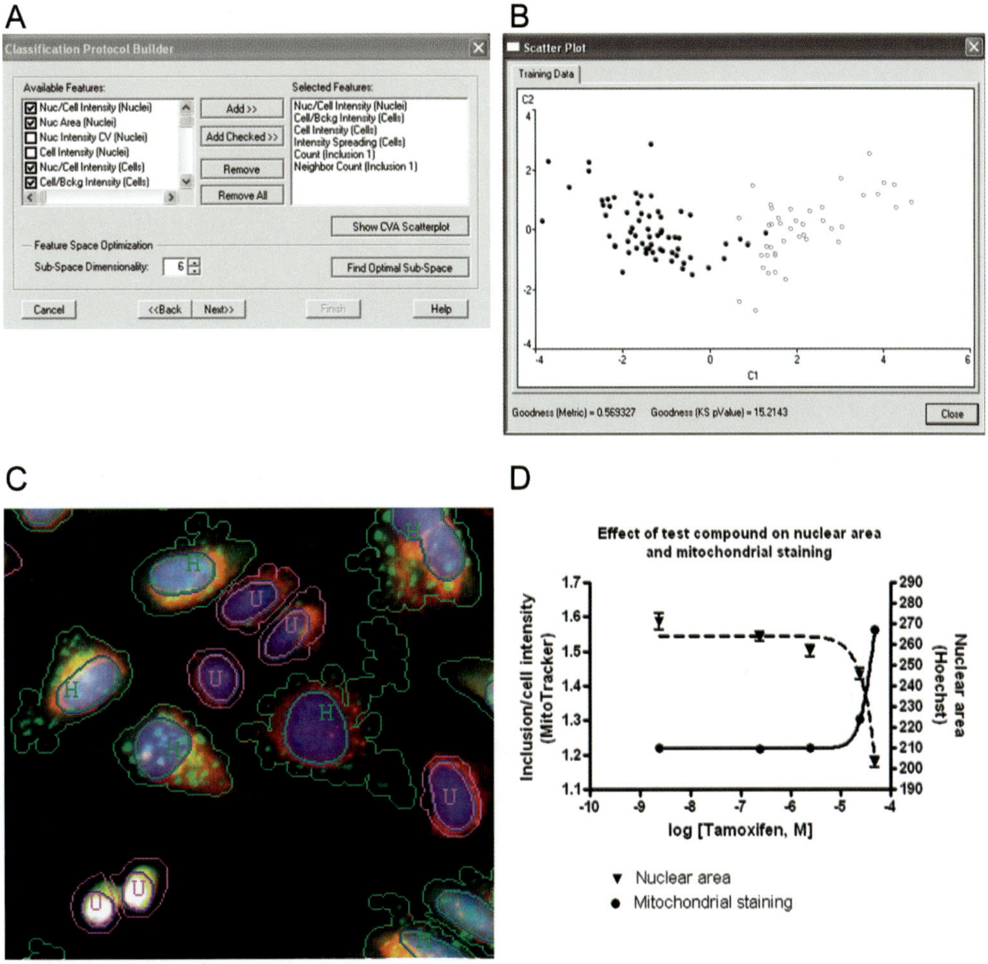

Figure 2. High-content analysis of the EGFP–FYVE domain profiling assay (see page 278). IN Cell Analyzer 1000 analysis software was used to analyze the assay described in *Protocol 5*. (*A*) The software facilitates design of a classification protocol, allowing the protocol developer to choose from a list of 39 available measures for classification. The Sub-Space Dimensionality field is shown here being used to reduce the number of parameters to the six most influential. (*B*) Color-coded data visualization scatter plot (printed here in gray scale) allowing assessment of how well a classification routine separates two control cell populations used to train the classifier. Filled circles represent healthy control cells; open circles represent cells exposed to a toxic compound. (*C*) After classification, cell assignments appear labeled and color-coded in the Image View window. The developer has chosen H/green and U/magenta to label the healthy and unhealthy classes, respectively. Nuclei are labeled with Hoechst (blue), mitochondria are stained with MitoTracker (red), and endosomes are labeled with EGFP–FYVE domain reporter (green). (*D*) Analysis results showing a dose-dependent effect of tamoxifen on nuclear size and mitochondrial staining intensity. Results are shown as mean ± SEM, with $n = 12$ replicate images (wells) per data point.

Chapter 13. High-content and high-throughput screening

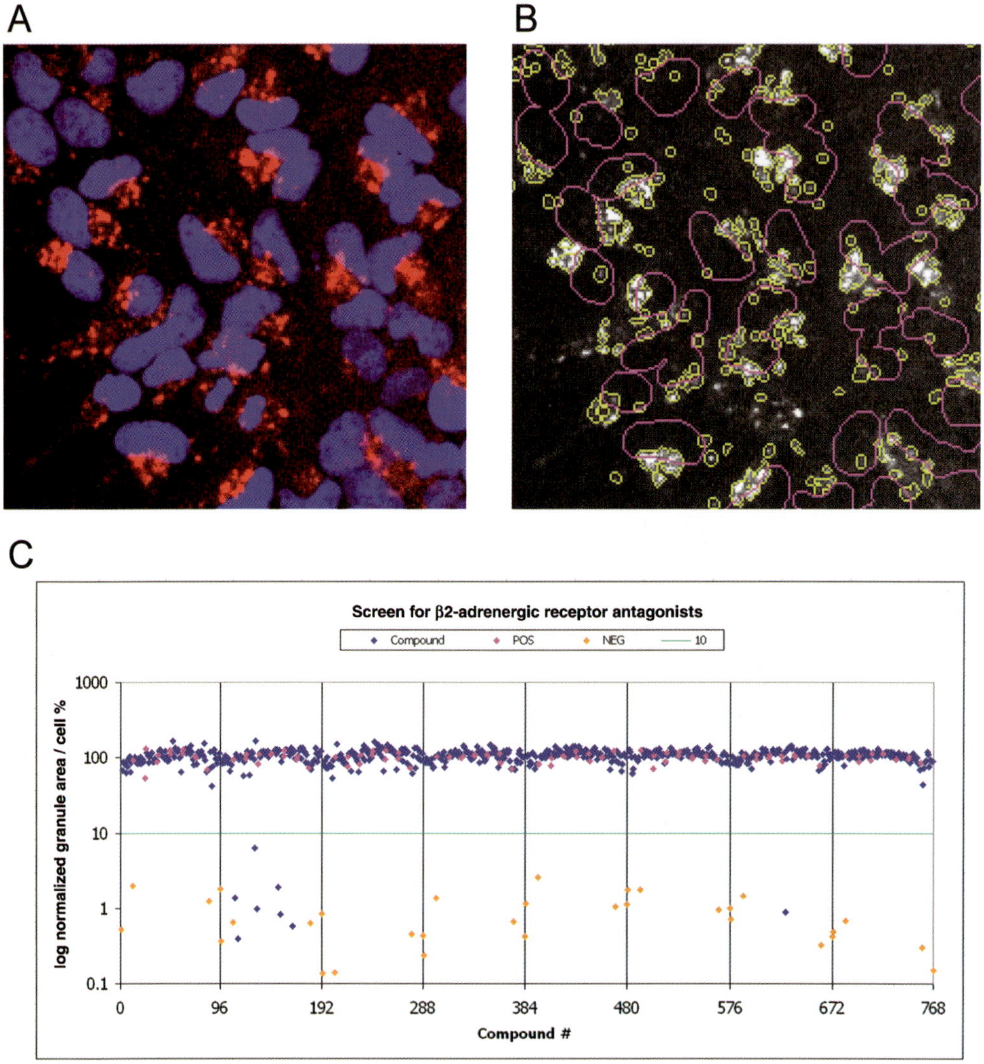

Figure 3. High-content screening for β2-adrenergic receptor antagonists (see page 284). When cells expressing β2-adrenergic receptor are treated with agonist, pH-sensitive CypHer5E dye becomes fluorescent at the sites of receptor internalization. (*A*) Cells imaged with the IN Cell Analyzer 1000 following treatment with agonist. Nuclei are stained with Hoechst (blue) and sites of receptor internalization are identified by CypHer fluorescence (red). (*B*) Bitmap overlay indicating analysis results for the image shown in (*A*). Nuclei are outlined in magenta; identified CypHer fluorescence is outlined in yellow. (*C*) Compound-induced responses from a screen of a 640-compound library are plotted as a percentage of the agonist-only control response. Potential antagonists (hits) were defined as those with a response of less than 10% of control values. All known antagonists in the library and no false positives were identified. Each data point represents the population-averaged result from a single image, with the following color-coding: blue, test compound; pink, positive control; orange, negative control.

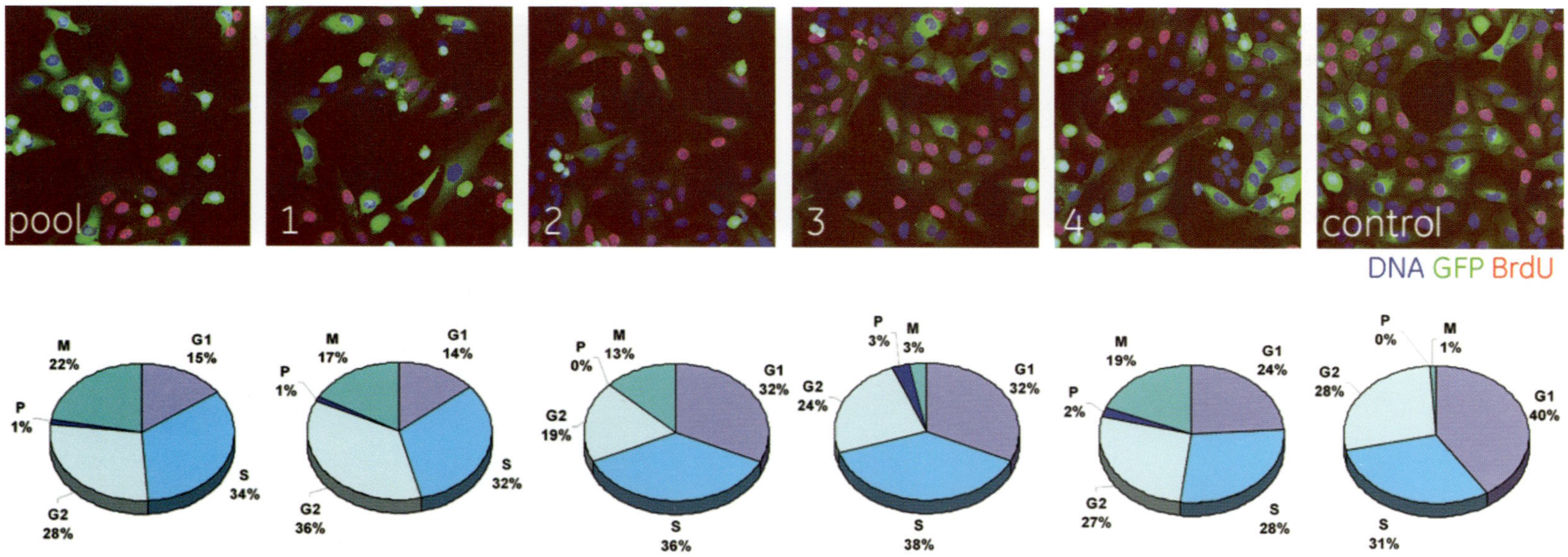

Figure 4. Analysis results from an siRNA study using *Protocol 4* with a G_2/M cell-cycle phase reporter cell line (see page 288). Images from wells treated singly (labelled 1–4) or in combination (labelled 'pool') with four siRNAs directed against different regions of the mRNA for polo-like kinase, which is a regulator of cell-cycle progression during mitosis. An EGFP fusion protein (green) reports cell-cycle phase; nuclei are stained with Hoechst (blue), and nuclei that have incorporated BrdU are stained with Cy5 (red). For each image, the corresponding analysis results are displayed as a pie chart that shows the proportion of cells in mitosis (M), G_1 phase, S phase, G_2 phase, and prophase (P). Changes in cell-cycle phase distribution can be assessed relative to the distribution determined from cells in the control well, which received a pool of scrambled siRNAs.

CHAPTER 1

An introduction to cell imaging

David Stephens

1. INTRODUCTION

Cell imaging has become a fundamental aspect of any molecular cell biology research project. This volume has been compiled with the aim of incorporating a wide range of cell imaging methods. It is aimed at those who are embarking on imaging projects, as well as relatively experienced microscopists who might wish to develop their experiments further. While it is not possible to detail every possible technique for imaging cells, this book attempts to provide working examples, troubleshooting, and tips relating to a wide range of the more commonly available technologies.

The methods described in this book are aimed at giving the reader some context for each imaging technology, as well as providing some basis for familiarizing oneself with these techniques. While microscopy has been used in the study of cells for centuries, it is only in the later part of the 20th century that it has become an integral part of so many research environments. This is, at least in part, due to the continued development of new hardware, with scanning and spinning disk confocal microscopes, cryo-ultramicrotomes, multi-dimensional live-cell imaging systems, and multi-photon excitation being just some examples (1). Regardless of the system that is used, one should always be able to understand and justify the use of a particular piece of equipment. In Chapter 2, Mark Jepson compares confocal and wide-field approaches; this, and indeed the other chapters in this book, are aimed at providing some guidance for the user in choosing between these commonly available systems. Why would you use the latest state-of-the-art scanning confocal rather than a simple wide-field microscope? What are the benefits of one technique over another? This becomes increasingly important as one progresses to more and more complex technologies. The chapters in this book will guide readers towards choosing the correct imaging system for their experiments. This includes choosing and detecting different types of probe (discussed by Timo Zimmermann in Chapter 5), selecting methods for functional analysis such as protein–protein interactions (Chapters 11 and 12 by Suhling, and Varma and Mayor, respectively, discussing the application of fluorescence lifetime imaging (FLIM) and fluorescence resonance energy transfer (FRET)), 'sectioning'

Cell Imaging: ***Methods Express*** (D. Stephens, ed.)

techniques including cryosectioning (for light and electron microscopy, as covered in Chapter 10 by Pombo and colleagues) and selective illumination (total internal reflection fluorescence (TIRF) microscopy, as discussed by David Zenisek in Chapter 8). Imaging is not limited to the acquisition of 2D pictures. Many microscopes are now equipped with optical sectioning capabilities that make 3D reconstruction relatively simple. One key technique in this area, for both confocal and wide-field imaging data, is deconvolution. This is essentially a mathematical method to reassign out-of-focus light to its original point source. In Chapter 9, José-Angel Conchello discusses the mathematical approaches to this complex problem. Many of the techniques detailed in this book are now widely available, notably within and around the many core cell imaging facilities that now exist. More specialized approaches such as FLIM and high-throughput screening are increasingly being applied in both academic and pharmaceutical laboratories.

2. FLUORESCENCE MICROSCOPY

Most of the techniques described in this book rely on the ability to detect cellular components selectively. This is most commonly achieved using fluorescent labels. As much as the advances in hardware, it is the development of fluorescent probes that has driven forward the application of microscopy. While there are many excellent examples of imaging of endogenous (autofluorescent) compounds within cells and tissues, the most common approach is to introduce fluorescently labeled molecules that can subsequently be specifically detected. Fluorescent lipids, sensors of ion concentration, DNA-binding dyes, and fluorescent antibodies have all been used for many years. The list of possible fluorophores is too vast to detail here, and many resources exist in this area, such as online databases of spectral data for commonly used dyes. It is essential that users have a good understanding of how to match fluorophores to the excitation sources and fluorescent filters that are available to them. Furthermore, one must consider the means of introduction of the fluorophore into the cell. Fluorescent protein-based fusions are typically expressed from plasmid DNA introduced into cells by transfection; other dyes are membrane permeable, while fluorescently labeled antibodies and proteins may have to be introduced by mechanical methods such as scrape loading or microinjection.

3. GENETICALLY ENCODED FLUOROPHORES

In the last few decades, the ability to express fluorescent molecules simply and efficiently inside cells has revolutionized the applications of cell imaging to the biomedical research field. Specifically, the discovery, cloning, and application of green fluorescent protein (GFP) has been central to this transformation of the way in which we conduct cell imaging experiments. GFP was initially purified from the Pacific Ocean jellyfish, *Aequorea victoria* (see *Fig. 1*), by Shimomura in 1962 (2). An elegant discussion of Shimomura's subsequent work with both aequorin and

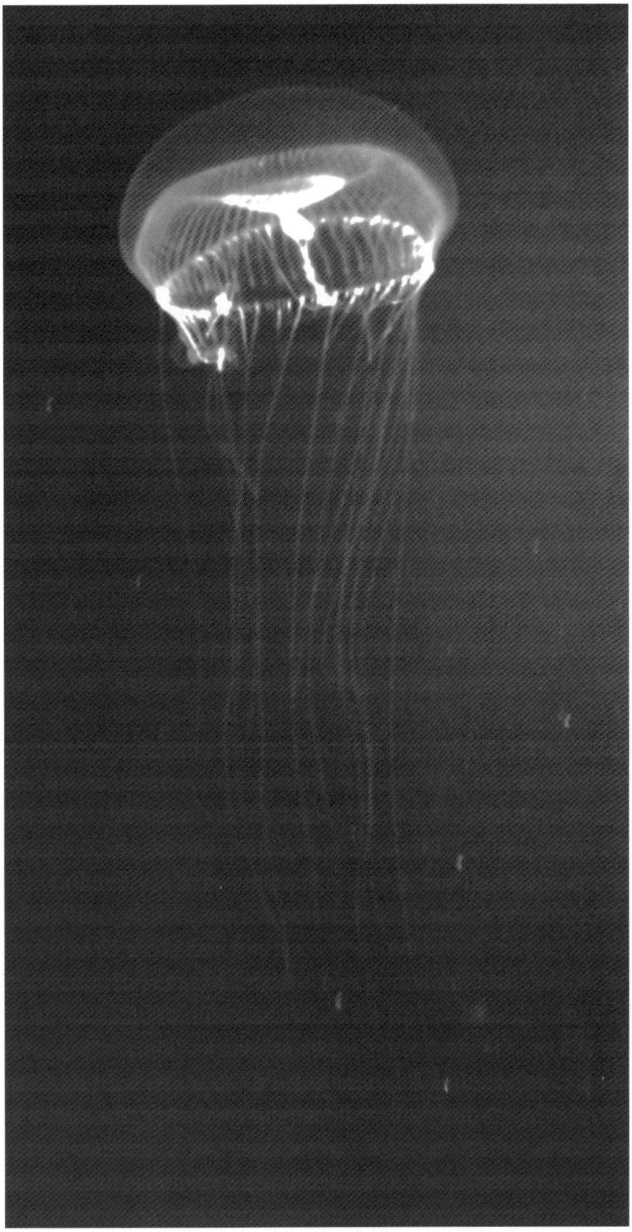

Figure 1. The jellyfish *Aequorea victoria* from which GFP was first purified and subsequently cloned (2).

GFP can be found in (3). Subsequently, the cDNA for GFP was cloned (4) and expressed in living organisms (5, 6). This led to its now widespread use as a reporter of gene expression and as a fusion tag to mark protein localization and dynamics. Subsequent mutagenesis of GFP has led to the development of brighter, more stable variants that are less susceptible to photobleaching, as well as multiple spectral forms of the protein such as cyan fluorescent protein (CFP) and yellow fluorescent protein (YFP). Furthermore, the discovery of other fluorescent proteins from reef corals (such as a red variant, DsRed, from *Discosoma* sp.; 7) has added to the array of genetically encoded fluorescent reporters available. The availability of multiple spectral variants of GFP allows the use of paired fluorescent proteins, which can be used for FRET experiments, as described in Chapter 12 by Varma and Mayor. Essentially, FRET occurs between two spectrally overlapping fluorophores. CFP and YFP are the most commonly used GFP derivates for FRET experiments. A key development has been to mutate these proteins to eliminate their inherent propensity to dimerize. Monomeric forms of multiple spectral variants of fluorescent proteins from both jellyfish and reef coral are now available. Monomeric forms are vital for applications in which there is a high local concentration of the fluorescent protein; this was nicely demonstrated for the clustering of proteins into lipid microdomains at the plasma membrane (8). Recently, directed evolution of fluorescent proteins using iterative somatic hypermutation was used to generate derivatives of a red fluorescent protein, which now span a rainbow of emission wavelengths (9, 10).

This volume also contains applications of GFP derivatives, such as photoactivatable and photoswitchable forms, that can be used as fluorescent highlighters. For example, while normally dim, photoactivation with near-UV light causes the conversion of photoactivatable GFP (PA–GFP; 11) into a bright state, allowing its use as a fluorescent highlighter. Kaede (12) is an example of a photoswitchable protein that can be converted from green-emitting to red-emitting states. Similarly, Dronpa (13) allows photoactivation from a normally dark state, but also undergoes further photoconversion back to the dark state. These photoswitchable proteins enable further analysis of protein dynamics using similar approaches to those used in photobleaching techniques such as fluorescence recovery after photobleaching (FRAP). FRAP and its associated methods are described in detail by John Presley in Chapter 6. A list of commonly available fluorescent proteins is maintained by Kurt Thorn at Harvard University and is available at http://www.thornlab.org/gfps.htm. The data from this resource, correct at the time of going to press, are reproduced in *Table 1*. Further lists of fluorescent proteins and details of other fluorophores, as well as a wealth of information concerning many aspects of microscopy, can be found at the excellent 'Multi-Probe Microscopy' website maintained by George McNamara (http://home.earthlink.net/~mpmicro/).

Further developments in fluorescent protein technology have included the generation of biosensors from fluorescent proteins and their derivatives (14). A key example of this is illustrated in Chapter 7 by Tepikin and colleagues in which the use of cameleons (15) and other fluorescent probes for imaging intracellular calcium concentrations is described. Essentially, cameleons are calcium-dependent biosensors based on FRET between CFP and YFP that have been covalently fused by the addition of a calcium-sensitive linker. Thus, as calcium binds, conformational changes alter the orientation of each fluorophore and thus affect FRET between them. Other developments in this area have included

Table 1. Fluorescent protein information

A list of available fluorescent proteins, with information about their photophysical properties. Reproduced and adapted with permission from the website of Dr. Kurt Thorn at the Bauer Center for Genomics Research, Harvard University, USA (http://www.thornlab.org/gfps.htm).
Notes (taken from the web site): This is not a complete list of existing fluorescent proteins. In general only those proteins that are commercially available or that are well characterized and represent a clear improvement over existing proteins are included. All values were taken from the manufacturer's data or the indicated papers. Brightness is the product of extinction coefficient and quantum yield, divided by 1000.

Protein	Ex	Em	Extinction coefficient	Quantum yield	Brightness	Aggregation	pKa	Source	Notes
Green proteins									
EGFP	488	507	56 000	0.6	33.6	Monomer		*	
Emerald	487	509	57 500	0.68	37.3	Monomer		(34)	EGFP derivative; fivefold brighter at 37°C
CoralHue Azami Green	492	505	72 300	0.67	48.4	Tetramer	<5.0	MBL International	From (35)
CoralHue Monomeric Azami Green	492	505	55 000	0.74	40.7	Monomer	5.8	MBL International	From (35)
CopGFP	482	502	70 000	0.6	42.0	Monomer		Evrogen	Faster folding, less pH sensitive than EGFP
AceGFP	480	505	50 000	0.55	27.5	Monomer		Evrogen	
ZsGreen1	493	505	35 600	0.63	22.4	Tetramer		Clontech	From (7)
Blue/UV proteins									
EBFP	380	440	29 000	0.31	9.0	Monomer		*	
GFPuv	395	509	30 000	0.79	23.7	Monomer		Clontech	
Sapphire	399	511	29 000	0.64	18.6	Monomer		(34)	Aka H9-40
T-Sapphire	399	511	44 000	0.6	26.4	Monomer		(36)	Faster folding than Sapphire
Cyan proteins									
ECFP	433	475	32 500	0.4	13.0	Monomer	4.7	*	
Cerulean	433	475	43 000	0.62	26.7	Monomer	4.7		Improved ECFP from (37)
AmCyan1	458	489	40 000	0.24	9.6	Tetramer		Clontech	More photostable than ECFP; from (7)

Table 1. Fluorescent protein information – cont'd

Protein	Ex	Em	Extinction coefficient	Quantum yield	Brightness	Aggregation	pKa	Source	Notes
CoralHue Midoriishi-Cyan	472	495	27 250	0.9	24.5	Dimer	6.6	MBL International	From (38)
Yellow proteins									
EYFP	513	527	83 400	0.61	50.9	Monomer		*	
Citrine	516	529	77 000	0.76	58.5	Monomer	5.7	(39)	EYFP derivative; less Cl, pH sensitive
Venus	515	528	92 200	0.57	52.5	Monomer	6.0	(40)	EYFP derivative; 30-fold brighter at 37°C
PhiYFP	525	537	130 000	0.4	52.0	Monomer		Evrogen	Less pH sensitive than EGFP
ZsYellow1	529	539	20 200	0.42	8.5	Tetramer		Clontech	From (7)
Orange proteins									
CoralHue Kusabira-Orange	548	561	73 700	0.45	33.2	Dimer	<5.0	MBL International	From (38)
CoralHue Monomeric Kusabira-Orange	548	559	51 600	0.6	31.0	Monomer	5.0	MBL International	From (38)
mOrange	548	562	71 000	0.69	49.0	Monomer	<6.5	(16)	~2.5 h maturation time
Red proteins									
tdimer2(12)	552	579	120 000	0.68	81.6	Monomer	4.8	(41)	Tandem dimer; functional monomer
mRFP1	584	607	44 000	0.25	11.0	Monomer	4.5	(41)	
DsRed-Express	557	579	31 000	0.42	13.0	Tetramer		Clontech	DsRed.T1 from (42)
DsRed2	563	582	43 800	0.55	24.1	Tetramer		Clontech	
DsRed-Monomer	556	586				Monomer		Clontech	
HcRed-Tandem	590	637	160 000	0.04	6.4	Monomer		Evrogen	Tandem dimer; functional monomer
HcRed1	588	618				Dimer		Clontech	HcRed-2A from (43)
AsRed2	576	592	56 200	0.05	2.8	Tetramer		Clontech	asFP595 from (44)?
eqFP611	559	611	78 000	0.45	35.1	Tetramer	<4.0	(45)	Tetramers dissociate at low concentration

mPlum	590	649		0.1		Monomer		(9)	~100 min maturation time
mRaspberry	598	625	86 000	0.15	12.9	Monomer		(9)	~55 min maturation time
mCherry	587	610	72 000	0.22	15.8	Monomer	<4.5	(10)	~15 min maturation time
mStrawberry	574	596	90 000	0.29	26.1	Monomer	<4.5	(10)	~50 min maturation time
mTangerine	568	585	38 000	0.3	11.4	Monomer	5.7	(10)	
tdTomato	554	581	138 000	0.69	95.2	Monomer	4.7	(10)	~1 h maturation time
Photoconvertible proteins									
CoralHue Kaede (green)	508	518	98 800	0.88	86.9	Tetramer	5.6	MBL International	Green to red photo-conversion, green state
CoralHue Kaede (red)	572	580	60 400	0.33	19.9	Tetramer	5.6	MBL International	Same protein as above, after photoconversion
KFP-Red	580	600	59 000	0.07	4.1	Tetramer		Evrogen	This data is for photoactivated state
PA-GFP	504	517	17 400	0.79	13.7	Monomer		(11)	This data is for photoactivated state
PS-CFP	402	468	34 000	0.16	5.44	Monomer	4.0	(46)	Before photoconversion
PS-CFP	490	511	27 000	0.19	5.13	Monomer	6.0	(46)	After photoconversion
mEosFP	505	516	67 200	0.64	43.0	Monomer	5.5	(47)	Before photoconversion
mEosFP	569	581	37 000	0.62	22.9	Monomer		(47)	After photoconversion
CoralHue Dronpa	503	518	95 000	0.85	80.7	Monomer		MBL International	From (13); photoactivated form

*Clontech no longer licenses enhanced GFP (EGFP) and its derivatives, but these are now such commonly available reagents that we have included them for completeness. The license for these vectors is held by GE Healthcare and they are available commercially in some vectors from Invitrogen.

Ex, excitation wavelength (nm); Em, emission wavelength (nm).

engineering of pH-sensitive fluorophores (e.g. the pHluorins; 16) and biosensors based on lipid binding or detection of activated states of signaling molecules (17).

Other genetically encoded fluorescent labels also exist, such as the FlAsH/ReAsH system for covalent labeling of genetically encoded tetracysteinyl peptides (18, 19). This system, developed by Roger Tsien, is now commercially available as the Lumio labeling system from Invitrogen. The ReAsH label has the advantage that it can be used for electron microscopy as well as light microscopy. There are some drawbacks to their use and they have not been widely applied in the same way as fluorescent protein technology.

4. BASIC NECESSITIES – CULTURING CELLS FOR IMAGING

These fluorophore developments have also, of course, been accompanied by continued and rapid hardware development (1). Perhaps the simplest of these is the now wide availability of different types of culture vessels suitable for cell imaging. Glass-bottomed tissue-culture dishes, chamber slides, and multi-well plates (from six to 384 wells) are widely available, as are chamber slides, perfusion chambers, and other suitable substrata for imaging. Critical parameters to evaluate when choosing these culture vessels are the optical clarity of the substrate, the thickness (one would probably wish for something matched to the standard coverslip thickness – typically 0.17 mm), the uniformity of the dish (when scanning across, small changes in level can make large differences to focusing), and the accessibility for perfusion/flow-dependent experiments. Cell imaging in the biotechnology and pharmaceutical industry frequently utilizes high-throughput and high-content screening applications; multi-well plates (typically 96- or 384-well formats) and chamber devices are, of course, particularly important here, as discussed in Chapter 13 by Elizabeth Roquemore. The advent of modern cell imaging means that high-content assays are available, which provide a much more information-rich method of analyzing the effects of potential drugs. These techniques are also being more widely used in academic and pure research sectors. Good examples include the screening of novel GFP-tagged cDNAs to provide functional annotation of genomic data (20, 21) and genomic profiling of early embryogenesis defects in *Caenorhabditis elegans* using RNA interference libraries and differential interference contrast microscopy (22).

5. LIMITS OF RESOLUTION AND DETECTION

Ever since the development of the first microscopes in the 17th century, scientists have been imaging cells and tissues. Anton van Leeuwenhoek (1632–1723) used his knowledge of lenses to build microscopes with which he viewed a wide variety of living and fixed specimens. He was the first to describe images of bacteria and yeast, as well as the microbes within a water droplet and the circulation of blood corpuscles. Subsequently, it was Robert Hooke (1635–1703), with the publication of his *Micrographia* in 1665 (23), who first coined the term 'cell' (although Hooke was not in fact looking at cells as we define them today, but at cell walls). This book also contained a diagram and description of Hooke's microscope, which was the forerunner to the modern microscope. *Fig. 2*(*A* and *B*) (also available in color section) shows this version of Hooke's microscope, alongside a modern-day counterpart, the scan head from a Leica TCS SP2 AOBS scanning confocal microscope (see *Fig. 2C*). Over 300 years of development have led us to the types of microscope that we typically find in research institutes, universities, and classrooms today (1).

For optimal use of the microscope it is important to understand the fundamental ways in which the image is formed by the system being used. This requires a basic knowledge of optics, notably of resolution and detection limits of the modern light microscope. It is essential to understand the difference between detection and resolution in microscopy. I would urge the reader to consult more detailed texts on this subject (24–26). The Rayleigh criterion defines the resolving power (*R*) of the light microscope. This is the distance (*D*) by which two point objects must be separated to be detected as discrete objects:

$$D = 0.61\lambda/\mathrm{NA}$$

where λ is the emission wavelength and NA is the numerical aperture of the lens.

Thus, if one considers a typical, well-configured, modern imaging system, using a common fluorophore such as GFP (emission wavelength 507 nm) and a 60× oil-immersion lens with NA = 1.4, the limit of resolution is:

$$(0.61 \times 507)/1.4 = 221 \text{ nm}$$

Fluorescence microscopy is restricted by the diffraction limit as described above. While there are now some possibilities for overcoming this limit (27), these are currently restricted in their availability and application to more general cell imaging. Resolution in cell imaging can, of course, be increased by moving from light microscopy techniques to electron microscopy. Indeed, much can be gained from correlating light and electron microscopy images of the same cells. A very useful example is provided in Chapter 10 by Pombo and colleagues, who exploit both fluorescence and electron microscopy to bridge this resolution gap. A more detailed discussion of resolution is largely beyond the scope of this text as it is assumed that readers will understand the basis of resolving power and the role of the digital detector. For more information, I would recommend the references (24)–(26). There are also a number of excellent online resources, foremost amongst which is the Molecular Expressions Microscopy Primer (http://www.microscopy.fsu.edu), which includes a number of interactive tutorials on a range of microscopy topics.

It is vital not to confuse this resolving power with detection limits. It is possible to image objects as small as single molecules using conventional light microscopy. This is simply a matter of having enough photons of light to detect. Fluorescence detection is now sensitive enough to allow single fluorophore detection, not only *in vitro* (28), but also in living cells (29). In addition to fluorescence detection, contrast techniques are extremely important for the detection of fine structure within cells; Chapter 4 by Lane and Stebbings covers

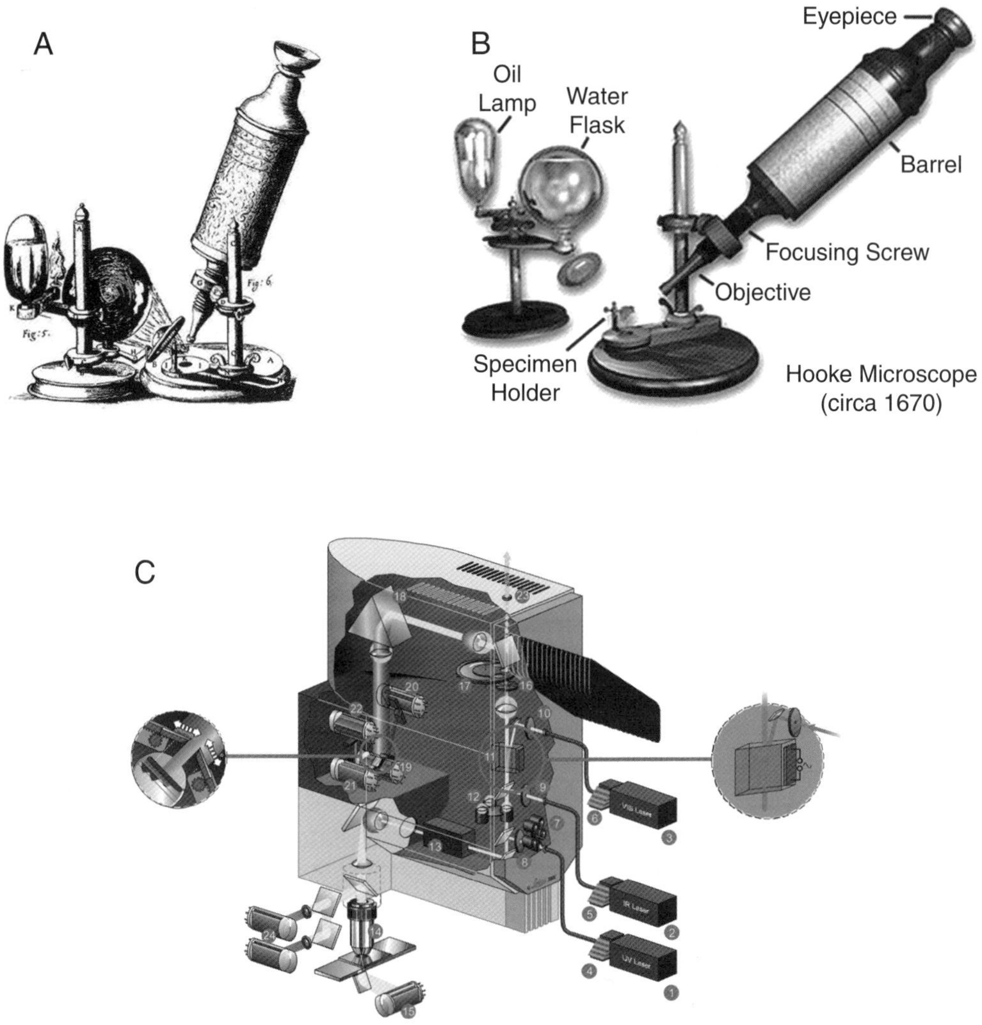

Figure 2. Microscopy from the 17th to the 21st centuries (see page xvii for color version). (*A*) Robert Hooke's microscope as detailed in *Micrographia*, published in 1665 (23). Image reproduced with permission of the Charles Deering McCormick Library of Special Collections at Northwestern University, USA. (*B*) A schematic of Hooke's microscope showing the relevant components. Image reproduced with permission from the Molecular Expressions Microscopy Primer (www.microscopy.fsu.edu/primer/museum/hooke.html). (*C*) Leica Microsystems' TCS AOBS confocal scan head, commercially available from 2002 (www.leica-microsystems.com). The numbered components are as follows: *1*, UV laser; *2*, IR laser; *3*, visible range laser; *4*, UV acousto-optical tunable filter; *5*, IR electro-optical modulator; *6*, visible range acousto-optical tunable filter; *7*, UV adaptation optics; *8*, UV excitation pinhole; *9*, IR excitation pinhole; *10*, VIS excitation pinhole; *11*, primary beam splitter; *12*, adjustable pupil illumination; *13*, 'K'-scanner with rotator; *14*, microscope and objective; *15*, transmitted light detector; *16*, confocal detection pinhole; *17*, analyzer wheel; *18*, spectrophotometer prism; *19*, photomultiplier channel 1; *20*, photomultiplier channel 2; *21*, photomultiplier channel 3; *22*, photomultiplier channel 4; *23*, external optical port; *24*, nondescanned reflected light detectors.

many transmitted light imaging techniques and details how one can use differential interference contrast microscopy to image single microtubules (diameter ~25 nm).

6. SELECTIVE IMAGING – OPTICAL AND MECHANICAL

Selective imaging of cells is facilitated in a number of ways. Ultramicrotomes provide a mechanical means of isolating and subsequently analyzing very thin sections of cells and tissues; confocal imaging techniques provide a nondestructive means of sectioning cells optically. Other optical techniques are also possible; one of the most elegant implementations is TIRF microscopy. Here, only the volume within 100–200 nm of the coverslip is imaged. In Chapter 8, David Zenisek describes the application of this technique that he and others have applied so successfully to the study of membrane trafficking events at the plasma membrane (30).

7. CHOICE OF SYSTEM

More developments are continually occurring in terms of microscope and detector technology. Where once wide-field microscopes were coupled to 35 mm film cameras, most systems now come equipped with some form of digital capture, typically a CCD (charge-coupled device) or PMT (photomultiplier tube). Recent developments in camera technology, typically with regard to speed and sensitivity, are well suited to cell imaging. Notably, the development of electron-multiplying CCDs has made it possible to image cells faster and with greater sensitivity than before. Furthermore, since their development, confocal microscopes have become almost commonplace in research laboratories worldwide. Electron microscopy has continually been an important technique in all aspects of cell imaging and is undergoing something of a renaissance at the present time. This is aided by the increasing need to obtain ultrastructural data that relate to light microscopy data such as time-lapse imaging.

Newer developments such as multi-photon microscopes, screening platforms, TIRF microscopy, and FLIM systems will doubtless become more and more commonplace in the coming years. Choice of system for a particular experiment therefore becomes paramount. This book assumes a certain level of knowledge in this area, but it is hoped that it will also spur the reader on to think about the best possible means of imaging their cells to obtain and process the data in the most appropriate manner. Clearly, the majority of projects require multiple techniques to address them. It is frequently sufficient to label fixed cells and undertake immunofluorescence analyses of cellular localization for example. However, increasingly, the level of understanding of a protein's function is greatly enhanced by live-cell imaging. Neither of these optical techniques provides sufficient resolution to define the ultrastructural localization of a protein and consequently electron microscopy must frequently be employed. If one is trying to image protein interactions, then FRET-based techniques may be required. Varma and Mayor discuss the application of FRET to cell imaging in Chapter 12. FRET can be monitored in many ways; FLIM is one of the more advanced methods for this, as discussed by Klaus Suhling in Chapter 11. It is with this diversity of techniques and equipment in mind that this book has been compiled. While many readers may not be familiar with many of the techniques described in this book, it is entirely possible that they will have the tools at their disposal to perform some of the more complex procedures described in this volume in the near future. Nevertheless, many of the chapters have been specifically written to apply to all forms of data, notably those on analysis and handling.

Clearly, reagents for cell imaging are relatively simple to generate on a rapid timescale. Many suitable antibodies are commercially available and most laboratories are happy to share reagents and expertise with others in academia. In all settings, acquisition of hardware is more challenging. Thus, it is necessary to make the best use of what is available. For example, relatively simple analyses using immunofluorescence labeling could be augmented in a number of ways. Time-lapse images can be simply obtained using a variety of different microscopes including wide-field, scanning, and spinning disk confocals. Specialist experiments typically develop from these. Principally, one needs to ask, what is the objective of my imaging experiment? Unless involved directly in the development of instrumentation, one's experiments should ideally be guided by biological questions as opposed to the availability of equipment. If this is ultrastructural resolution, then electron microscopy will probably be required. If one needs 3D spatial information, then 3D reconstruction, now frequently coupled with deconvolution, is the best approach. Specialist techniques such as TIRF might be applied to image-specific regions of a cell, or photobleaching experiments could be considered to measure protein dynamics. It is important to note that many of the techniques described in this book are in fact complementary to one another.

8. IMAGE PROCESSING AND PRESENTATION

Image processing is a cornerstone of all imaging experiments. The complexity of image acquisition is matched by the diverse set of techniques available for image processing. The latest array of digital detection and manipulation tools available means that one must gain some understanding of the way in which electronic detectors function and programs (such as Adobe PHOTOSHOP) handle and process images. While some information to this end is provided in this book, the reader is referred to more detailed texts that clearly explain many of the basic principles in more detail (24–26). A full understanding of image acquisition and processing is required for accurate cell imaging. One of the most important and frequently overlooked aspects of modern cell imaging is image quantification and analysis; Miura and Rietdorf discuss various approaches available in Chapter 3. It is important to note that many journals now require full disclosure, not only of the hardware used for imaging, but also of the image processing methods that have been used (see, for example, 31 and 32).

9. A WORKED EXAMPLE

Fig. 3 shows an example image taken from a data set that was acquired using several of the approaches described in this book. Here, we have used multidimensional wide-field imaging to acquire 3D stacks of HeLa cells stably expressing GFP–histone H2B, which had been grown on glass-bottomed dishes (MatTek). This experiment was designed for a specific purpose; we required fast time-lapse imaging to view the congression of chromosomes during metaphase. While the entire cell cycle of a HeLa cell lasts for about 18 h, chromosome condensation and alignment on the metaphase plate occur within approximately 30 min. For this imaging application, our available means of confocal scanning was too slow and did not provide sufficient light throughput. Cells were illuminated using a monochromator, which provides even light across the imaging

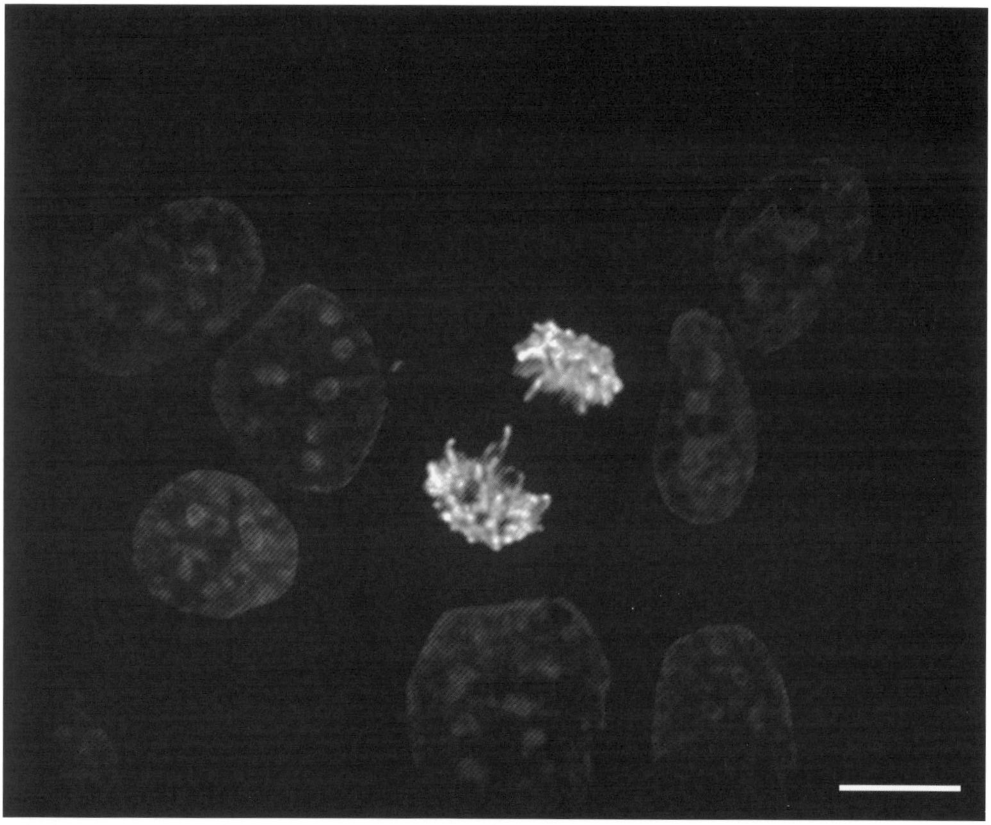

Figure 3.
Image of HeLa cells stably expressing GFP–histone H2B (33) acquired using a TILL Photonics multi-dimensional wide-field time-lapse imaging system. Cells were imaged using 50 ms exposures at ten *z*-planes each separated by 500 nm. Images were then deconvolved using iterative deconvolution with a measured point spread function using VOLOCITY software (Improvision). Bar, 10 μm.

field and relatively low intensity. This maintains cells in a healthy state such that they will go through at least two complete cycles of mitosis (unpublished observations). In order to maintain cell growth, the entire microscope stage was enclosed in a Perspex box (Solent Scientific) and heated to 37°C. Images were acquired using the lowest exposure time possible in order to give a good signal-to-noise ratio (in this case 50 ms). Images were taken at multiple focal planes (500 nm apart) using a Piezo-controlled Pifoc device coupled to the objective lens (a 60× water-immersion lens with NA = 1.2). We have previously determined that acquiring ten slices at this interval provides sufficient information for deconvolution, while minimizing exposure of the cells to light. After acquisition, these data were then deconvolved using Improvision's VOLOCITY software. We used a proprietary iterative algorithm with a measured point spread function.

A single still image is shown in *Fig. 3*. This is an 'extended focus' (also known as 'maximum intensity projection') image in which all planes have been compressed to a single 2D image. In the laboratory, we are acquiring these images over time to provide 4D data sets. This allows us to examine the precise spatio-temporal coordination of chromosome alignment during mitosis. We have now developed these experiments to include automated *x-y* scanning to include more cells in each experimental run, thus maximizing data acquisition. These experiments could also be allied to automated object recognition to define the stage of the cell cycle at a particular time point. It would also be possible to include any number of additional hardware configurations in these experiments. This experiment is shown as an example of the integration of multiple approaches described in this book. One must always consider multiple parameters for these experiments, and the final experiment is often a trade-off between sensitivity, resolution, and speed. More detail concerning these considerations can be found in the next chapter.

10. FURTHER RESOURCES

The axiom 'there is no substitute for experience' is particularly relevant for cell imaging. Central to any of the techniques detailed in this book is training. It is important that researchers in all areas of science and at all career stages exploit the many opportunities that exist for training in microscopy methods. Every year many courses are run by professional societies such as the Microscopy Society of America (www.microscopy.org) or the Royal Microscopical Society (www.rms.org.uk), or by funding agencies such as the European Molecular Biology Organization (www.embo.org). Many specialist courses also exist such as the hugely successful ones run by the University of Columbia (www.3dcourse.ubc.ca), Cold Spring Harbor Laboratory (http://meetings.cshl.edu), and the Marine Biology Laboratory at Woods Hole, USA (www.mbl.edu). I would urge readers to make use of these training opportunities, as well as to exploit the possibility of formal collaborative visits to established laboratories in related areas. There is no substitute for hands-on experience in any of the techniques described here.

11. REFERENCES

★★ 1. **Stephens DJ & Allan VJ** (2003) *Science,* **300**, 82–86. – *Review providing a brief description of diverse imaging techniques for light microscopy including scanning and spinning disk confocal, and TIRF.*

★★ 2. **Shimomura O** (2005) *J. Microsc.* **217**, 1–15. – *An interesting retrospective on the discovery of the photoprotein aequorin and GFP.*

3. **Shimomura O, Johnson FH & Saiga Y** (1962) *J. Cell. Comp. Physiol.* **59**, 223–239.
4. **Prasher DC, Eckenrode VK, Ward WW, Prendergast FG & Cormier MJ** (1992) *Gene,* **111**, 229–233.
5. **Inouye S & Tsuji FI** (1994) *FEBS Lett.* **351**, 211–214.
6. **Chalfie M, Tu Y, Euskirchen G, Ward WW & Prasher DC** (1994) *Science,* **263**, 802–805.
7. **Matz MV, Fradkov AF, Labas YA, *et al.*** (1999) *Nat. Biotechnol.* **17**, 969–973.

★ 8. **Zacharias DA, Violin JD, Newton AC & Tsien RY** (2002) *Science,* **296**, 913–916. – *Contains details of the generation of monomeric fluorescent proteins and their necessity for FRET experiments.*

★ 9. **Wang L, Jackson WC, Steinbach PA & Tsien RY** (2004) *Proc. Natl. Acad. Sci. U. S. A.* **101**, 16745–16749. – *Description of the generation of multiple fluorescent protein variants spanning a range of excitation and emission wavelengths generated using an iterative somatic hypermutation approach.*

10. **Shaner NC, Campbell RE, Steinbach PA, Giepmans BN, Palmer AE & Tsien RY** (2004) *Nat. Biotechnol.* **22**, 1567–1572.
11. **Patterson GH & Lippincott-Schwartz J** (2002) *Science,* **297**, 1873–1877.
12. **Ando R, Hama H, Yamamoto-Hino M, Mizuno H & Miyawaki A** (2002) *Proc. Natl. Acad. Sci. U. S. A.* **99**, 12651–12656.
13. **Ando R, Mizuno H & Miyawaki A** (2004) *Science,* **306**, 1370–1373.
14. **Zhang J, Campbell RE, Ting AY & Tsien RY** (2002) *Nat. Rev. Mol. Cell Biol.* **3**, 906–918.
15. **Miyawaki A, Llopis J, Heim R, *et al.*** (1997) *Nature,* **388**, 882–887.
16. **Miesenbock G, De Angelis DA & Rothman JE** (1998) *Nature,* **394**, 192–195.
17. **Mochizuki N, Yamashita S, Kurokawa K, *et al.*** (2001) *Nature,* **411**, 1065–1068.
18. **Griffin BA, Adams SR, Jones J & Tsien RY** (2000) *Methods Enzymol.* **327**, 565–578.
19. **Adams SR, Campbell RE, Gross LA, *et al.*** (2002) *J. Am. Chem. Soc.* **124**, 6063–6076.
20. **Liebel U, Starkuviene V, Erfle H, *et al.*** (2003) *FEBS Lett.* **554**, 394–398.
21. **Simpson JC & Pepperkok R** (2003) *Genome Biol.* **4**, 240.
22. **Sonnichsen B, Koski LB, Walsh A, *et al.*** (2005) *Nature,* **434**, 462–469.
23. **Hooke R** (1665) *Micrographia.* J. Martyn and J. Allestry. Reprinted in 1961 by Dover Books, New York.

★★ 24. **Inoue S & Spring KR** (1997) *Video Microscopy: the Fundamentals,* 2nd edn. Plenum, New York. – *A classic text dealing with the basic principles of microscopy, detection, and processing of images.*

★★ 25. **Pawley JB** (1995) *The Handbook of Biological Confocal Microscopy,* 2nd edn. Plenum Press, New York. – *Excellent description of the basic principles of confocal microscopy – a true handbook.*

★★ 26. **Murphy DB** (2001) *Fundamentals of Light Microscopy and Electronic Imaging.* Wiley-Liss, New York. – *An excellent guide to the basic principles of microscopy, detection, and processing of images.*

27. **Klar TA, Engel E & Hell SW** (2001) *Phys. Rev. E Stat. Nonlin. Soft Matter Phys.* **64**, 066613.
28. **Ishijima A & Yanagida T** (2001) *Trends Biochem. Sci.* **26**, 438–444.
29. **Mashanov GI, Tacon D, Knight AE, Peckham M & Molloy JE** (2003) *Methods,* **29**, 142–152.
30. **Merrifield CJ, Perrais D & Zenisek D** (2005) *Cell,* **121**, 593–606.
31. **Rossner M & Yamada KM** (2004) *J. Cell Biol.* **166**, 11–15.
32. **Rossner M** (2002) *J. Cell Biol.* **158**, 1151.
33. **Kanda T, Sullivan KF & Wahl GM** (1998) *Current Biology,* **8**, 377–385.
34. **Cubitt AB, Woollenweber LA & Heim R** (1999) *Methods Cell. Biol.* **58**, 19–30.

35. **Karasawa S, Araki T, Yamamoto-Hino M & Miyawaki A** (2003) *J Biol Chem*, **278**, 34167–34171.
36. **Zapata-Hommer O & Griesbeck O** (2003) *BMC Biotechnol.* **3**, 5.
37. **Rizzo MA, Springer GH, Granada B & Piston DW** (2004) *Nat. Biotechnol.* **22**, 445–449.
38. **Karasawa S, Araki T, Nagai T, Mizuno H & Miyawaki A** (2004) *Biochem. J.* **381**, 307–312.
39. **Griesbeck O, Baird GS, Campbell RE, Zacharias DA & Tsien RY** (2001) *J. Biol. Chem.* **276**, 29188–29194.
40. **Nagai T, Ibata K, Park ES, Kubota M, Mikoshiba K & Miyawaki A** (2002) *Nat. Biotechnol.* **20**, 87–90.
41. **Campbell RE, Tour O, Palmer AE, *et al.*** (2002) *Proc. Natl. Acad. Sci. U. S. A.* **99**, 7877–7882.
42. **Bevis BJ & Glick BS** (2002) *Nat. Biotechnol.* **20**, 83–87. Erratum **20**, 1159.
43. **Gurskaya NG, Fradkov AF, Terskikh A, *et al.*** (2001) *FEBS Lett.* **507**, 16–20.
44. **Lukyanov KA, Fradkov AF, Gurskaya NG, *et al.*** (2000) *J. Biol. Chem.* **275**, 25879–25882.
45. **Wiedenmann J, Schenk A, Rocker C, Girod A, Spindler KD & Nienhaus GU** (2002) *Proc. Natl. Acad. Sci. U. S. A.* **99**, 11646–11651.
46. **Chudakov DM, Verkhusha VV, Staroverov DB, Souslova EA, Lukyanov S & Lukyanov KA** (2004) *Nat. Biotechnol.* **22**, 1435–1439.
47. **Wiedenmann J, Ivanchenko S, Oswald F, *et al.*** (2004) *Proc. Natl. Acad. Sci. U. S. A.* **101**, 15905–15910.

CHAPTER 2

Confocal or wide-field? A guide to selecting appropriate methods for cell imaging

Mark A. Jepson

1. INTRODUCTION

Those of us who are involved in running central microscope facilities are called on to advise potential users which type of imaging system is most suitable for a particular application. It may also be the case that we are not asked this question as often as we should be. Some researchers have a preconceived idea that they need to use a confocal microscope to get the 'best pictures' and it is not always easy to convince them that other options should be considered. In fact, the choice of which imaging system to use for a particular application is not always straightforward, as they have broadly overlapping capabilities. This means that there are few, if any, firm rules. In this chapter, we will discuss some of the pros and cons of different types of imaging systems, but it is also important to bear in mind that most of us are faced with a rather limited choice of which imaging systems we can use for our research. Few, if any, institutions provide access to the entire range of imaging systems that would be required to cater optimally for every application and every type of specimen. Most of us therefore have to make some compromises to address our research requirements as best we can with the equipment available. Here, we will discuss factors that affect the performance of the more commonly encountered classes of imaging systems for different applications and different types of sample. The aim is to provide information that will be helpful to newcomers to microscopy wishing to make a more informed choice of which imaging system to use and also to help existing microscope users to optimize imaging systems for specific applications. The advice contained is, by necessity, fairly general and somewhat biased towards the imaging systems that I have most experience with. Some comparisons performed with confocal and wide-field imaging systems in the Cell Imaging Facility at the University of Bristol

Cell Imaging: ***Methods Express*** (D. Stephens, ed.)

(http://www.bristol.ac.uk/biochemistry/mrccif.html) are included as examples of the types of system that most researchers will have access to and, where possible, more thorough, peer-reviewed comparisons of imaging system performances are discussed.

I have largely steered away from discussing more technical aspects of the imaging technology and computational methods, as I want the advice in this chapter to be as accessible as possible to microscope users without a thorough background in mathematics and optics. It is undoubtedly the case that a deeper understanding of optical properties and other technical aspects of microscope systems can be enormously helpful in enabling users to optimize the quality and reliability of their data. That said, many microscope users manage to get started – and often get by for quite some time – with a more limited technical understanding of the systems they use. I have tried to give due consideration to those aspects of the way the microscope system works that are most important for users to understand and have attempted to explain their relevance in a straightforward way. This will be oversimplistic for some readers seeking a more thoroughly understand of these imaging systems and I would refer such readers to other sources for fuller explanations (1–3). I would also encourage readers to consult their local imaging experts to get the best available information on the capabilities of the imaging systems available to them. Discussion of the capabilities of available equipment is heavily biased towards what are currently the most commonly encountered facilities, but imaging technology is constantly changing. I have briefly discussed newer developments in imaging technology that are becoming available at the present time and encourage regular reconsideration of the capabilities of imaging equipment as their availability changes and research requirements shift. All too often, microscope users will stick with one system because of familiarity rather than through proper consideration of its ability to fulfill their requirements. Finally, it is worth pointing out that there is extensive background material available on the internet and at least one on-line discussion group that includes some of the most knowledgeable scientists in the area of fluorescence imaging[1].

The fluorescence imaging facilities of most research establishments are likely to include wide-field microscopy (WFM) and confocal laser scanning microscopy (CLSM) systems, which are widely regarded as the most flexible types of imaging system. For these reasons, WFM and CLSM, and specifically point-scanning CLSM, are the primary focus of this article. Other types of confocal microscope will also be discussed briefly. Many users will also have access to different classes of imaging system that may have distinct advantages in some aspects and/or for some applications. For example, CLSM and WFM systems within the University of Bristol Cell Imaging Facility are complemented by additional imaging capabilities within the laboratories of some of our users (including spinning disk confocal, total internal reflection fluorescence (TIRF), advanced wide-field deconvolution, multi-photon (MP), electron microscopy and atomic force microscopy systems).

[1]http://listserv.acsu.buffalo.edu/archives/confocal.html

This still only covers part of the repertoire of cell imaging techniques. It is beyond the scope of this article to describe any of these alternative imaging technologies in any detail, although some are covered more extensively in other chapters of this book and other sources (1, 4–8).

In concentrating on CLSM and WFM, we are discussing systems somewhere near two extremes in fluorescence imaging capabilities, with CLSM providing the best axial resolution[2] (ability to discriminate objects accurately in different focal planes), but at the cost of sensitivity, speed, greatly increased risk of photodamage, and price. The other principle advantage that CLSM has over WFM is that its use of point scanning makes it inherently applicable to techniques that require selective illumination of defined areas (photobleaching, photoactivation, uncaging). Although the addition of lasers to WFM systems and other types of confocal microscope can also enable such techniques on these systems, the flexibility of point-scanning systems gives them a clear advantage. In some circumstances, imaging of multiple fluorophores can also be achieved optimally by CLSM in a more straightforward way than with typical WFM systems. However, the overriding advantages of WFM in other respects mean that the most straightforward advice we can give to users is that, if the added resolution or functionality that CLSM provides is not required, WFM is most likely a better option. Much of the rest of this chapter will expand on these points and outline how these imaging systems perform in different applications.

2. METHODS AND APPROACHES

2.1. Comparison of CLSM and WFM

As we have already seen, the principle advantage of CLSM for standard imaging applications is its ability to provide enhanced axial resolution – its 'optical sectioning' capability. This comes at a high price, in the form of reduced speed and reduced sensitivity, as well as in monetary terms. Put simply, the choice between using CLSM and WFM (for straightforward imaging, at least) comes down to how much the enhanced axial resolution is really needed and how much we can afford the price we must pay to achieve it. The responses to these questions will depend on specimen properties, experimental requirements, and the biological question being addressed. For example, the 'price' to be paid in terms of sensitivity and speed may easily be affordable when cells are fixed, stained, and protected by antifade mountant, but are often harder to bear in live-cell imaging applications. Furthermore, the ability to acquire images with improved axial resolution will be much more valuable for thicker specimens with complex fluorescent labeling distributed over a broad depth than for thin, simple specimens. It is also important

[2]While CLSM provides the best axial resolution available to most researchers there are higher-resolution systems available for a more limited set of applications, and to a smaller proportion of researchers – these will be discussed later.

to remember that not all biological questions require the best axial resolution and, for some applications, there are advantages in detecting fluorescence from more than a narrow depth of focus.

As already stated, the main advantage of CLSM over WFM is in its ability to 'optically section' material. CLSM achieves this by excluding fluorescence emanating from levels above and below the focal plane ('out-of-focus' fluorescence) with an aperture through which light must pass to be detected. Of course, moving from WFM to CLSM is not the only way of answering a requirement for increased axial resolution or optical sectioning. Later we will discuss how deconvolution can be applied to WFM images to enhance the 'resolution' or 'spatial accuracy' of images[3]. The efficacy of deconvolution in improving spatial accuracy of WFM images is, however, dependent on how much out-of-focus fluorescence contaminates in-focus information, the accuracy of image stacks, and the speed with which they can be acquired. WFM/deconvolution also suffers from the drawback that it requires post-acquisition processing, so there is a significant gap between acquiring images and knowing whether the end result is satisfactory[4]. *Table 1* summarizes some other techniques that can be used to localize fluorescence precisely in limited depth within specimens, including both physical sectioning and optical sectioning techniques. More detailed discussion of some of these techniques is provided in other chapters and elsewhere (1, 4–8).

In addition to the differing capabilities of CLSM and WFM with respect to key imaging parameters, in general their cost also differs widely. CLSM systems are relatively expensive (though less so than some other types of imaging system), and for most people their use also involves significant access charges, reflecting the relatively high maintenance costs resulting, in turn, from their use of sophisticated electronics and expensive hardware, including lasers with finite lifetimes. The costs of WFM systems start much lower and running costs (and hence access charges) are also generally much lower than for CLSM. Many microscope users will therefore find a distinct advantage to using WFM in preserving hard-earned research funds. It is, however, quite possible to spend similar, or even higher, amounts on WFM as on CLSM, since high-end WFM/deconvolution systems with the most extensive capabilities can be significantly more expensive than lower-end CLSM systems. Not surprisingly, the relative costs of these systems generally reflect their comparative capabilities.

2.2. Optimizing imaging parameters: the need for compromise

Thus far, we have considered the capabilities of CLSM and WFM in rather simplistic terms and focused on their main differences with reference to sensitivity, speed,

[3]Although the term 'resolution' is more properly reserved for optical discrimination of objects without mathematical processing, for convenience we will use the term 'axial resolution' in comparing all techniques that enhance the ability to discriminate objects at different focal planes.
[4]The time required to deconvolve image stacks is becoming progressively shorter as algorithms and computers improve, and for those systems already supported by large arrays of computers the time interval no longer provides a serious drawback.

Table 1. Alternative approaches for acquiring fluorescence information from limited focal planes

General approach	Specific technique	Comments	Review references
Physical sectioning	Ultramicrotome sectioning	Labeling of ultrathin cryosections or sectioning of fluorescent material. No out-of-focus fluorescence present. High spatial resolution and straightforward application to correlative light/electron microscopy. Impossible to obtain dynamic information and difficult to acquire 3D information.	8; Chapter 10
Limit detection to single plane	CLSM	Point-scanning or line-scanning systems. Confocal aperture excludes out-of-focus fluorescence. High resolution but relatively low sensitivity and (mostly) low speed. High-intensity excitation (laser) can cause photodamage and limits use for some live-cell applications. Point-scanning systems are most common. Very applicable to ROI scanning applications (FRAP, FLIP, etc.).	1–3, 7
	Nipkow/Yokogawa (spinning) disk	Spinning disks with thousands of pinholes and (in newer systems) microlenses. Image the entire field simultaneously. At best, axial resolution is not as good as CLSM and is limited to certain lenses. Usually not as flexible as CLSM for multiple fluorophores, bleaching, etc. Popular for time-lapse imaging of GFP, etc. due to speed and reduced photodamage.	1, 7, 19
	Structured illumination (e.g. Optigrid, Thales Optem; ApoTome, Carl Zeiss)	WFM imaging of specimen on to which grid pattern is projected and moved. Images are combined and processed with a high-resolution result. Relatively low cost. No laser. Limited use for broader imaging applications. The need for several images to be captured and processed makes it not ideal for live-cell imaging and it is reportedly prone to artifacts.	21–23
Limit excitation to single focal plane	MP	Fluorophore excitation by two or more photons of longer wavelength (lower energy) than conventional excitation wavelength. Excitation is limited to the point of focus, so no need for confocal aperture. Longer wavelength light gives better tissue penetration and often less photodamage. High-intensity pulsed lasers (high cost). Lower axial resolution than CLSM. Becoming more user friendly and so more widely available.	1, 3, 6, 18
	TIRF	Nonstandard microscope set-up with high-NA lenses allows total internal reflection of laser light. An evanescent wave penetrates around 100 nm into the specimen, allowing selective excitation of fluorophores immediately adjacent to the coverslip. Used for specialized	Chapter 8

		applications, e.g. imaging events at plasma membrane.	
Computational image restoration	Deconvolution	Mathematical processing of image stacks (usually WFM) to calculate the correct location of fluorescent material from its contribution to images at all focal planes. Improved spatial accuracy of WFM images with attendant advantages of WFM in sensitivity and lack of photodamage. Limited effectiveness in some samples due to difficulties in predicting PSF, inconsistencies in acquired images, or excessive mixing of in-focus and out-of-focus light.	24; Chapter 9
Advanced imaging techniques	Scanning near-field optical microscopy	Combination of scanning probe microscopy and optical microscopy. In theory, allows acquisition of data on single fluorescent molecules and events. A complex biophysics tool out of the realm of most biologists at present, although biological applications at the single-molecule level were recently reported.	25–27
	4Pi	Exceeds the resolution limit of conventional microscopy with a double-objective imaging system linked to CLSM. Four- to sevenfold higher axial resolution than CLSM/MP (~100 nm). Only applicable to a limited depth (tens of microns). Technically demanding. Requires precisely matched lenses and very precise laser alignment. Commercially available, but expensive.	4

and resolution. Before considering in more detail how the performance of microscope systems can be configured to favor certain aspects of their performance, it is worth reiterating how the main imaging parameters are interrelated and how compromises between them underscore all types of fluorescence imaging.

It is helpful to consider the capabilities of fluorescence imaging systems with reference to the most important parameters that users may wish to optimize for a particular application. These aspects of the performance of microscope systems are largely interdependent and are often represented as a triangle (see *Fig. 1*), having, at its points, acquisition speed (temporal resolution), sensitivity (or signal-to-noise ratio, SNR), and resolution (or image quality). Within the boundaries of the triangle, moving closer towards one point takes one further away from the others, such that trade-offs must be made to optimize any of these three imaging parameters[5]. For example, decreasing image resolution may allow faster image acquisition, or speed may be increased without losing resolution if a resulting loss

[5]A web-based 'virtual confocal microscope' (http://micro.magnet.fsu.edu/primer/virtual/confocal/index.html) can be used to compare the imaging capabilities of WFM and CLSM and interactively demonstrates the effects of adjustments in imaging parameters on CLSM images.

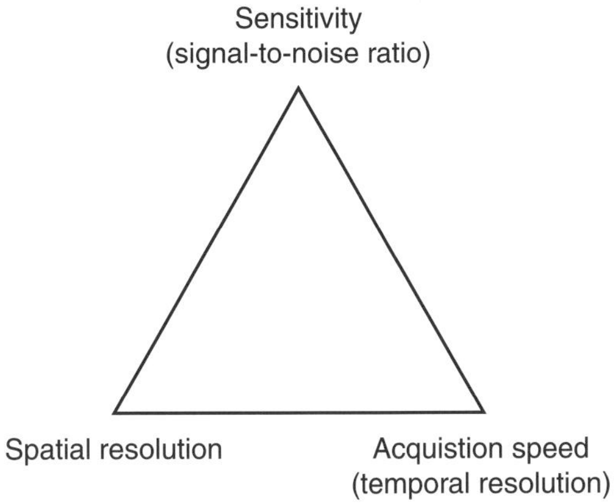

Figure 1. The 'imaging triangle'.
The desirable capabilities of fluorescence imaging systems can be considered as three largely interdependent parameters, which are here represented as a triangle. Attempts to optimize any one of these three parameters (acquisition speed, sensitivity, and resolution) will tend to move one further away from at least one of the others. To a large extent, this 'rule' applies both when attempting to optimize the performance of an imaging system and when comparing different types of imaging system.

of sensitivity can be tolerated. Conversely, optimization of SNR is generally achieved at the expense of speed and/or by compromising resolution. These issues need to be considered when using any imaging system that allows such parameters to be configured, and we will return to them later when discussing compromises that can be made to optimize one or other imaging parameter. To a large extent, the 'imaging triangle' analogy also holds true when comparing different types of imaging system. Since no single system offers the best performance in speed, sensitivity, and resolution, the choice of which imaging system to use for a particular purpose is guided by similar considerations. CLSM and WFM are somewhere near two extremes in terms of their imaging capabilities – CLSM offers greatly enhanced resolution, but largely at the expense of speed and sensitivity, while WFM performs much better in terms of speed and sensitivity, but provides images with lower resolution.

These generalizations only give part of the picture and we need to consider the details of how these systems are used for different applications to understand better their strengths and weaknesses and how they can be configured to make compromises between sensitivity, speed, and resolution.

2.2.1. Speed and sensitivity

The speed with which images are acquired can be of critical importance for some live-cell imaging applications, as the resulting temporal resolution may limit the available biological information. For other applications, such as fixed cells and slow time courses, it is clearly less critical, but may still have implications for the time required to obtain data. For example, acquisition of high-quality data through the entire volume of a cell might require 30 min or more of CLSM time, while the same volume could be imaged in a minute or so with standard WFM systems. The sensitivity of imaging systems, often measured as SNR, is intimately related to speed of image acquisition. It is self-evident that greater efficiency of detection will allow adequate signal to be acquired in a shorter time. The sensitivity of imaging systems also affects the ability to acquire images of live cells without inducing photodamage; if a lower level of fluorescence is required to generate acceptable signal, then the intensity and/or duration of excitation can be decreased to limit damage. As we will discuss later, sensitivity and speed are also related to image quality, since high levels of contrast (SNR) are required to achieve optimal resolution, increasing detector sensitivity ('gain') will usually result in noisier images, and since resolution can be compromised to increase temporal resolution. We will first consider in more detail the relative speeds and sensitivities of CLSM and WFM.

Standard CLSM has a scan speed of 1–2 s at conventional resolution (e.g. 512 × 512 pixels), which is rather slow when compared with WFM systems, which acquire images of the same sample at a rate at least tenfold faster with mid-range charge-coupled device (CCD) cameras – and faster still with more advanced hardware such as the 'electron multiplier' cameras. The absolute temporal resolution of CCD-based systems depends not only on the speed at which an image is acquired (i.e. detector sensitivity), but also on the chip readout rate, which can become the limiting factor for live-cell imaging. Many cameras have readout rates of less than 10 frames/s at their highest resolution, although higher-end (and usually more expensive) cameras will often have much higher readout rates.

It is often possible to increase frame rates during CLSM, e.g. by increasing the scan speed, decreasing *x*-*y* resolution, or by limiting the field scanned. These are useful compromises, but are, of course, not always practical if the highest resolution and SNR is required over a large scan field. Furthermore, imaging with point-scanning devices will always generate data in which there is a time difference between different parts of the scanned field. This is an important consideration in applications where movement is expected (most live-cell imaging applications) and can be reduced by limiting the dimensions of the field scanned (e.g. decreasing the number of lines scanned), or by optimizing scan speed (e.g. using one of the more specialized fast scanning devices on the market, which will be discussed later). While these options reduce the time delay between distinct parts of the image, the only way to eliminate it is by using an imaging system that detects fluorescence simultaneously across the entire field, e.g. WFM, or camera-based confocals, such as the spinning disk confocal systems (e.g. the Yokogawa system).

Often in WFM and CLSM there will be a need to acquire stacks of images at different levels of focus to obtain volume data, and the speed (and accuracy) of focusing can thus limit data acquisition. Higher-end WFM systems will have Piezo drives on individual lenses or to drive the microscope stage. Other focus devices may be at least an order of magnitude slower, so will limit acquisition of volume data over time courses (4D data acquisition). This limitation is, of course, less critical when imaging fixed cells where a single stack of images only is required, although the accuracy of focusing remains important and the time taken to acquire image stacks may also have implications if time and money are in short supply. The time taken to acquire stacks of images during live-cell imaging is perhaps most critical where deconvolution is applied to the image stacks, since any movement of fluorescent material during acquisition of the volume will introduce errors when the deconvolution algorithm is applied. The possibility of movement-based deconvolution artifacts cannot be eliminated entirely, but will be reduced by decreasing the time taken to acquire image stacks, e.g. using Piezo drives, the fastest available wavelength switching, and the most sensitive cameras. The potentially damaging effects of movement during stack acquisition can also be decreased by limiting the number of images acquired, so many researchers will choose to acquire suboptimal stacks for 4D imaging, as they balance the requirements for spatial and temporal resolution. These considerations are difficult to provide firm advice on, as the biological information required will vary widely between experiments.

While speed of focusing is a major factor when considering the time taken by CLSM and WFM to acquire data throughout a volume, different factors come into play when considering how quickly and conveniently the different imaging systems can acquire high-quality images of single optical sections. With CLSM, it is possible to acquire very high-resolution images of a single optical section in one go, or, if a thicker depth of section is required, by combining (projecting) a small number of images captured at different levels of focus. Obtaining comparable images by WFM may, depending on the thickness and complexity of the specimen, be possible only by acquiring images throughout the entire depth of the specimen and beyond, and then applying a deconvolution algorithm to the resulting data set (see *Fig. 2*). This approach does not always work reliably due to specimen properties as discussed later[6]. Thus, where the requirements are limited to defined focal planes, CLSM will generally provide the required images much more quickly and more reliably than WFM. For these reasons, CLSM tends to be preferred by many for imaging fixed cells with good levels of fluorescence. For live-cell imaging and specimens where fluorescent signal is weak, our choice more often moves towards WFM due to its greater sensitivity (see below).

Another factor that comes into play while considering acquisition speed is the ability to image, and accurately separate, different fluorophores. Despite its relatively slow imaging speed, point-scanning CLSM can provide some advantages over WFM systems in this respect. Most WFM systems will have a single CCD camera, necessitating switching between fluorescent filters to acquire images of more than one fluorophore. In some cases, this will require movement of filter blocks (containing the excitation filter, dichroic mirror, and emission filter) in between acquiring images of each fluorophore. While this option may provide the best separation of fluorophores (minimizing the possibility that the emission of one fluorophore contaminates the image of another), the slow

[6]To obtain images comparable with CLSM using our WFM/deconvolution system, we have sometimes needed to obtain several image stacks and adjust the deconvolution parameters, as we have yet to find a foolproof way of acquiring and processing data sets.

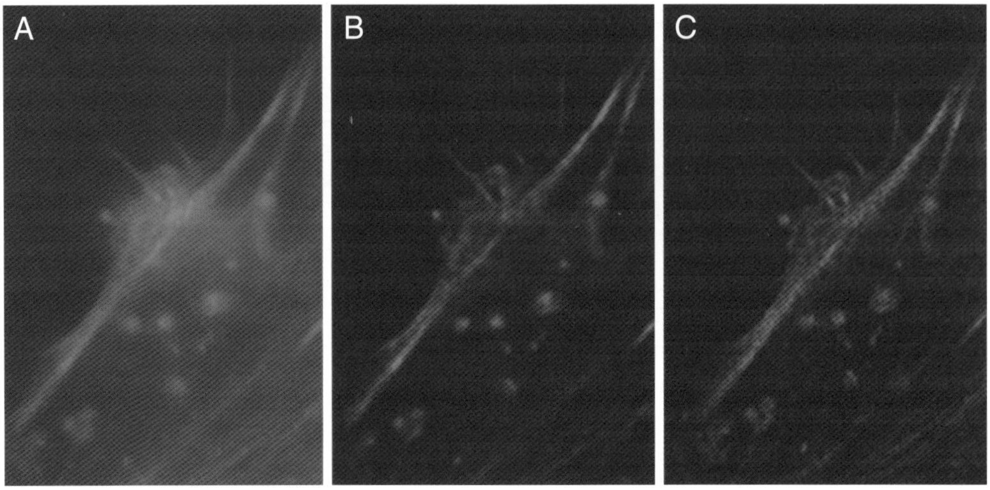

Figure 2. Comparison of WFM, WFM/deconvolution, and CLSM images of a HeLa cell stained with TRITC-phalloidin to localize F-actin.
Images were acquired from the entire cell at defined focal increments as described fully in *Protocols 1–3*. (*A*) The raw WFM image obtained at a focal plane within the cell is 'hazy' due to out-of-focus fluorescence. (*B*) After deconvolution using VOLOCITY software (and a calculated PSF), a prominent increase in clarity is observed due to removal of out-of-focus information and increased spatial accuracy within this optical section. (*C*) A CLSM image (acquired with the Leica SP2 AOBS system) of the same cell and same plane of focus as shown in (*A*) and (*C*) – or as near as it was possible to determine. The CLSM image shows similar 'resolution' to the deconvolved WFM image in (*B*) apart from an arguably clearer delineation of some very fine filamentous structures and a noticeably greater variation in signal intensity ('speckling') within actin structures, which may be indicative of the greater degree of noise expected in CLSM imaging. Field of view approx. 20×30 μm.

movement of filter blocks introduces delays that are unacceptable for many applications. Faster switching between fluorophores can be achieved using multiple band-pass beam splitters and emission filters, so that only switching between excitation filters is needed. This option, although faster[7], is less efficient in terms of fluorophore separation, so that the possibility of 'bleed-through' between images of different fluorophores needs to be tested. If such cross-contamination of fluorophore images is minimal, it can often be ignored or, if necessary, can be corrected by image processing. However fast the switching can be made, WFM systems with a single detection channel (i.e. most WFM systems) will always give a finite time gap between images of two or more fluorophores[8].

[7]Using a monochromator (generally faster) or filter wheels of varying speed and sophistication. In the case of filter wheels, the number and size of filters and their relative position will affect switching speed. Switching between adjacent filters will be quicker than if they are further apart.
[8]A minority of WFM systems are now equipped with optical field-splitting systems, allowing emission at two wavelengths to be acquired side by side on the same CCD chip, or with special CCD chips that allow simultaneous detection of different colors on the same chip. These options potentially allow simultaneous acquisition of more than one fluorophore, but if two excitation wavelengths are simultaneously used will be very prone to bleed-through between fluorophores.

The impact of the interval between images of different fluorophores will depend on how long it is (and how repeatable the interval is) and also on the nature of the imaging application. The interval will be more significant for fast time courses where movement of fluorescently labeled objects between images of different fluorophores may adversely affect data interpretation. The impact will also increase with the number of different fluorophores imaged, and will be exacerbated further if phase contrast or differential interference contrast images are also required. On the other hand, most CLSMs have more than one detection channel, so can in principle acquire images of two or three fluorophores simultaneously. While in some cases simultaneous imaging of more than one fluorophore can be achieved with excellent separation, in others there will be problems with bleed-through of fluorescent signal between channels, so that sequential image acquisition is often preferable. Most CLSM systems (at least those manufactured over the past few years) have an advantage over WFM in that excitation wavelength can be controlled extremely rapidly using acousto-optic tunable filters (AOTFs). Control of the AOTF allows continuous switching between different laser lines on a millisecond timescale, thereby enabling 'line-by-line sequential imaging' in which separate images of two or three fluorophores and, if required, transmitted light images are acquired with essentially no time gap. Thus, the lower imaging rate of CLSM compared with WFM can, to some extent, be negated if multiple fluorophores need to be imaged.

While the previous discussion might imply that CLSM should be favored for live-cell imaging of multiple fluorophores, it is not all good news. As already mentioned, there are other factors that limit the applicability of CLSM for live-cell imaging – its relatively low sensitivity and the requirement for intense laser excitation. CLSM lacks sensitivity largely because light is discarded (excluded at the pinhole) to allow optical sectioning, giving a low light efficiency. Although the efficiency of CLSM systems is constantly being improved, e.g. by modifications to optics and the introduction of more sensitive detectors (discussed in more detail later), the fact remains that the confocal aperture throws away a lot of photons. WFM systems are inherently more light efficient, as these systems essentially detect all the light, with the consequent drawback that, depending on the sample properties, the resultant image can include a large amount of out-of-focus light. If required, and if the data allow, deconvolution can be employed to enhance spatial accuracy of the resulting images, utilizing out-of-focus light to contribute to the resulting images rather than discarding it. Other parameters affect sensitivity (SNR). Point-scanning devices have a very short dwell time such that the detected signal at each pixel contains very few photons (a major reason why confocal images can be inherently noisy). Slowing scan speed (and thereby increasing dwell time) will boost signal, but, of course, decrease acquisition speed even further and potentially cause more photodamage.

In all imaging systems, light detected at more than one point (pixel) can be pooled to increase sensitivity (at the cost of resolution). Binning pixels in CCD cameras or decreasing *x-y* resolution of CLSM images (e.g. to 256×256 pixels) can be employed to achieve this. Amplification of detected signal in WFM and CLSM by increasing detector 'gain' will improve the ability to detect low-level signals, but will also increase noise and potentially compromise image quality. Perhaps the simplest way to increase CLSM sensitivity is to increase the size of the confocal pinhole and thereby increase the proportion of light detected (including more out-of-focus light) at the cost of somewhat reduced spatial resolution. Most confocal microscopes allow the pinhole to be adjusted in this way and we will discuss its effects in more detail later.

2.2.2. Resolution

The ability to discriminate between (resolve) closely separated objects is of fundamental importance in optimizing the information available in fluorescent images. The resolution achievable by CLSM is vastly superior to WFM in the *z*-axis direction (axial resolution) and, in theory at least, slightly better in *x* and *y* (lateral resolution) (1–3). In practice, data acquisition, especially volume data, often falls far short of optimum resolution. This is not always a significant drawback; it is down to the user to consider how important, practical, or desirable the optimization of resolution is for their own applications. To achieve this and to avoid oversampling, it is necessary to understand what the resolution limits are[9]. This is not the place to provide a detailed account of resolution limits or sampling theory, which can be found elsewhere (1, 2), but some brief consideration of these is important to understand how they affect image acquisition by CLSM and WFM.

Before acquiring 3D or even 2D data, it is important to consider the level of resolution required and the sampling frequency needed to achieve this. The most commonly encountered rule governing sampling theory is the Nyquist criterion, which specifies the sampling interval required to locate features faithfully based on the resolution limit[10]. According to the Nyquist theorem, a sampling frequency equal to twice the spatial resolution is sufficient for optimal sampling, but for practical reasons a figure of 2.3× is generally accepted for digital imaging. Thus, optimal resolution should be achieved by a pixel size of approximately 43% of the theoretical resolution limit. Resolution is proportional to emission wavelength and inversely proportional to numerical aperture (NA) of the objective lens (or NA^2 in the case of axial resolution). In aiming to satisfy the Nyquist criterion, it is also important to bear in mind that the resolution limit ultimately depends on contrast, such that optimal spatial separation is dependent on satisfactory SNR, again illustrating the interdependence of these parameters.

WFM images include information from in-focus and out-of-focus material, and the latter can contribute the vast majority of light detected at a single level of focus when fluorescent material is distributed throughout relatively thick

[9]As an example, using CLSM with optimum resolution and a 1.4 numerical aperture oil-immersion objective, the theoretical resolution limits for enhanced green fluorescent protein are approximately 150 nm (*x*-*y*, lateral) and 500 nm (*z*, axial).

[10]CLSM systems usually provide guidance on optimal *z*-steps to achieve 'Nyquist sampling' and will in some cases automatically set optimum step size for fully sampling a volume. Optimization of lateral sampling frequency is more often overlooked.

specimens. CLSM excludes out-of-focus fluorescence by virtue of the aperture (pinhole) through which light must pass to be detected and is thus able to discriminate objects at different focal planes much more reliably than WFM. Most CLSM systems have default confocal aperture settings (usually adjusted for objective lens properties and wavelength of fluorescence) where the aperture is closed down to an effective optimum size, equivalent to 1 Airy unit (1–3), at which optical sectioning is close to its best and beyond which the adverse effects of signal reduction will outweigh any small resolution benefit. In our experience, users rarely adjust the confocal aperture setting and, while there is often no need to, it is worth considering the enhancement in signal that can be gained by opening the aperture. This strategy will result in a decrease in axial resolution, but this is often not significant and is outweighed by the advantage provided by increased sensitivity and the reduction in photodamage that arises from the ability to decrease laser intensity. Also, for many applications, optimum resolution is not actually required and can even be a disadvantage. For example, in live-cell imaging applications, it may be advantageous to acquire fluorescence from a thicker optical section to include more information and avoid disappearance of entities moving out of the focal plane. This is one reason why live-cell imaging with WFM and other confocal imaging systems with less-effective optical sectioning (e.g. Yokogawa spinning disk confocals) often gives results preferable to CLSM[11].

Acquisition of data by CLSM through a 3D volume requires imaging at defined focal steps. Using CLSM with high-NA lenses (e.g. 1.3–1.4 NA) can provide axial resolution around 400–500 nm (depending on the fluorophore emission wavelength) so that optimal data collection acquisition, i.e. acquiring all fluorescence data in a defined volume, requires image stacks to be obtained at narrow intervals (e.g. 160–200 nm). Many confocal microscopes will automatically calculate the step size required for optimal sampling, while allowing the user to override this setting. This is important, since optimal sampling often requires the acquisition of too many images to be practical (i.e. expending too much time, incurring photodamage, and resulting in impractically large files). For these reasons, image stacks are more often acquired with focus increments larger than optimum. It is important to consider the effect of this on the data collected, since, under conditions affording fine axial resolution, information will be 'missed' in the intervals between the suboptimally acquired images and cannot be retrieved. This is much less likely to be the case for WFM image stacks, where the low axial resolution will include more information from levels of focus between the focus steps, which can potentially be reassigned to its point of origin by deconvolution. I will not attempt to prescribe a 'correct' way of working with these potential

[11]Comparisons of CLSM with other optical sectioning techniques (such as the Yokogawa spinning disk confocals) sometimes overlook the lower axial resolution of these alternatives, which are arguably more comparable to CLSM with increased confocal aperture. Sensitivity comparisons can thus be biased towards lower-resolution systems. However, even when the suboptimal resolution is taken into account, the higher detector sensitivity and light efficiency of spinning disk confocals are generally regarded as providing improved performance in terms of sensitivity and tolerance to illumination compared with CLSM (see later).

limitations, as each user's application will have different requirements. Personally, I usually choose a slightly compromised step size for most imaging with high-NA lenses (usually 500 nm, or larger for very thick specimens) unless I intend to produce the best-quality 3D reconstruction from the data. If this is the case, I try to reduce the focus interval to 150–160 nm. On the other hand, it is important to remember that axial resolution is inversely proportional to NA^2, so that for low-NA lenses (e.g. lower-magnification lenses) the axial resolution limit may be increased to several microns so that choosing a focus step size of 500 nm results in significant oversampling, i.e. needlessly acquiring images that will contain no additional information. Suboptimal sampling is also often applied to live-cell imaging applications where a smaller number of sections and/or more open pinhole are often preferred to enable acquisition of data without photodamage and loss of information due to movement of fluorescent entities. Care needs to be taken to consider the specific requirements of each application and to adjust the imaging parameters appropriately to optimize acquisition of biological information. This will more often than not require some degree of compromise, balancing the need to acquire enough images to sample the entire volume with the potential loss of information due to photodamage and loss of temporal resolution due to prolonged acquisition time.

The point illumination used in CLSMs provides a small (about 30%) enhancement in lateral (*x*-*y*) resolution compared with WFM – at least in theory (1, 2). Although this increased lateral resolution is not always achievable in practice, we should consider how images would need to be acquired to approach this resolution limit. As with axial resolution, Nyquist sampling frequency is the key to this and requires that pixel size should be less than half (approx. 43%) of the lateral resolution limit. Thus, if the lateral resolution limit is around 130–190 nm, as it is with the highest-NA objective lenses and commonly used fluorophores, it is necessary to acquire images where each pixel measures fluorescence from a specimen area of between 57 × 57 nm and 83 × 83 nm, depending on the fluorophore emission wavelength. Using standard CLSM systems, pixel size can be adjusted by altering the size of the scanned field (i.e. zooming) or by changing the pixel density of the image – for example, scanning at 1024 × 1024 pixels instead of the default 512 × 512 on the Leica confocals. The inherent ability of CLSM to 'zoom' by scanning smaller areas provides great flexibility over the magnification achieved with individual objective lenses. For example, many users will prefer to use a 63× lens and zoom as required, rather than to use a 100× lens when higher magnification or smaller pixel size is required. On our Leica confocals, achieving Nyquist sampling in *x*-*y* with our most commonly used objective lenses (63×) and 512 × 512 pixel density requires a zoom factor of anything between 4 and 8, depending on the fluorophore to be imaged and which confocal is used (the scanned field varies in size between systems). Data on pixel size is constantly displayed, so it is relatively simple to adjust zoom size and/or pixel density to achieve optimum values. Of course, in practice many users will not optimize the pixel size in this way, and in some cases the application will not allow the Nyquist criterion to be fulfilled. For each application, users should consider what pixel resolution and zoom factor to select, as well as whether data is required from large areas, whether sufficient signal is present to allow higher resolution, and whether file size and/or laser intensity is a limiting factor. Most users will choose to keep pixel resolution and the zoom factor constant for sets of experiments, so that the resulting image sets are directly comparable.

As with the step size chosen for acquisition of image stacks, it is not possible to provide set rules for pixel size. For example, using relatively high-magnification objective lenses (40× and 63×), for practical purposes I will often undersample in *x* and *y* by selecting a pixel size of 200 nm or more – unless I need to resolve very fine structures or to acquire optimal images for 3D reconstruction when I will take care to reduce the pixel size to 60–80 nm depending on the objective lens and fluorophores used.

WFM, on the other hand, has a more restricted pixel size and density depending on chip size and the objective lens used. In some cases, the pixel size required for optimal sampling is only achieved with 100× objective lenses - although the loss of resolution in 63× or 40× lenses is unlikely to be very noticeable for most applications. There is much less flexibility over magnification with WFM systems, since changing overall image size usually requires changing objective lens. It is possible to decrease pixel resolution to boost signal and/or increase speed by binning (pooling groups of CCD pixels), and for many applications this will not adversely affect the results. Pixel size increases linearly with binning (2 × 2, 4 × 4, etc.), and for some applications the loss in lateral resolution will be significant. Our WFM systems with Hamamatsu cameras (1344 × 1024 pixels) provide a pixel size of approximately 100 × 100 nm using a 63× objective lens and without binning[12]. This allows Nyquist sampling to be comfortably achieved with longer-wavelength fluorophores (TRITC, Cy3) but not quite reached with shorter wavelength dyes such as DAPI and green fluorescent protein (GFP). In practice, few applications will be sensitive to this failure to achieve the theoretical resolution limit, but any further increase in effective pixel size (e.g. binning) will result in a more noticeable resolution shortfall in many cases.

Of course, the preceding discussion of resolution limits has referred to theoretical optima. In real life, these limits are often unachievable, although images of carefully prepared, and relatively thin, specimens should get very close to these theoretical limits. Many will be using microscopes to image living specimens or deeper tissues. The complex nature of cellular materials means that the passage of light is affected in ways that cannot be accurately modeled, resulting in image quality falling well short of theoretical resolution limits. Imaging systems are not affected equally by this potential decrease in spatial accuracy as the focal plane moves further from the coverslip. CLSM is generally more tolerant to increasing working distance than WFM systems, where out-of-focus light becomes increasingly problematic. Absorption and scattering of light

[12]Our Hamamatsu ORCA II ER camera has a CCD chip with 6.45 × 6.45 μm pixels. Image formed by a 63× objective lens undergoes no further magnification, so the area of specimen sampled by each pixel is approximately 0.102 × 0.102 μm, obtained by dividing 6.45 by 63. The resulting pixel size of images can be verified by capturing images of a micrometer slide or hemocytometer.

also contribute to the reduction in image quality with depth, and these factors become less prominent with increasing excitation wavelength. This is one of the main reasons why MP imaging is often preferred for deeper penetration of tissues[13].

The loss of resolution occurring deeper within samples is exacerbated by mismatches in refractive index (RI) between immersion fluid, coverslips, and media. For this reason, water-immersion lenses are preferred by many for live-cell imaging, as they allow deeper imaging without loss of image quality (longer effective working distance). One of the advantages of water over oil as an objective lens immersion medium is that temperature has little effect on its RI – the same is not true for oil. However, water-immersion lenses, like dry lenses, are sensitive to the thickness of the coverslip so usually have correction collars that should be adjusted for the precise thickness of coverslip used[14].

2.3. Flexibility of point-scanning systems for specialized imaging applications

Point-scanning CLSMs allow adjustments to the size of the scanned field, as we have already discussed with respect to their variable magnification (zoom capability). Another advantage of this scan control is that a single line can be repeatedly scanned while the objective lens or microscope stage is moved through different planes of focus, generating an image of a 'section' perpendicular to the conventional field of view (an *x-z* section). This option is unavailable to WFM system users and is achieved most effectively with those CLSM systems that allow rapid focus control. Parking the beam on a user-defined spot can also be used for point bleaching or uncaging in a small region of the cell.

With the introduction of AOTFs to CLSM systems, the control over which parts of the sample are subjected to laser illumination was taken to another level. AOTF switching allows any user-defined regions of interest (ROIs) to be selectively illuminated. This capability makes CLSM a popular choice for applications requiring selective bleaching such as fluorescence recovery after photobleaching (FRAP), fluorescence loss in photobleaching (FLIP) and related techniques, which are widely used to monitor the mobility of fluorescent molecules in living cells (see Chapter 6) (9). In recent years the advent of GFP variants that change fluorescent properties upon laser illumination, including photoactivatable and photoswitchable fluorescent protein variants (10–12), has further extended the use of ROI scanning to additional applications for monitoring movement of fluorescently tagged proteins.

[13]Reduced photodamage due to the longer-wavelength excitation and the elimination of excitation outside of the plane of focus are other reasons why MP imaging is preferred by many users for *in vivo* imaging.

[14]Unfortunately one cannot assume that coverslips are the exact thickness their packaging claims, so they should be measured or the coverslip correction collar ring adjusted to give optimal resolution (i.e. the visually equivalent effect of defocusing above and below the plane of focus). It is also helpful to check that the coverslip correction collar on the objective lens is calibrated correctly. This is not always the case.

The speed of switching between scanning parameters required, e.g. for 'bleaching' and 'recovery' modes, limits the quality of data from CLSM photobleaching and photoactivation studies. For example, the relatively slow switching between different 'zoom' settings delays the start of acquisition of 'recovery' images if zooming (with a consequent increase in effective laser intensity) is used to bleach a limited field. Where AOTF-controlled switching of laser intensity can be applied to effect bleaching (or photoactivation) within defined ROIs, the delay between 'bleaching' and 'recovery' phases (or between 'activation' and 'monitoring' phases) is reduced, as scanning can proceed without any additional delay related to switching zoom factor. Further refinement of these specialized CLSM applications is ongoing. For example, we use AOTF-controlled Leica CLSM, which now offers a 'Fly-mode FRAP' option in which the AOTF continually switches from high intensity (for bleaching) to low intensity (to record recovery) between the forward and fly-back phases of each scanned line. This effectively eliminates the time interval between 'bleaching and 'recovery' phases of FRAP experiments and has similar advantages for other applications requiring selective photobleaching or photoactivation.

Of course, other imaging systems (e.g. WFM, spinning disk confocals, etc.) can be adapted for selective photobleaching or photoactivation by the addition of extra lasers, which can be focused on parts of the imaged field, but these adaptations are available to a minority of users. Such approaches will also tend to be less flexible as experimental requirements change or for different users compared with the ROI scanning ability of point-scanning CLSM. Thus, where CLSM does not give insurmountable problems for the live-cell imaging requirements of the experiment (due to photodamage, low sensitivity, etc.), this is likely to remain the preferred instrument for most users requiring these specialized applications.

2.4. Multi-color imaging

In the early years, commercial CLSM systems could exploit what was then a very limited choice of lasers[15], so the range of fluorophores that could be imaged was much more limited than used for WFM applications. WFM systems, on the other hand, most often use mercury or xenon arc lamps, which produce a broad range of excitation wavelengths that can be selected using appropriate excitation filters so that the range of fluorophores is consequently broad[16]. The distinction between early CLSM and WFM in availability of excitation wavelengths has essentially disappeared over the years as CLSM systems now exploit the much more extensive

[15]At first, most people used argon lasers with 488 and 514 nm lines; argon/krypton lasers with 488, 568, and 647 nm lines then became popular.

[16]The most commonly used lamps are mercury, which emit several high-intensity peaks including some in the UV range and much lower intensity elsewhere in the spectrum, and xenon, which emit more evenly over the visible and infrared spectrum but less in the UV range. Mercury lamps tend to be preferred if UV is required, despite their uneven emission at other wavelengths. Newer mercury/xenon sources provide a compromise between the two, with more even emission spectrum as well as strong UV peaks.

range of lasers that have subsequently been developed. Today, CLSM systems are usually equipped with high-power argon lasers producing a variety of blue lines together with several helium/neon, diode and/or other lasers producing a broad range of distinct excitation lines and allowing fluorophores from UV excitation to far-red emission to be exploited in CLSM imaging studies[17].

As we have already discussed, the ability to switch rapidly between excitation wavelengths using AOTFs, and the presence of several detectors, provides CLSM with some advantages for simultaneous imaging of multiple fluorophores with reliable separation into distinct channels. In recent years, the introduction of spectral detection on CLSM systems has allowed detailed fluorescence properties to be monitored accurately and has additionally made it easier to distinguish closely related fluorophores. CLSM with spectral detection capabilities offers options for spectral unmixing[18], which applies mathematical algorithms to sets of images acquired at discrete wavelengths to localize separately each of the fluorophores present, or separate fluorophore signal from autofluorescent background (see Chapter 5) (13). An alternative method for distinguishing fluorophores with distinctive but largely overlapping spectra is provided by the Leica acousto-optic beam splitter (AOBS) CLSM system. The replacement of traditional dichroic mirrors with the AOBS allows precise reflection of excitation lines and detection of emitted light from fluorophores including a larger proportion of the emission spectra. This, together with the precise detection of selected wavelengths provided by spectrophotometer-based detection, allows good, although often imperfect, separation of fluorophore combinations such as GFP and yellow fluorescent protein (YFP), which are too similar to distinguish by conventional WFM. The remaining bleed-through between images of such fluorophores can usually be eliminated by simple mathematical correction of percentage crosstalk between channels.

2.5. Improving spatial accuracy of WFM images by deconvolution

As we have seen, while WFM systems are undoubtedly fast and sensitive, they are outperformed by CLSM in resolution, and most markedly in optical sectioning, which is essentially nonexistent in WFM, since light from a point source contributes to images at all focal planes. This is not always a significant problem, and we, along with countless others, have used standard WFM to show dynamic processes effectively in living cells. The failure of WFM to discriminate fine structures at different focal planes has led to the widespread application of deconvolution algorithms to enhance the spatial accuracy (especially axial resolution) of WFM images. Details of deconvolution techniques and their

[17]For example, the combination of lasers on the four CLSM systems in our Cell Imaging Facility allows the following excitation wavelengths to be individually selected: 351, 364, 405, 430, 458, 476, 488, 496, 514, 543, 561, 568, 594, 633, and 647 nm.

[18]Spectral unmixing has been available on some CLSM systems for several years and is now becoming available on WFM systems with the advent of CCD cameras with integral liquid crystal tuneable filters for multi-spectral imaging (Cambridge Research & Instrumentation). It remains to be seen how much impact this technology will have on WFM imaging.

application are discussed more fully in Chapter 9. Here, we will restrict ourselves to discussing general aspects of the approach and to considering the pros and cons of WFM/deconvolution compared with CLSM.

Some so-called 'deconvolution' programs remove what is assumed to be haze from out-of-focus fluorescence based on the contrast in single images. These 'deblurring' or 'no-neighbors' methods may superficially improve image quality, but cannot reliably improve spatial accuracy of volume data, so have limited use in high-quality imaging studies. The more effective deconvolution methods employ algorithms developed over a number of years to reconstruct volume data, usually with reference to measured or calculated point spread functions (PSFs) that predict how a point source contributes fluorescence in cones above and below its physical position. Various commercial and freeware algorithms are available that, by forming increasingly effective (iterative) guesses of the correct location of fluorescent objects within a stack of images, can derive a restored image set, which in many cases rivals or improves on data acquired by CLSM. Most importantly, they can do this on data that was acquired faster than most CLSMs, with superior light efficiency, and consequently with less photodamage or bleaching. This type of approach is used by many researchers to achieve live-cell imaging of fluorescent samples (e.g. GFP fusion proteins expressed in cells), while avoiding the high light intensities that are required for CLSM. While this may sound a perfect approach, there are limitations to its effectiveness with certain samples; its benefits over CLSM are most clearly seen with relatively thin specimens and/or with well-defined fluorescent structures and situations where photon efficiency is most critical (e.g. some live-cell imaging applications).

The deconvolution software currently available differs in complexity and adaptability. The simpler options may work adequately with simple data sets, while the strength of the more sophisticated approaches will be more evident when applied to more complex fluorescence distributions in deeper specimens. For many users, deconvolution software is a 'black box' in which volume data can be miraculously improved, or sometimes not. Undoubtedly, a fuller understanding of the mathematical processes involved can help users better exploit the capabilities of deconvolution software, identify potential problems, recognize the types of data that lend themselves to deconvolution, and optimize image acquisition and PSF measurement. Deconvolution is a fast-moving area, and the continual refinement of algorithms is likely to increase their reliability for 'difficult' data sets as well as their speed. For those of us using deconvolution systems that are not supported by extensive computing resources, there is a significant gap between obtaining the data and seeing the end result following deconvolution. This has important practical implications for the progress of research relying on WFM/deconvolution. We will return to the drawbacks and potential pitfalls of WFM/deconvolution after discussing studies that have directly compared WFM/deconvolution with CLSM and provide some guidelines about the type of sample that is best suited to each approach.

There are few published comparisons of the performance of WFM vs. CLSM. One such paper is that of Swedlow *et al.* (14), which provided an elegant comparison of the abilities of WFM (with deconvolution) and CLSM to image the low content of YFP–tubulin in subcellular structures of *Toxoplasma gondii*. Tubulin-containing structures were imaged adequately in these small organisms (approx. 7×2 μm) by WFM, but not CLSM. The problems with CLSM imaging were prominent in these experiments because of a low SNR (due to intensity fluctuations). However, CLSM clearly outperformed WFM in imaging a much thicker specimen with widespread fluorescent labeling (a phalloidin-labeled quail embryo). It should be noted that these studies employed a high-end WFM/deconvolution system (Applied Precision DeltaVision) with robust deconvolution algorithm and system stability, and that the sensitivity of at least some CLSM systems has improved since these studies were performed. Nevertheless, the observations remain pertinent to most users' experiences with these techniques. Swedlow *et al.* suggested some 'rules of thumb' that WFM with deconvolution was likely to outperform CLSM when:

1. Fluorescence is restricted to a thin layer – whatever the thickness of the specimen.
2. The ratio of the vertical resolution limit (in practice, approx. 0.5 μm) to sample thickness is greater than the tolerance of the deconvolution algorithm, i.e. the lowest acceptable SNR (in practice, 1% or more), assuming that fluorescence is evenly distributed and absorbance is minimal.

A minority of specimens fulfills the first criterion. Examples include bacteria, yeast, and subnuclear structures where impressive results with WFM/deconvolution attest to the power of this type of imaging (15–17). The second point might suggest that most cells grown in culture may be better imaged with WFM/deconvolution than CLSM (14) since they will generally have a thickness of less than 100 times the vertical resolution limit (i.e. are <50 μm thick). However, in practice the performance of WFM/deconvolution is highly dependent on specimen properties due to heterogeneity in fluorescence, absorbance, and PSF. These factors will result in deconvolution working more reliably in some specimens than others and, perhaps more importantly, in some parts of specimens better than others.

The images in *Fig. 2* illustrate our experiences with deconvolution using VOLOCITY software to process image stacks acquired with our WFM system. This is a fairly typical acquisition system of relatively low cost and lacking the sophistication of some dedicated WFM/deconvolution systems on the market. It comprises a high-quality microscope, objective lenses, a CCD camera, and acquisition software, but, of particular relevance to the acquisition of volume data for deconvolution, has only a simple motorized focus control and is equipped with a standard mercury lamp-house without any additional stabilization of illumination or means to measure and correct for intensity fluctuations[19]. Despite these limitations our acquisition and deconvolution of WFM images of F-actin (labeled with TRITC–phalloidin and exhibiting a complex distribution) resulted in images comparable to, and arguably preferable to, CLSM images of the same cell (see *Fig. 2*). Certainly deconvolution provided a huge advantage over the raw image in

[19]Further details of image acquisition and processing are given in *Protocols 1–3*.

precisely localizing actin in the cell. An important caveat to this illustration of the power of deconvolution is that not all parts of this image stack produced such good results after deconvolution. For example, fluorescence distribution at the top of the cells was less clear in WFM/deconvolution images than after CLSM, so that the full volume of data was better represented in the CLSM image stack. I would stress that this may have been due to inadequacies in the collection of images, but nevertheless it serves to illustrate the point that we have not yet found a foolproof way of acquiring and deconvolving WFM images using our systems. Rather, we tend to use CLSM for imaging of fixed cells where labeling is usually relatively intense and has a complex distribution pattern. The full power of deconvolution is seen better with other types of sample and may also be more reliably demonstrated with dedicated WFM/deconvolution systems that tightly control image acquisition.

2.6. Potential problems and pitfalls in WFM/deconvolution

As we have already mentioned, the effectiveness of WFM/deconvolution is sample dependent and there is a requirement for sophisticated algorithms to tackle deconvolution where fluorescence is distributed in a complex pattern throughout the specimen. Although the algorithms and required computing power are developing apace, WFM/deconvolution will always be susceptible to problems arising from inconsistent properties of the input data and the unpredictability of PSF behavior in complex samples. For example, in our attempts to deconvolve WFM images, we have sometimes seen severe artifacts that were rejected because they were obvious (although their sources were not). It is of course possible, maybe even likely, that less-severe artifacts could be overlooked, potentially compromising data interpretation. WFM/deconvolution is thus acutely dependent on the quality of the input data and the accuracy with which the PSF can be modeled.

The first potential problem with WFM/deconvolution is the consistency of the acquired data. Uneven illumination intensity, inaccurate focus increments, and RI mismatches are all potential sources of flaws in the images collected for deconvolution. The effect of such 'flaws' in individual images is potentially more damaging to WFM/deconvolution than other types of imaging, since one flawed image can potentially 'corrupt' the entire volume of data, as the content of each image is used to calculate every image in the stack. For these reasons, the high-end WFM/deconvolution systems may encompass highly accurate focus drives, stable light sources, monitoring of illumination intensity (to allow correction for fluctuations during processing), and painstaking selection of coverslips and immersion oil. Precise control of temperature may also improve the quality of image stacks by improving the accuracy of focus increments. We have already discussed the importance of minimizing the possibility of movement-based artifacts in WFM/deconvolution. It is impossible to eliminate the possibility that something moves during stack acquisition and introduces errors in the results of deconvolution because its contribution to out-of-focus planes changes mid-stack. To reduce the chance of this happening, it is necessary to maximize the speed at which the image stack is acquired and possibly by collecting suboptimal stacks (i.e. fewer images than ideal for sampling) to enable this. It may be preferable to acquire a 'suboptimal' stack quickly, rather than an 'optimal' stack more slowly.

The next factor affecting effectiveness of deconvolution is the accuracy of PSF estimation. The PSF used in deconvolution algorithms can only ever be an approximation, since the behavior of light will vary in different parts of the sample. In most cases, the PSF is measured by acquiring stacks of images of subresolution beads (see Chapter 9) or calculated from properties of the sample and objective lens. Measurement of PSF can be problematic, since the location of the fluorescent microsphere used to measure it will differ from at least some parts of the sample it will be applied to – unless we are interested only in imaging fluorescent beads on coverslips! The distance from the coverslip affects the PSF of an object, as does its environment (RI varies among different subcellular compartments). Thus, even if the PSF of the bead can be accurately measured inside a cell – which is rarely done – the resultant measured PSF will not be a precise representation of the behavior of light in all parts of the (heterogeneous) cell. The difficulty in accurately measuring a PSF and extrapolating this to PSFs within the sample is one reason why some prefer to use a calculated PSF instead. This approach can give very good results (see *Fig. 2*). An advantage of using calculated PSFs is that measuring a PSF can potentially introduce its own errors (due to flaws in the bead image stack), which may then be transposed to each image stack deconvolved with the inaccurate PSF. Carefully measured PSFs are, nevertheless, favored by many experts. Algorithms are developing to cope better with irregularities of PSF (see Chapter 9). For example, blind deconvolution algorithms circumvent the requirement for either measured or calculated PSFs, as the image stack itself is used to derive the PSF during the deconvolution process. Such approaches can provide excellent results (see Chapter 9).

As discussed earlier, mixing of in-focus and out-of-focus light provides the main challenge for deconvolution algorithms. Even relatively simple deconvolution software can be expected to work well in situations where there is minimal interference from out-of-focus fluorescent material, e.g. where fluorescence is limited to a narrow depth of field and especially where fluorescence is located in limited numbers of discrete structures (vesicles, microtubules, etc.). Deconvolution software will struggle more where fluorescent material is present throughout a thick volume. The more sophisticated algorithms can cope with relatively complex and extended distributions of fluorescence, but only up to a point. To illustrate the problem, consider a cell containing numerous fluorescent structures that are unevenly distributed and have varying fluorescence intensities. A dimly fluorescent structure in the middle of the cell may be readily imaged by WFM/deconvolution if there are few similar structures above and below it, since its signal will be relatively 'uncontaminated' by out-of-focus fluorescence. Elsewhere in the cell, an identical structure could be situated a few microns above or below a large cluster of fluorescent structures of greater brightness and will consequently be much harder to image by WFM/deconvolution due to its fluorescent signal effectively being 'swamped' by the large amount of out-of-focus fluorescence. It is especially difficult to locate low-level fluorescence reliably where there is higher-intensity material in out-of-focus planes, because it is necessary to acquire all images without detector saturation so as not to confound the deconvolution algorithm. Such problems are not present using CLSM where pixel saturation is rarely a problem; given a broad range of fluorescence intensities it is quite acceptable to increase sensitivity to enable sufficient signal to be acquired to image dim objects, even if this would allow the brighter objects to saturate the detector if they were present in the optical section being imaged. This is not to say that deconvolution of complex structures is impossible, but it is worth bearing in mind that limits are reached beyond which even the best algorithms begin to fall down.

Deconvolution of thicker specimens is made more challenging by the fact that, in addition to the potential presence of large amounts of out-of-focus fluorescence (14), PSFs are less accurately modeled as the light path becomes more complex (e.g. with differing RI values), and as light scattering and absorption increase. Although these factors affect all types of imaging, CLSM (14), and to a greater extent MP imaging (6, 18), tolerate the potential problems of imaging deeper within tissues and multicellular complexes better than WFM/deconvolution (14). Application of deconvolution to thicker specimens is also affected by the requirement for large numbers of images effectively to reconstruct the entire volume. This increases the possibility of inconsistencies in the data set compromising deconvolution of the stack, e.g. introducing artifacts due to illumination fluctuations or movement (see above).

Despite the recognition of potential pitfalls in the application of WFM/deconvolution, it is clear that for some applications, and when carefully done, deconvolution provides an excellent tool with which to improve the spatial accuracy of WFM images (see Chapter 9 for a fuller discussion of deconvolution). However, the potential difficulties associated with WFM/deconvolution do tend to limit its use to circumstances where CLSM has significant drawbacks, e.g. because of low signal, a need for rapid imaging, and/or where there is a high risk of laser-induced photodamage.

2.7. Imaging systems with enhanced performance

This chapter has focused primarily on comparing the performance of standard CLSM and WFM systems with respect to key imaging parameters: speed, sensitivity, and resolution. Hopefully, this has also indicated ways in which the performance of these systems can be optimized to favor one or more of these parameters. We have deliberately restricted ourselves to the classes of imaging systems that will be available to most readers, but as the field is rapidly evolving, many other types of imaging system will become more readily available. In this section, we will turn our attention to alternative approaches to fluorescence imaging and to some recent advances and ongoing developments in imaging technology that aim to address some of the limitations we have discussed. This is necessarily a 'snapshot' of where we are in 2005, and some of these technologies may have become more widely available, or disappeared without trace, by the time you pick up this book. As well as the improvements in confocal and related instruments, it is also worth reiterating that WFM is itself undergoing advances, some of which have been mentioned elsewhere (e.g. increased camera sensitivity, increased light stability, improved speed and reproducibility of focus drives, enhanced performance and speed of deconvolution algorithms approaching 'real-time' deconvolution, multi-spectral imaging, adaptation for photobleaching applications, etc.).

As we have already seen, a major disadvantage of standard CLSM systems for live-cell imaging is their relatively slow frame rate compared with WFM. Manufacturers have addressed this limitation recently by introducing faster systems, which can, in some cases, now capture confocal images at several hundred frames per second. These use very fast scanning mirrors (e.g. Piezo-driven) and, in some cases, line scanning instead of the conventional point scanning[20]. The decreased pixel dwell time of fast CLSMs and the potentially large numbers of images collected mean that such systems are prone to low SNR and increased photodamage of specimens unless the increased scan rate is accompanied by increased sensitivity of signal detection and/or light efficiency of the systems. Fortunately, there have also been advances here, with redesigned beam splitters (e.g. Leica AOBS and Zeiss 5 *Live*) transmitting a high percentage of emitted fluorescence compared with conventional beam splitters, improved coating on optical elements for higher transmission, and the introduction of more sensitive detectors (photomultiplier tubes (PMTs) replaced or augmented with avalanche photodiodes or CCDs).

Another type of confocal imaging system has been around for a few years offering increased speed and decreased photodamage. 'Spinning disk' confocals achieve confocal imaging with a pair of disks containing patterns of thousands of pinholes; only light from an 'in-focus' plane passes through the corresponding pinholes to reach a CCD camera. The multi-point scanning enables simultaneous imaging of the field, a distinct advantage over point-scanning CLSMs for imaging rapid movement of fluorescent entities. The first generation had relatively low light efficiency, but the introduction of microlenses to focus laser light on to the excitation pinholes (in the Yokogawa confocal head) resulted in a substantial increase in light efficiency and hence usability. Many users prefer the Yokogawa spinning disk systems (as supplied by a number of companies, including Perkin Elmer), especially for straightforward live-cell imaging for which they produce excellent results. Notably, the frame rates achievable with these spinning disk systems are much greater than those with conventional point-scanning confocal microscopes. These systems are often used in conjunction with deconvolution, in part because axial resolution (even with the best-matched objective lens) falls

[20]Examples of fast-scanning CLSM systems include the Leica RS (resonant scanner – a point-scanning instrument), the recently introduced Zeiss 5 *Live* system (which uses line scanning and a novel optical path), and the new Nikon LiveScan Swept Field Confocal Microscope, which optionally uses fast point or slit scanning and lines of pinholes for multi-point scanning. Line-scanning systems have a somewhat decreased axial resolution compared with point-scanning systems (1).

short of CLSM. Other disadvantages of the spinning disk systems compared with CLSM include their relative lack of flexibility. The fixed pinhole size means they are ideally matched to one high-magnification lens and work much less well with others, and the field that is scanned is fixed so they offer only one magnification for each lens, no ROI scanning (so limited adaptability for applications requiring selective photobleaching or photoactivation), and no *x-z* imaging capability. More recent systems allow imaging of more than one fluorophore with multiple laser lines and transmitted light imaging, but only by switching between settings and hence introducing a time delay between each image, which can be avoided when using conventional point-scanning CLSM. As with the WFM/deconvolution vs. CLSM comparison mentioned earlier, there are few published comparisons of spinning disk and CLSM systems. Wang *et al.* recently provided such a comparison (19) in which the Yokogawa system was shown to be much more light efficient and to induce much less photobleaching compared with CLSM, although, as with other studies, the axial resolution was found to be somewhat inferior with the best-matched lens (100×) and markedly worse with lower-magnification lenses.

MP imaging is yet another type of imaging system that is preferred by many for imaging tissues and living cells and for *in vivo* imaging due to a combination of enhanced light penetration and reduced photodamage (1, 6, 18). Both of these advantages arise in part from the use of near-infrared light, which penetrates cellular material better than visible light and induces less photodamage due to its lower energy. MP imaging relies on fluorophores being excited by the simultaneous arrival of more than one photon with energy lower than light of its conventional excitation wavelength. The extreme improbability of such coincident excitation occurring effectively limits excitation (and consequently any chance of photodamage) to the point of focus. The absence of an out-of-focus fluorescent signal negates the need for a confocal aperture, so that all signal can be collected and, optionally, with nondescanned detectors, thereby increasing sensitivity. Optimum resolution falls somewhat short of that achievable with CLSM, by a factor of approximately 2 (depending on the properties of the fluorophore) (1, 3). MP excitation requires a large supply of photons at the focal point to increase the likelihood of coincident interaction with fluorophores – hence high-energy pulsed lasers are required. The high cost of such lasers and the technical difficulties associated with operating the first-generation MP systems limited their use, but in recent years they have become much more widely available.

The final class of imaging system we will consider is those that provide enhanced resolution, although in limited fields. In TIRF microscopy, a nonstandard microscope set-up with an unusually high-NA objective lens is used to produce an evanescent wave that penetrates a short distance beyond the coverslip, usually penetrating less than 100 nm into the specimen (see Chapter 8). TIRF enables selective excitation of fluorophores in a very limited field immediately adjacent to the coverslip so it can be used to image, at very high resolution, events at or very close to the plasma membrane of cells attached to the coverslip. The range of uses for TIRF is rather limited, especially when compared with CLSM and WFM, but it remains peerless for certain specialist applications, include monitoring single events in exocytosis and other membrane-trafficking events probed with GFP.

Another class of microscope has effectively broken the resolution limit of conventional light microscopes to offer greatly enhanced axial resolution. 4Pi microscopy uses a special phase- and wavefront-corrected double-objective imaging system (4) linked to a CLSM system. These systems provide axial resolution of around 100 nm within a depth of field of less than 100 μm and have been used to provide extremely fine images of subcellular compartments, microorganisms, etc. (4). 4Pi microscopes have only recently become commercially available (Leica) and their high cost and technical complexity will probably restrict their availability to specialist laboratories, for the time being at least.

2.8. Concluding remarks

A broad range of fluorescence imaging technology is now available to biomedical scientists, each having their own strengths and weaknesses for particular applications. The choice of system to use is not always straightforward and is likely to be limited by availability, as few, if any, research establishments will be equipped with every type of imaging system. In this chapter, I have discussed some of the principle features of the commonly available fluorescence imaging systems. It is perhaps helpful to conclude by reiterating our personal preferences, which are guided by the main differences among the imaging systems we have available to us.

In our Cell Imaging Facility, we currently have four Leica CLSM systems of varying vintage, two WFM systems (Leica microscopes, Hamamatsu cameras, Improvision software, LUDL or Prior shutters and filter wheels), and VOLOCITY software for deconvolution[21]. For my own research, I prefer to use CLSM for imaging fixed cells because of the enhanced resolution it offers. The advantage of CLSM is prominent in my own research, which often involves imaging relatively thick cells with complex fluorescence distribution, e.g. phalloidin staining of polarized epithelial cells. Our experience with WFM/deconvolution with such samples shows that it can provide a very good alternative to CLSM, but for our applications we have seen no clear advantage to this approach for fixed-cell imaging and, since we are well equipped with CLSM systems, this remains our preferred option. For more straightforward fluorescence imaging of relatively thin samples, WFM provides very good results and is cheaper, quicker, and more sensitive than CLSM – a major benefit if staining is on the low side. We use WFM for most of our live-cell imaging applications (mostly GFP and phase contrast imaging of cells with 10–20 s time intervals). When we instigated these studies, we saw major advantages in WFM over CLSM in terms of sensitivity for this type of imaging, which does not require the highest axial resolution. We usually acquire images at two or three focal planes (separated by 1.5 μm) to maximize information, but do not need 4D imaging for these studies. Others use WFM systems with faster focus devices, and sometimes deconvolution, to acquire 4D data sets with highest light efficiency. Increased sensitivity of CLSM systems over

[21]Further details of the equipment in the University of Bristol Cell Imaging Facility is available at http://www.bristol.ac.uk/biochemistry/mrccif.html

the years has improved their potential for live-cell imaging, but, while many users of our Cell Imaging Facility do use them for this purpose, we have not extensively compared their capabilities with WFM for our own studies. For practical reasons, we have also not tested the potential of the Yokogawa spinning disk system for our live-cell imaging studies, but some researchers in Bristol have now largely moved their live-cell confocal imaging to such systems, as they found distinct advantages in speed of image acquisition and avoidance of photodamage. As we alluded to earlier, this type of system has some disadvantages compared with CLSM in terms of resolution and flexibility for different applications, but, at least for relatively simple live-cell imaging of GFP-tagged proteins, has found many fans. Our CLSM systems come into their own when more complex live-cell applications are needed, since they are readily adaptable for different fluorophores, simultaneous transmitted light imaging, and selective photobleaching and photoswitching (due to the inherent advantage of point-scanning devices for ROI illumination). Although other types of imaging system (including WFM) can be adapted for such applications, the in-built capabilities of CLSM systems makes them a more obvious choice.

At the present time, CLSM and WFM are the most widely used and most flexible imaging systems. As other types of imaging system with distinctive capabilities (e.g. MP, TIRF, etc.) become more widely available, they open up new possibilities for fluorescence imaging-based research. It is interesting to speculate what a book such as this will focus on in 10 years' time and how the newly emerging imaging systems with enhanced resolution capabilities will impact on research. What is certain is that the power of fluorescence imaging to address questions across a broad range of research areas will continue to drive the development and wider use of imaging techniques. Understanding the capabilities of the available imaging systems will remain of critical importance to exploiting the new opportunities these will bring to different research fields.

3. RECOMMENDED PROTOCOLS

Protocol 1

Sample preparation for WFM and CLSM image aquisition

Method

1. Grow HeLa cells on glass coverslips and infect with enteropathogenic *Escherichia coli.*
2. Fix cells in 2% paraformaldehyde and permeabilization in 0.1% Triton X-100.
3. Label cells with TRITC–phalloidin to localize F-actin and with an anti-HA monoclonal antibody and Alexa 488-conjugated secondary antibody.
4. Mount cells on standard microscope slides in Vectashield (Vector Laboratories), a glycerol-based antifade mountant incorporating DAPI to label DNA (not shown).

Protocol 2

WFM and deconvolution

Equipment and Reagents

- Imaging system: Leica DMIRB inverted microscope with Hamamatsu ORCA ER (12-bit CCD) camera and Orbit (Improvision/Prior Scientific hardware) filter wheel, shutters and motorized focus controlled by Improvision OPENLAB 4 software (further details available at http://www.bristol.ac.uk/biochemistry/mrccif.html)
- Alexa 488- and TRITC-conjugated antibodies (Molecular Probes)

Method

1. Manually select a triple dichroic and emission filter set (460/20, 520/35, 600/40), which is mounted in a filter cube (Chroma).
2. Manually select a 63× oil-immersion objective lens (PLApo 1.32 NA) and add a small drop of immersion oil (Leica) to the coverslip.
3. Focus on cells, having selected the appropriate filter for TRITC excitation (555/25 nm filter; Chroma) using the Orbit filter wheel controller.
4. Switch excitation using the Orbit controller (484/15 nm filter; Chroma) and examine Alexa 488 labeling.
5. Locate a suitable area for imaging and the brightest level of focus within the specimen. At this stage it is worth taking care to ensure that the image is clear (high resolution) by eye. It is fairly easy to do this with this specimen, as we are familiar with the staining and clarity of image that should be obtained. An alternative is to use commercial slides that contain cells labeled with multiple fluorophores (e.g. from Molecular Probes/Invitrogen or other suppliers). If the image is not as clear as it should be, it is worth checking the coverslip correction collar (if applicable) or cleaning the coverslip and replacing the oil – insufficient oil or contamination of the coverslip or oil are common causes of poor image quality.
6. Select the maximum camera resolution (1344 × 1024; i.e. no binning) and 12-bit depth using OPENLAB 4 software. At this resolution, each pixel relates to a specimen area of 100 × 100 nm and thus fulfils Nyquist sampling criteria for both TRITC and Alexa 488 with this lens (theoretical optima approx. 115 and 102 nm, respectively).
7. Capture images of TRITC and Alexa 488 and adjust the exposure times and (minimally) camera gain to select suitable exposure times for each fluorophore (ideally the image will have a large dynamic range, without saturation, and with low background). Selecting the color 'look-up table' (LUT) showing the adjustable range of intensities as red (top of the range) and blue (bottom of the range) is useful for this adjustment of camera parameters. Using this LUT, it becomes instantly apparent where the image approaches 0 and 4095 (the limits of the 12-bit range).
8. Manually focus through the specimen with the camera in 'live' mode and with TRITC excitation selected to check that brighter objects are not present at other levels that would saturate the camera. The exposure time will need to be decreased if this is the case.
9. Repeat the 'saturation check' (step 8) for any other fluorophores to be imaged (in this case, Alexa 488). The avoidance of 'out-of-range' pixels is most important when applying deconvolution to image stacks, as any inaccuracies in measured intensity will compromise the operation of deconvolution algorithms. If deconvolution is not used, it may sometimes be necessary to allow some features to saturate in order to image dimmer ones adequately.

10. Manually focus the sample to a suitable start point above the specimen. For deconvolution, this may need to be a few microns beyond the limit of the specimen to optimize deconvolution of the first images in the stack. If deconvolution is not required, an image stack need only cover the actual limits of cells, or possibly just one level of focus if that is all the experiment requires.
11. Select the software option to link the shutter to exposure. This limits the illumination of the specimen as much as possible during acquisition of image stacks and avoids the need to include shutter control commands in automation.
12. Open an existing OPENLAB automation previously used to acquire similar image stacks of TRITC and Alexa 488.
13. Modify the automation by typing in the appropriate exposure times for TRITC and Alexa 488 as previously determined; input the required focus steps (in this case 200 nm) to be taken after acquiring each pair of images and set the number of image pairs to be acquired (in this case 51, giving a 10 μm depth). The automation starts at the current point of focus and moves focus-user-defined increments by controlling the Orbit focus drive (Prior ProScan II motorized focus), a cost-effective option, but by no means the fastest focus drive available.
14. After running the automation, image stacks are saved in Improvision's LIF format; these are transferred to an off-line image processing workstation and imported directly into Improvision VOLOCITY software. This software allows 3D rendering, as well as deconvolution, in separate modules. The VOLOCITY software uses a proprietary constrained iterative deconvolution algorithm based on the maximum entropy algorithm originally derived by Skilling & Bryan (20).
15. Deconvolution is applied to the image stack using PSFs calculated for each fluorophore wavelength and objective lens used. Deconvolution parameters are set at 95% confidence (generally involving 13–15 iterations). In our experience, these options give acceptable results for this type of image stack, but may need adapting for alternative experiments. Measured PSFs may give more reliable results in some cases, but have not been extensively examined by us.

Protocol 3

CLSM

Equipment

- Imaging system: Leica TCS SP2 AOBS CLSM attached to a Leica DM IRE2 inverted epifluorescence microscope. Of the five lasers on this system, we use the argon laser (488 nm) and green helium neon laser (543 nm) with detection set at 500–535 nm for Alexa 488 and 555–700 nm for TRITC (further details of the imaging system are available at http://www.bristol.ac.uk/biochemistry/mrccif.html)

Method

1. Manually select a 63× oil-immersion objective lens (PLApoBL 1.4 NA). This is one of the range of lenses that are corrected for chromatic aberration in the blue part of the spectrum.
2. Locate a suitable area for imaging (in this case matched to previously acquired WFM images). At this stage it is worth taking care to ensure that the image is clear (high resolution) by eye. It is easy to do this if one is familiar with the staining and clarity of image that should be obtained. With thicker specimens, it can be much harder to tell whether the image is clear due to the level of out-of-focus fluorescence, so it may be worth checking with a pre-prepared, familiar, and clear specimen. An alternative is to use commercial slides that contain cells labeled with multiple fluorophores (e.g. from Molecular Probes/Invitrogen or other suppliers). If the image is not as clear as it should be, it is worth checking the coverslip correction collar (if applicable) or cleaning the coverslip and replacing the oil – insufficient oil or contamination of the coverslip or oil are common causes of poor image quality.
3. In order to match the pixel size of WFM images collected with an equivalent lens (100 × 100 nm), the scan size is set at the default size of 512 × 512 and zoom factor adjusted to 4.65 (pixel size is displayed, so it is easy to adjust the zoom to the appropriate level interactively). Zoom in on the same area as that imaged by WFM, adjusting the field of view (by panning) as necessary. Although the pixel size of 100 × 100 nm meets the requirements of Nyquist approaches for WFM, CLSM could (in theory, at least) facilitate an increase in resolution of approximately 1.3-fold such that this pixel size may be slightly larger than the theoretical optima.
4. The pinhole (confocal aperture) should be left at the default setting (equivalent to 1 Airy unit) to optimize resolution.
5. Averaging is set at four frames to reduce noise. Further averaging would potentially improve this further at the expense of increasing time taken to acquire stacks, as well as increasing the light dose received and the consequent risk of photobleaching.
6. Laser power (AOTF setting) is selected to allow images of these well-stained cells to be acquired at a PMT setting of 500–600 V, above which more 'noise' is apparent. In the case of fixed (and antifade-mounted) specimens, the laser power is less critical than for live-cell imaging, where it is often necessary to use higher PMT voltages to enable laser power, and hence photobleaching/damage, to be minimized. As with WFM, we use a color LUT (called GlowOver/Under) that clearly identifies pixels with intensities of 0 or 255 (the extremes of the 8-bit range) by displaying them as green or blue, respectively. In the case of CLSM, there is less advantage to be gained by using 12-bit imaging, since, in contrast to WFM, the number of photons detected for each pixel is usually rather small.
7. With the TRITC setting selected, the top and bottom extremes of the cell are located and selected. This allows a stack of images to be acquired at user-defined intervals within the

selected limits. Of course, it is not always necessary to obtain image stacks; a single image is sometimes all that is needed, or a small number of sections if a limited part of the cell depth contains the required information. In this case, the full depth amounts to an 8 μm stack and step size is set at 163 nm intervals. These narrow focus increments are selected to optimize data acquisition (Nyquist sampling) and, in this case, to allow identification of the sections closest to those obtained from the equivalent WFM stack. More often, larger step sizes are selected (e.g. 500 nm).

8. After adjusting the AOTF setting and PMT voltage for each fluorophore, these settings are saved to allow sequential imaging of the two fluorophores. This approach has advantages over acquiring images simultaneously, as it usually avoids the cross-talk between images of the two fluorophores that usually occurs when labeling with green fluorophore is relatively intense.
9. Select sequential imaging and import the previously defined TRITC and Alexa 488 settings into the sequential imaging dialogue box to allow switching between the two excitation wavelengths. In this case, the Alexa 488 and TRITC stacks are acquired using 'line-by-line' sequential imaging where AOTF is controlled to switch rapidly between excitation lines for every line scanned, such that the fluorophores are imaged at millisecond intervals and appear simultaneously on the display panel. Other parameters such as PMT settings, the wavelength range detected, pinhole size, etc. can be adjusted between fluorophore settings only in the other, slower, sequential modes, i.e. 'between frames' and 'between stacks'.
10. The image sequence is then acquired and saved as a stack of TIFF files (the default save option), which can simply be exported to other software.

Acknowledgements

I would like to thank Isa Martinez-Argudo for preparing the specimen imaged in *Fig. 2*, Alan Leard for deconvolving WFM images and assistance with writing the protocols, and David Stephens for helpful discussions. Work in my laboratory is supported by the BBSRC and MRC. The University of Bristol Cell Imaging Facility was established with funding from the MRC.

4. REFERENCES

★★ 1. **Pawley J** (ed.) (1995) *Handbook of Biological Confocal Microscopy*, 2nd edn. Plenum Press, New York. – *An excellent and thorough discussion of confocal microscopy.*
★★ 2. **Sheppard CJR & Shotton DM** (1997) *Confocal Laser Scanning Microscopy*. BIOS Scientific Publishers, Oxford, UK. – *A detailed discussion of the technique of confocal microscopy.*
3. **White NS & Errington RJ** (2005) *Adv. Drug Deliv. Rev.* **57**, 17–42.
4. **Egner A & Hell SW** (2005) *Trends Cell Biol.* **15**, 207–215.
5. **Gustafsson MG** (1999) *Curr. Opin. Struct. Biol.* **9**, 627–634.
6. **Konig K** (2000) *J. Microsc.* **200**, 83–104.
★★ 7. **Stephens DJ & Allan VJ** (2003) *Science*, **300**, 82–86. – *A review covering a diverse range of imaging techniques, providing a good overview of their application.*
8. **Takizawa T & Robinson JM** (2003) *J. Histochem. Cytochem.* **51**, 707–714.
9. **Lippincott-Schwartz J, Altan-Bonnet N & Patterson GH** (2003) *Nat. Cell Biol.* **Suppl.**, S7–S14.
10. **Ando R, Hama H, Yamamoto-Hino M, Mizuno H & Miyawaki A** (2002) *Proc. Natl. Acad. Sci. U. S. A.* **99**, 12651–12656.

11. **Chudakov DM, Verkhusha VV, Staroverov DB, Souslova EA, Lukyanov S & Lukyanov KA** (2004) *Nat. Biotechnol.* **22**, 1435–1439.
12. **Patterson GH & Lippincott-Schwartz J** (2002) *Science*, **297**, 1873–1877.
13. **Zimmermann T, Rietdorf J & Pepperkok R** (2003) *FEBS Lett.* **546**, 87–92.
★★ 14. **Swedlow JR, Hu K, Andrews PD, Roos DS & Murray JM** (2002) *Proc. Natl. Acad. Sci. U. S. A.* **99**, 2014–2019. – *A thorough comparison of confocal and wide-field approaches, highlighting the benefits and disadvantages of each.*
15. **Jones LJ, Carballido-Lopez R & Errington J** (2001) *Cell*, **104**, 913–922.
16. **Pardo M & Nurse P** (2005) *J. Cell Sci.* **118**, 1705–1714.
17. **Platani M, Goldberg I, Swedlow JR & Lamond AI** (2000) *J. Cell Biol.* **151**, 1561–1574.
18. **Niggli E & Egger M** (2004) *Front. Biosci.* **9**, 1598–1610.
★★ 19. **Wang E, Babbey CM & Dunn KW** (2005) *J. Microsc.* **218**, 148–159. – *A detailed comparison of scanning and spinning disk confocal microscopes.*
20. **Skilling J & Bryan RK** (1984) *Mon. Not. R. Astr. Soc.* **211**, 111–124.
21. **Gustafsson MG** (2000) *J. Microsc.* **198**, 82–87.
22. **Neil M, Juskaitis R & Wilson T** (1997) *Opt. Lett.* **22**, 1905–1907.
23. **Schaefer LH, Schuster D & Schaffer J** (2004) *J. Microsc.* **216**, 165–174.
24. **Wallace W, Schaefer LH & Swedlow JR** (2001) *BioTechniques*, **31**, 1076–1078, 1080, 1082 passim.
25. **Hausmann M, Liebe B, Perner B, Jerratsch M, Greulich KO & Scherthan H** (2003) *Micron*, **34**, 441–447.
26. **Hoppener C, Siebrasse JP, Peters R, Kubitscheck U & Naber A** (2005) *Biophys. J.* **88**, 3681–3688.
27. **Kim J, Muramatsu H, Lee H & Kawai T** (2003) *FEBS Lett.* **555**, 611–615.

CHAPTER 3
Image quantification and analysis

Kota Miura and Jens Rietdorf

1. INTRODUCTION

Recent advancements in biomedical microscopy are largely due to the advent of techniques that specifically label molecules and probes inside living samples using fluorescent proteins (1). While the aim of microscopists in earlier times was restricted to resolving structural details in transmitted or reflected light illumination and to obtaining contrast from unstained specimens, questions asked in research these days additionally concern the localization, concentration, and dynamic behavior of specifically labeled proteins, organelles, or cell types.

Concomitantly, the need to analyze and extract numerical information from microscope images has steadily increased. New digital image processing techniques and the availability of cheap and fast personal computers have facilitated these trends. Many modern methods have been developed to measure molecular dynamics and interactions of molecules (e.g. fluorescence recovery after photobleaching and fluorescence resonance energy transfer). Since many of these methods have been described in some detail elsewhere in the literature (2, 3) and in this book, we will only briefly introduce the basic methods and principles of how to extract shape, position, and intensity information from digital images and will then focus on methods for the analysis of particle movement inside living specimens.

At different levels, movement is an essential function in biological systems. Various methods are available to measure movement in digital image sequences. Each method has both advantages and disadvantages. The best choice of method will depend on the character of the movement that the researcher is dealing with. We will briefly describe the principle of tracking and discuss which algorithms are suited/not suited to certain applications. Throughout the chapter, IMAGEJ (4) will be referred to as the standard imaging software, since it is freely available via the internet, is platform independent, and is widely used by biological researchers.

Cell Imaging: ***Methods Express*** (D. Stephens, ed.)

2. METHODS AND APPROACHES

2.1. Densitometry

Densitometry, i.e. the determination of intensity, attempts to measure the apparent amount of specific label at a certain position inside the sample. For a number of reasons, we cannot determine the concentration of a fluorescent label as precisely with a light microscope as with a fluorimeter. In the light microscope, we typically deal with minute measurement volumes in the order of 0.1–10 fl as determined by the focal volume of the microscope (5, 6). For physiological dye concentrations, we may expect the number of fluorophores to be in the order of ten or less inside the focal volume of a confocal microscope, which in principle will emit photons in all directions. For highly efficient objective lenses, we can at best expect 25% of the emitted photons to be collected by the objective, and by the time these photons are translated into countable photoelectrons, the overall efficiency of the detection device will roughly be in the order of 1%. In confocal microscopes, bright voxels may only be represented by approximately 10–20 photoelectrons, resulting in a high level of uncertainty in the measurement (7). In practice, there is a long list of factors that can degrade the measured signal (7, 8). For example, complications may result from misalignments of the instrument (9). In addition, the label may cause alterations in the specimen and, vice versa, the specific environment inside the specimen may cause alterations to the label (10). The bottom line is that, even though many measurement uncertainties caused by the properties of the device can be quantified and sometimes corrected for (11), it is impossible to do so for the uncertainties resulting from the physico-chemical properties of the specimen. Even for the measurement of relative intensities, it is necessary to align the instrument carefully (9) and to make sure that illumination and detection is even across the field of inspection.

Calibration slides with a defined fluorescein polymer layer thickness allow the estimation of illumination intensity distribution and a separation of this illumination intensity distribution from detection efficiency distribution (12). Other calibration kits – for example InSpeck (Invitrogen), a collection of beads with different staining densities – can help to make densitometric measurements more reliable and reproducible.

2.1.1. Segmentation

In many situations, we want to define the boundaries of regions of interest to separate objects locally from what is referred to as the background. This task is called 'segmentation' and is a prerequisite for quantitative analysis of digital images. There are various ways to segment images and it is far beyond the scope of this chapter to give an exhaustive overview. We will briefly discuss intensity-based thresholding and edge tracing as the most widely used approaches. For a more detailed discussion of the topic, see (2) and (13). Obviously, the efficiency of the segmentation process depends critically on the quality of the images, and although there are plenty of image processing methods to improve segmentation of noisy and deteriorated images, it is easier and more efficient to optimize the acquisition process than to improve the segmentation.

2.1.2. Segmentation by intensity thresholding and edge tracing

To set a threshold manually, a histogram, i.e. a plot of the abundance of all pixel intensity values in the image, is typically used. Even after careful shading correction (see *Protocol 1*), noise and measurement artifacts can falsely cause areas in the image to appear to be of different intensity, making it hard to select an entire object with only one threshold. Two strategies are taken to deal with the problem: (i) pre-filtering of the image to denoise or 'despeckle' it before thresholding. Median, salt-and-pepper, or low-pass filtering methods are often used in this process. Deconvolution algorithms can also be used to reduce noise (14); (ii) after thresholding, parts of the image that have been miscategorized to either regions of interest or background can be eliminated by binary image operations that take neighboring pixel categories into account. These are referred to as closing and opening operations. The concern with setting a threshold manually is the introduction of bias by the user. There are many algorithms described to set the threshold automatically (13).

Another segmentation method is edge tracing and is based on finding areas of maximal intensity differences that represent object boundaries. This method is used if the object does not have an even signal intensity, as for example in pictures obtained by phase or differential interference contrast imaging. The contour line is either set manually or using edge-enhancing methods before thresholding. Simple filters to enhance edges are called 'Sobel' and 'Laplace' filters (available in IMAGEJ; 4).

If there are many objects in an image, they may touch each other after thresholding. These objects can be separated by the use of a 'watershed' algorithm. Some procedures are nicely illustrated in the IMAGEJ documentation web pages (15).

Ultimately, if there are no convincing ways to segment the target object numerically, the user can trace the object manually (see *Protocol 6*).

Once objects are segmented, a number of parameters can be derived automatically as described in *Protocol 3*.

2.2. Morphometry

With its telecentric optics, the light microscope is in principle well suited to measuring distance, shape, and volume of well-resolved objects accurately, as described in *Protocol 3*. However, small objects of subresolution size will appear inflated and blurred under a fluorescence microscope. For example, the size and volume of most of the small cellular organelles (mitochondria, plastids, Golgi cisternae, etc.) cannot be measured accurately. Microtubules, 25 nm in diameter, will appear almost ten times wider.

A reasonably accurate morphometric measurement is only possible if the structures are fully resolved (16) and the optical elements of the microscope are in good alignment (9). The capability to resolve small structures is limited not only by the optical properties of the microscope, but also in most biological samples by the available contrast (16).

2.2.1. Position

Position measurements can be achieved either manually or with very high precision through calculation of the centroid of the segmented object, fitting the Gaussian curves, or matching template patterns. Details will be explained in section 2.3.3.

2.2.2. Length and distance

Most automated systems contain software tools that allow length and distance measurements. On camera-based systems, the size of a single pixel can be estimated from dividing the pixel size of the camera – typically provided in the data sheet of the camera – by the total magnification of the microscope optics. It is, however, more practical and more precise to calibrate the system by taking a reference picture of a scale (for example, a stage micrometer or a standard cell-counting chamber) and calculating the exact pixel size from it (also see *Protocol 2*).

For unresolved structures, only the measurement of an apparent length is possible. For example, the outline of a cell may, depending on the overall resolution of the image, appear as a smoothly curved line or very rough with many indentations and protrusions, which will result in vastly different estimates of its length, and consequently also of surface measurements.

2.2.3. Area, volume, and surfaces

The volume of a well-resolved 3D structure can be measured by counting the voxels occupied by the signal of the structure. Measurements of the surface area of 3D objects in a confocal image require fitting a surface to the selected object and calculating its area. This operation can be performed by some software packages (for example, IMARIS and IMAGEPRO+).

2.3. Measurement of movement

A time series of digital images, usually called a 'stack', contains temporal dynamics of position and intensity, and kinetics can be obtained from these data. There are three types of dynamics: (i) the position does not change, but intensity changes over time; (ii) the position changes, but the intensity does not change; (iii) both position and intensity change over time. A typical example of type (i) is the measurement of cargo transport dynamics in vesicle trafficking (17). Transition of protein localization from the endoplasmic reticulum to the Golgi and then to the plasma membrane was measured over time by measuring the signal intensity in each statically positioned compartment. This technique has evolved to various sophisticated methods based on the same principle – measurement of signal intensity at a constant position. Type (ii) corresponds to the measurement of movement, or object tracking, and an example is the single-particle tracking of membrane surface proteins (18). An example of type (iii) is the measurement of chemotaxis-related protein accumulation dynamics during *Dictyostelium* cell migration (19). Since analysis of type (i) dynamics will be covered elsewhere in this book (for example, in the photobleaching methods described in Chapter 6), we will focus on the analysis of type (ii) dynamics. Analysis of type (iii) dynamics is more specific and advanced, and readers are referred to other literature (20).

2.3.1. Principles of object tracking

In any tracking method, the ultimate goal is to obtain the position coordinates of a target object in each image frame, so that the movement of the target object can be represented as changes in the position coordinates. Velocity and movement direction can then be calculated from the resulting coordinate series.

The process of tracking has two major steps. In the first step, the object must be segmented. This then enables us to calculate the coordinates of the object position, such as its centroid. There are various ways to carry out segmentation (see section 2.1.1.). In the second step, successive positions of the object must be linked to obtain a series of coordinates of that object. We call this process 'position linking'. Position linking is also complicated when there are too many similar objects. In this case, it becomes impossible to identify the object in the next time point among the multiple candidates. Position linking also becomes difficult when successive positions of the object are too far apart. This happens when the time interval between frames is too long. In this case, multiple candidates may appear in the next time point, even though similar objects are only sparsely present in the image frame. If the target object is single and unique, linking of coordinates to successive time points has none of these problems. In this chapter, we will assume that the user has acquired images with a sufficient temporal and spatial resolution, so that the successive points can be linked by the nearest-neighbor principle, i.e. linking an object in a frame to the nearest object in the next frame. For a further description of position linking, readers are referred to other literature (20).

2.3.2. Manual tracking

In the simplest case, tracking can be done manually. The user can read out the coordinate position of the target object directly from the imaging software. (In the case of IMAGEJ, use the selection tool and move the cross-hair pointer over the object. The position coordinates of the pointer will be indicated in the status bar of the IMAGEJ window.) The user can list the coordinates for each position in standard spreadsheet software such as Microsoft EXCEL for further analysis. An IMAGEJ plug-in is also freely available to assist this simple way of tracking (see *Protocol 4*). An obvious disadvantage of manual tracking is that mouse clicking by the user could be imprecise in its position. An estimation of the measurement errors can be obtained by tracking the same object several times. The resulting error can then be indicated together with the results.

2.3.3. Automatic tracking

Automatic tracking reduces the workload of the researcher, enabling them to deal with a huge amount of data. Statistical treatments can then be more reliable. In addition, the results can be regarded as more objective than manual tracking. Automatic tracking is an optional function that can be added on to some imaging software. However, these readily available functions are not adaptable for all analyses, since the characteristics of target objects in biological research vary greatly. In particular, when the target object changes shape over time, further difficulties arise. For these reasons, automatic tracking often requires custom programs. In IMAGEJ, Pascal-like macro programming is possible and there are many programming functions that enable such automated segmentation and position measurements.

Basic tracking programs can be coded relatively easily in the form of script. The main part of the program will be segmentation of the target object (see section 2.1.1). The target object position is then measured and recorded. This main part will be iterated for every frame in the image stack. If there are multiple target objects in each frame, a position-linking algorithm must be considered (see section 2.3.1).

Centroid method

The centroid is the average of all pixel coordinates inside the segmented object and is the most commonly used feature for representing the object position. The centroid coordinate $p_c(x, y)$ can be calculated as

$$p_c(x,y) = \left(\frac{\sum x_i}{n_x}, \frac{\sum y_j}{n_y} \right) \Bigg|_{x,y \in \Re} \quad \text{(Equation 1)}$$

where $\Re$ is the region surrounded by the contour and n is the total number of pixels within that region. In other words, all x coordinates of the pixels inside the object are added and averaged. The same happens for all y coordinates. The derived averages for x and y coordinates represent the centroid coordinates of the object. The position of the object can be measured for every frame of the sequence manually. In IMAGEJ, a centroid calculation is available as the 'Particle Analysis' function (see *Protocols 3* and *5*).

Gaussian fitting method

For spherical signals such as fluorescence beads or fluorescently labeled subresolution particles, the signal intensity distribution can be fitted to a standard 2D Gaussian curve (21–23):

$$I(x,y) = z_0 + z_n \exp\left\{ -\frac{(x - X_n)^2 + (y - Y_n)^2}{W_n^2} \right\} \quad \text{(Equation 2)}$$

where $I(x, y)$ is the intensity distribution of an image, z_0 is the background intensity, and z_n is the height of the peak. W_n is the width of the curve that peaks at (X_n, Y_n). This peak position is the signal position (see *Fig. 1*, also available in color section). Although this fitting method is restricted to spherical or oval objects, it yields the most precise measurements, even with a very low signal-to-noise ratio (24). Another advantage of the Gaussian fitting method is that the results are in subpixel resolution. Such high resolution can, for example, enable the analysis of molecular diffusion at nanometer resolution. For further discussion on localization accuracy in position measurements, see (25–27).

Pattern matching method

In this method, a kernel containing a template pattern of the object is compared with different positions within the image in order to find the position with the highest similarity to the kernel. A cross-correlation function, $C(x, y)$, is usually used to evaluate the resemblance of the template to different parts of the image (28):

$$C(x,y)=\sum_{i=0}^{n-1}\sum_{j=0}^{m-1} I(x+i,y+j)\{K(i,j)-\bar{K}\} \qquad \text{(Equation 3)}$$

where $I(x, y)$ is the intensity distribution of the image frame and $K(i, j)$ is an $n \times m$ pixels kernel that contains the template image pattern. $\bar{K}$ is the mean intensity of the kernel. $C(x, y)$ will be maximal at the position where the pattern best matches. In actual application, the template pattern is sampled in the kth image frame $I_k(x, y)$ by the user manually or by semi-automatic segmentation. The cross-correlation value $C(x, y)$ between this template kernel and the consecutive $k+1$th image frame $I_{k+1}(x, y)$ can then be calculated (see *Fig. 2*). Gelles *et al.* (28) introduced the following formula for obtaining the peak position (x_c, y_c):

$$x_c=\frac{\sum x\{C(x,y)-T\}}{\sum\{C(x,y)-T\}} \qquad \text{(Equation 4)}$$

$$y_c=\frac{\sum y\{C(x,y)-T\}}{\sum\{C(x,y)-T\}} \qquad \text{(Equation 5)}$$

where T is the threshold value and negative values are discarded. (x_c, y_c) corresponds to the centroid of the magnitude of correlation that is thresholded by T.

Since the cross-correlation function (Equation 3) tends to give higher values at brighter regions rather than at regions of similar shape, a normalized form of cross-correlation can also be used:

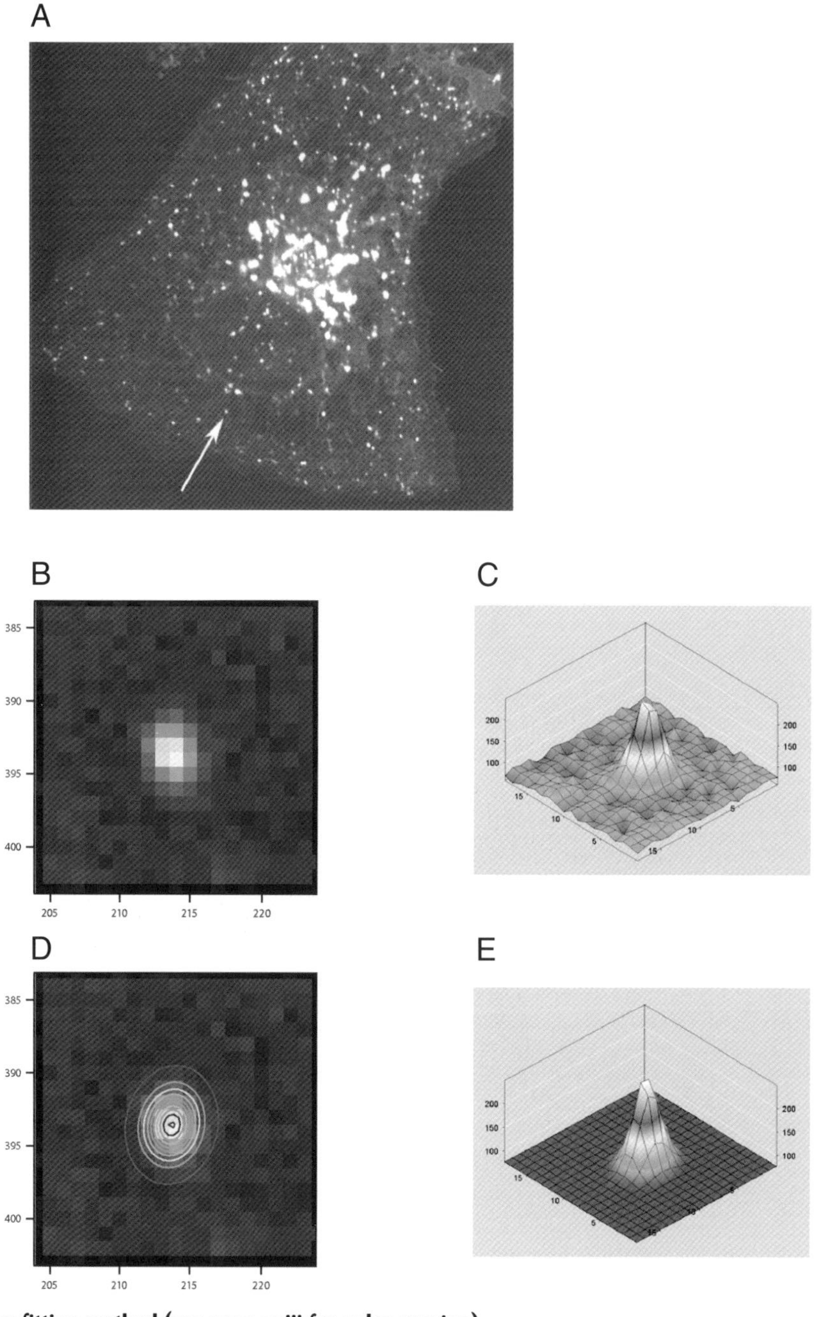

Figure 1. Gaussian fitting method (see page xviii for color version).
In (*A*), a Vero cell expressing vesicular stomatitis virus G protein fused to yellow fluorescent protein is shown, a model protein frequently used for studying intracellular vesicle transport. The arrow indicates a vesicle, which is shown enlarged (*B*) and by 3D plotting (*C*). In (*C*), signal intensity was converted to the height in the *z*-axis. Fitting of this vesicle image to a 2D Gaussian equation in IGOR PRO resulted in images (*D*) and (*E*), showing the contour and 3D plot of the fitted curve, respectively. Note how well the fitting works, as one can see by the 3D shape of the peak. From the fitted parameters of the 2D Gaussian curve, the position coordinates of the vesicle was determined as (213.7, 393.7).

$$C_N(x,y)=\frac{\sum_{i=0}^{n-1}\sum_{j=0}^{m-1}[I(x+i,y+j)-\bar{I}]\{K(i,j)-\bar{K}\}}{M_{Ix,y}\cdot M_k} \quad \text{(Equation 6)}$$

$$M_{Ix,y}=\sqrt{\sum_{i=0}^{n-1}\sum_{j=0}^{m-1}[I_{x+i,y+j}]^2} \quad M_k=\sqrt{\sum_{i=0}^{n-1}\sum_{j=0}^{m-1}[K_{i,j}]^2}$$

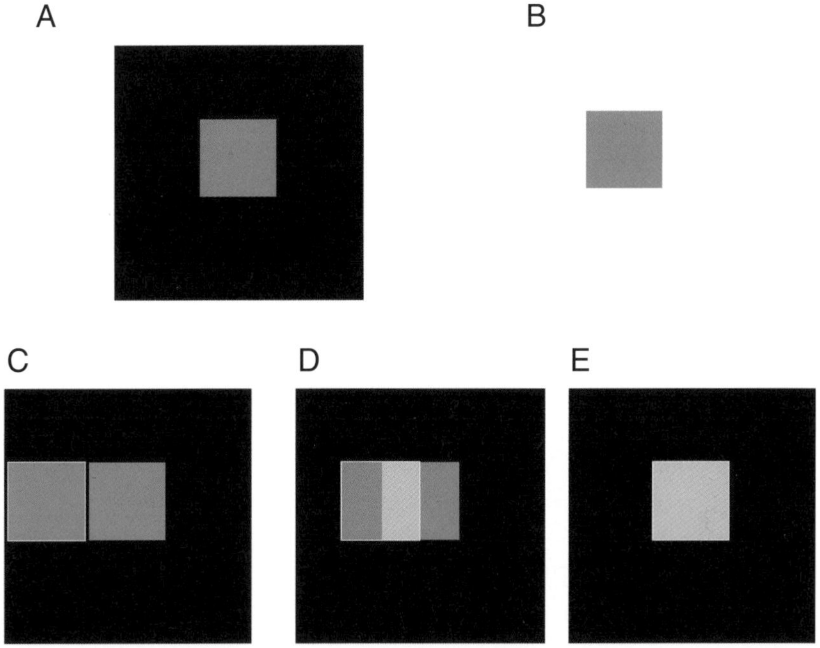

Figure 2. Principle of image cross-correlation calculation.
To obtain the position of a square object in an image frame (*A*; *I*(*x*, *y*)) by the cross-correlation method, a kernel containing a template image (*B*; *K*(*i*, *j*)) is first prepared. The cross-correlation value between the template pattern and various positions within the image frame (*A*) is then calculated (*C*–*E*). As an example, we assume that the square object is 4 × 4 pixels (total area = 16 pixels), the background intensity is 0, and the object has an arbitrary signal intensity of 10. When there is no overlap between the template kernel and the image frame (*C*), the cross-correlation value is 0, since:

10 [kernel intensity] × 0 [image frame intensity] × 16 [pixels] = 0.

When half of the area (8 pixels) overlaps (*D*), the cross-correlation value is:

(10 [kernel intensity] × 10 [image frame intensity] × 8 [pixels]) + (10 [kernel intensity] × 0 [image frame intensity] × 8 [pixels]) = 800.

When there is a perfect overlap, the cross-correlation value is:

10 [kernel intensity] × 10 [image frame intensity] × 16 [pixels] = 1600.

By searching the position coordinates of the template kernel where the cross-correlation value is the highest, the position of the object within the image frame can be determined.

where $\bar{I}$ is the mean intensity of a portion of the image overlapping the kernel, and $M_{Ix,y}$ and M_k are the root mean square values of the kernel and the corresponding portion of the image, respectively. The precision of the measurement is high (24), but object tracking will fail when the shape of objects changes radically between frames.

The cross-correlation method has been used extensively in single-particle tracking (SPT). SPT is a technique developed for measuring the mobility of membrane-bound proteins and the movement of motor proteins with nanometer precision (28–31) and has recently been reviewed (32). In these studies, a single protein was attached to a very small gold particle or labeled with a fluorophore and its movement was analyzed by video microscopy. Theoretical examinations showed that different modes of protein movement could be discriminated with nanometer resolution by measuring the mean square displacement of the labeled proteins (33). Various types of membrane protein motions, such as immobile, directed, confined, tethered, normal diffusion, and anomalous diffusion, were resolved, revealing the kinetics of the membrane protein mobility (34, 35). An automatic tracking program for multiple proteins has been developed by Ghosh & Webb (36) and used for measuring the movement of actin patches in yeast (37).

2.3.4. Summarizing data

Tracking of a moving object results in a list of position coordinates. This list can be saved as a text file and imported to spreadsheet software such as Microsoft EXCEL. The instantaneous velocity is derived by calculating the distance between consecutive time points. For example, if the position of an object at time point t is (x_t, y_t) and the object moves to a position (x_{t+1}, y_{t+1}) in the next time point $t + \Delta t$, then the instantaneous velocity v is:

$$v = \frac{\sqrt{(x_{t+1} - x_t)^2 - (y_{t+1} - y_t)^2}}{\Delta t} \qquad \text{(Equation 7)}$$

For each frame, the instantaneous velocity can be calculated. One way to summarize the data is to average the resulting instantaneous velocities and append its standard deviation. In most cases in biology, velocity changes with time, and this dynamic can be studied by plotting instantaneous velocity vs. time on a graph. Such plotting reveals characteristics of the movement, such as acceleration kinetics or periodicity.

Movements within organisms can be both random and directed. To make a clear distinction between these different types of movement, mean square displacement (MSD) plotting is a powerful method. Although this method has been extensively used in studying molecular diffusion within the plasma membrane, it can also be used at other levels such as bacterial swimming or cell movement within tissues (34, 38–41).

The most basic equation that describes the diffusion is:

$$<r^2> = 4D\tau \qquad \text{(Equation 8)}^1$$

where $<r^2>$ is the MSD, D is the diffusion coefficient, and τ is the time scale. MSD is calculated by squaring the net distance a molecule moves during the period of time τ. For example, if an object moves from (0, 0) to (x, y) during a period of τ, the square displacement will be:

$$r^2 = x^2 + y^2$$

or

$$r = \sqrt{x^2 + y^2} = \sqrt{4D\tau} \qquad \text{(Equation 9)}$$

If we have many samples, then we can average them and the MSD will be:

$$<r^2> = <x^2 + y^2> \qquad \text{(Equation 10)}$$

Equation 8 tells us that MSD ($<r^2>$) is proportional to τ, which means that when MSD is plotted against various τ, the plot will be a straight line with a slope $4D$ (see *Fig. 3a*). If the mobility of the target object is lower, then the slope becomes less steep (see *Fig. 3d*).

The mobility of molecules is not always a pure diffusion. In the case where there is constant background flow, such as laminar flow in the medium, molecules drift in a certain direction. This is called diffusion with drifts, or 'biased diffusion'. *Fig. 3*(*b*) is a typical MSD curve of this type. Since the flow causes the movement $v\tau$, where v is the flow rate, the displacement r will be:

$$r = \sqrt{4D\tau} + v\tau \qquad \text{(Equation 11)}$$

When MSD ($<r^2>$) is plotted against time, $v\tau$ causes an upward curvature of the graph.

Another mode of movement is 'constrained diffusion'. This happens when the diffusion is limited within a space. Consider a molecule diffusing in a bounded space. The molecule can diffuse normally until it hits the boundary. In such a case, the MSD curve displays a plateau such as that shown in *Fig. 3*(*c*). When τ is small, MSD is similar to pure diffusion, but as τ becomes larger, MSD becomes attenuated, since the displacement is hindered at a defined distance. Constrained diffusion is often observed with membrane proteins (35).

[1]This equation is for 2D mobility. In case of 3D mobility, $<r^2> = 6D\tau$.

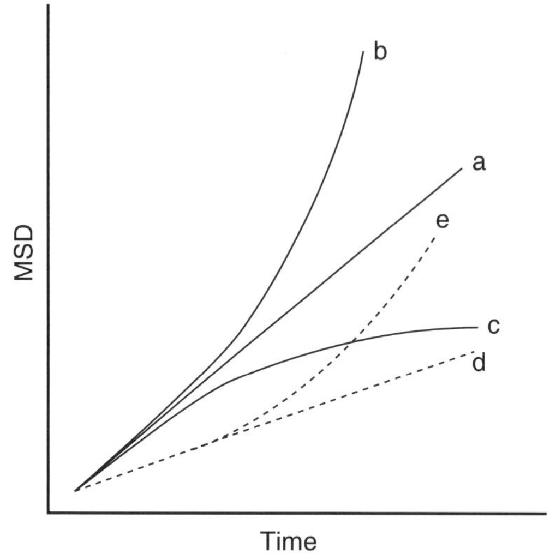

Figure 3. MSD plotting and different types of movement.
(*a*) Pure random walk. (*b*) Biased random walk (diffusion with drifts). (*c*) Constrained random walk. (*d*) Pure random walk, but with a lower diffusion coefficient. In the case of molecules, this could be due to larger molecule size. In the case of cells, this could be due to reduced migration activity. (*e*) As in (*b*), but with a lower diffusion coefficient.

2.3.5. Coding the automatic tracking programs

To create computer programs for automated tracking, we have used the IMAGEJ macro programming language and the IGOR PRO procedure programming language (WaveMetrics). Both programming languages are easier to use than general programming languages such as C or Java, and have sufficient capability for coding complicated image-processing algorithms. IMAGEJ has many built-in image-processing functions for macro programming (see *Protocol 9*). IGOR PRO is numerical calculation software similar to MATHEMATICA or MATLAB used by engineers. Either 2D- or 3D-image time series can be imported and treated in the program as 3D or 4D matrices. This enables many complex numerical calculations on the image stack. A wide range of mathematical functions is available for programming. A useful function of IGOR PRO is the 'History' window. Each menu command executed by the user will be listed in the History window, so the user can copy those commands and use them directly during programming.

We have added an appendix to show the actual algorithms for automated tracking by the Gaussian fitting method (see *Appendix*, section 6.1) and pattern matching method (see *Appendix*, section 6.2).

2.3.6. Choice of method

The choice of tracking method is dependent on the character of the movement of interest. There is no one tracking method that is superior to others. Manual tracking is easy to achieve and is the most convincing way to start out the tracking. During manual tracking, the researcher can also discover many details that could be informative for developing computer programs for automated tracking. The disadvantage of manual tracking is that it is labor-intensive. For deriving massive amounts of data, automated tracking is necessary.

Thresholding works well when the object has a clear signal against the background. For images with lower signal-to-noise ratios, the Gaussian fitting method or the pattern matching method tracks the target object more reliably than thresholding. The Gaussian fitting method results in subpixel resolution position coordinates, but the shape of the target object is limited to spheres and ovals. The pattern matching method results in position coordinates with less precision compared with Gaussian fitting, but the shape of the target object can vary. It is also possible to update the template image kernel for every time frame, which enables the tracking of target objects that change their shape over time.

2.3.7. Imaging software with tracking functions

Several popular commercial and noncommercial imaging software packages are equipped with tracking functions. For example, in METAMORPH (Molecular Devices), object tracking is available. In IMAGEJ, manual tracking plug-in and semi-automatic tracking plug-ins based on thresholding are freely available via the internet (4). For cell migration tracking, DIAS has been used extensively by researchers studying the model organism *Dictyostelium discoideum* (Soll Technologies). IMARIS (Bitplane) is software for 3D image visualization and has an add-on function, IMARISTRACK, which is equipped with 3D tracking tools, for both manual and automatic tracking based on threshold segmentation.

3. RECOMMENDED PROTOCOLS

Protocol 1

Shading correction

Equipment

- Fluorescence microscope
- Calibration slide (12)

Method

1. Using the same optics as for the sample measurement, focus and take an image from the calibration slide. Make sure noise is low in the correction image by taking an average of several frames.
2. Using IMAGEJ, divide each pixel of the correction image by the mean value of all pixels in the image to derive a 'deviation factor image'.
3. Make sure that the result is not affected by truncation, clipping, or rough rounding (the '32-bit results' option is recommended for this purpose).
4. Multiply the image of the object of interest by the 'deviation factor image'.

Protocol 2

Estimating pixel size

Equipment

- Microscope
- Calibration scale

Method

1. Using the same optics as for the image measurement, focus and take a transmission light image from the calibration scale.
2. Make sure the detector is not saturated, i.e. avoid having areas of pixels containing the maximum possible intensity value (for example, 255 in an 8-bit image).
3. Align the scale parallel to the *x*- or *y*-axis of the image.
4. Save the image (TIFF format).
5. Load the image into IMAGEJ. Optionally switch the look-up table to 'Spectrum' to emphasize bright pixels.
6. Using the 'Line tool' of IMAGEJ, select a line perpendicular to the scale in the image, ideally crossing several grid lines of the scale. For beads, make a line selection crossing the brightest point of the bead and spanning twice the diameter of the bead.
7. From the 'Analyze' Menu of IMAGEJ, select 'Plot Profile'. From the resulting graph, read the *x*-distance from peak to peak by moving the curser in the graph. The real distance of the grid lines, indicated on the scale, is now divided by the number of pixels between two peaks to derive the distance per pixel.

The accuracy of the distance measurements depends on the pixel resolution of the image. If the pixel resolution is low, the measured distance could be larger or smaller than the actual distance. Averaging several measurements of different parts of the scale or different beads will improve the accuracy.

Protocol 3

Measurement of shape and intensity parameters

Equipment

- Fluorescence microscope

Method

1. Make sure that the optical elements of the microscope are in good alignment.
2. Take an image without saturating the detector.
3. Load the image into IMAGEJ.
4. Correct for shading correction (see *Protocol 1*).
5. Set the threshold by choosing 'Adjust...Threshold' in the IMAGEJ menu.
6. In the IMAGEJ menu, choose 'Measure...Set Measurements'. Make sure the checkbox 'Limit to Threshold' is checked.
7. Select the parameters of interest from the list.
8. In the IMAGEJ menu, choose 'Measure'. The following parameters are measured and displayed in the results window:

   Area, i.e. the area of the selection in square pixels. The area is in calibrated units, such as square millimeters, if 'Analyze/Set Scale' was used to calibrate the image spatially (see *Protocol 2*).

   Min & Max Gray Level, Mean Gray Value, Standard Deviation, Modal Gray Value. The modal gray value is the most frequently occurring gray value within the selection.

   Centroid. The center point of the selection (see Centroid method in section 2.3.3).

   Center of Mass, i.e. the brightness-weighted average of the x and y coordinates of all pixels in the selection.

   Perimeter, i.e. the length of the outside boundary of the selection.

   Bounding Rectangle, i.e. the smallest rectangle enclosing the selection.

   Circularity. 4π(area/perimeter^2). A perfectly round object would return a circularity measure of 1.

   Feret's Diameter, i.e. the longest distance between any two points along the selection boundary.

Protocol 4

Simple tracking by clicking

Method

Manual tracking can be assisted by a freely downloadable IMAGEJ plug-in, 'Manual Tracking' (42). This plug-in enables the user to record *x-y* coordinates of the position where the user clicked using the mouse in each frame within a stack. The download website has detailed instructions on how to use this macro, as well as worked examples of how the data will be presented and the numerical information that it generates.

Protocol 5

The 'Particle Analysis' function in IMAGEJ

Method

The 'Particle Analysis' function in IMAGEJ is useful if the object can be segmented by thresholding (see section 2.1.2). Calculation using the 'Particle Analysis' function measures many parameters of many objects (particles) within the image frame automatically including the centroid, maximum, minimum, average, and standard deviation of the intensity, circularity, and perimeter. A specific range of object sizes (in the pixel area) can be set in the option window for defining the object size. The size-setting capability is useful for cutting off background noise, which may otherwise be considered as target objects.

Protocol 6

Manual edge tracing

Method

Manual edge tracing to segment the target object can be done using the 'Segmented Line Selection' tool or the 'Freehand Line Selection' tool in IMAGEJ. The user first draws the boundary as a closed region of interest. The boundary is then 'filled' by foreground color (e.g. in case of black background, white color). The closed contour of the object traced in this way can be easily recognized numerically by thresholding and the 'Particle Analysis' function (see *Protocol 5*). A slightly more efficient way is to use the NEURONJ plug-in of IMAGEJ (43). Although the name of this plug-in sounds like a specific tool for analyzing nerve cells, the plug-in generally assists the user in tracing the edge of target objects more easily than doing it in a fully manual way. It has a function similar to the magnetic lasso tool in Adobe PHOTOSHOP, called the 'Local Snapping' function. During tracing of the object edge, this function automatically snaps the selection line on to the object edge while the user is tracing.

Protocol 7

The 'Wand Tool' in IMAGEJ

Method

The 'Wand Tool' function of IMAGEJ is also a useful tool for tracing object edges and measuring the parameters of the object, if the object has uniform pixel intensity. By clicking anywhere within the object, this tool does an automatic tracing of uniform pixel values. In combination with the 'Measure' function, the wand tool enables analysis of objects with complicated shapes (also see *Protocol 3*).

Protocol 8

Automatic tracking using IMAGEJ plug-ins

Method

Two automatic object-tracking plug-ins are available in IMAGEJ. 'Object Tracker' uses the IMAGEJ native function 'Particle Analysis' to segment the object by intensity thresholding and linking the successive positions by the nearest-neighbor principle (44). The 'MultiTracker' plug-in is an extension of the 'Object Tracker' plug-in and can be used for tracking multiple objects (45). For both plug-ins, the user must first manually label the object to be tracked.

Protocol 9

Macro recording in IMAGEJ

Method

A useful function for the programming is the IMAGEJ 'Command Recorder'. Each menu command executed by the user will generate a macro function[2] and will be listed in the recorder window. These commands can be copied and pasted on to the macro program that the user is coding.

[2]'run([command])' is generated by the recorder. [command] contains words that correspond to one of the menu commands in IMAGEJ.

4. TROUBLESHOOTING

4.1. Probable sources of error in measurements

In the following, we will list measurement errors that are frequently encountered by biologists and explain their causes and possible solutions.

4.1.1. Aberrations

Measurement imprecision can arise from two types of aberration – visual artifacts, and as a result of defects in the objective lens in the microscope system.

Spherical aberration seriously affects image sharpness and intensity. Since waves passing near the periphery of the lens are refracted to a greater degree, light passing through different parts of the lens will end up at different focal points, causing spherical aberration. Recent high-quality microscopes are equipped with improvements in the lens material and architecture to suppress spherical aberration. However, even with such spherical aberration-corrected optics, users can still cause the artifact by utilizing an incorrect set-up for the specimen, such as using unmatched mounting medium and immersion medium. In addition, when using high-magnification, high-numerical-aperture dry objectives, the user must use the correct thickness for the coverslip (generally 0.17 mm). An objective equipped with a correction collar enables the user to correct for different coverslip thickness by adjusting the distance between the lens and the coverslip.

The lens can also cause chromatic aberration. This is due to variations in the refractive indices over the range of frequencies within visible light. Blue light is refracted to the greatest extent, followed by green and red light. Dispersion causes different image sizes and focal points for different colors. Measurements of length, area, and volume then become unreliable in different colors. There are two types of chromatic dispersion: (1) axial chromatic aberration causes different focal points for different colors and results in a white point ringed with color, and (2) lateral chromatic aberration is caused by off-axis ray fluxes being dispersed according to the wavelength, and results in a series of overlapping images varying in both size and color. Different-quality objectives differ in how well they are corrected for chromatic aberration. Between the achromatic and apochromatic type correction, there are also objectives known as semi-apochromats, or fluorites. The fluorites cost less, but correct almost as well as the apochromats.

4.1.2. Reproducibility

In addition to the intrinsic problems of using microscopes as measuring devices, there can also be a strong heterogeneity of responses from the same instrument, and variations from instrument to instrument, making it a challenging task to reproduce intensity measurements from day to day and place to place. It is only recently that the need for certain specifications and standards for performance have been recognized to be an urgent need (46).

4.1.3. Possible errors in morphometry

Misrepresentations can result from inhomogeneities of the sample, embedding medium, and air, since light has to pass through them. Distances measured inside the sample will only be precise if the microscope is in good alignment and if the embedding medium is of the specified refractive index that the objective requires. A biological specimen composed of a mixture of water, proteins, and lipids, when imaged with an oil-immersion objective, will appear compressed in the axial direction compared with the horizontal direction of the object, while using an air-immersion objective with the same sample will result in elongation of the object. These misestimations of length or distance will consequently result in a misestimation of derived sizes such as area and volume. Depending on the distance the light travels in the 'wrong' medium and the numerical aperture of the objective, spherical aberrations (see section 4.1.1) can severely degrade image brightness and quality.

Volume, surface, and area measurements are available as standard features in many commercial and noncommercial software packages. However, rarely will the outline of a biological structure follow a straight line, and thus the measurement will produce an apparent length (see section 2.2.2), which will strongly affect surface estimations. Area and volume measurements will mostly depend on accurate settings of the threshold (see section 2.1.2).

4.1.4. Errors in measurement of time series and solutions

Bleaching is a typical problem when imaging fluorescently labeled proteins. The fluorescence intensity decreases during image acquisition due to illuminating light. Since the images in the latter frames become darker than those at the beginning of the sequence, the thresholding approach may not segment the target object correctly in the latter frames. A similar problem may arise when there is a frame with brighter or darker average intensity. In fluorescence imaging, this could happen due to either accidental timing failure of the camera shutter or fluctuation of the irradiation light. In both cases, the average intensity of each image frame can be corrected against the first frame to equalize the intensity level. One of us has written an intensity correction function as a macro program in IMAGEJ. The macro 'Bleaching Correction' can be downloaded from the internet (47).

Acknowledgements

We thank M. Rosa Ng for going through the text for grammatical corrections.

5. REFERENCES

1. **Miyawaki A, Nagai, T & Mizuno, H** (2005) In *Microscopy Techniques*, pp. 1–17. Edited by J Rietdorf. Springer-Verlag, Heidelberg.
2. **Russ J** (1998) *The Image Processing Handbook*, 3rd edn. CRC Press, Boca Raton, FL.

3. **Darzynkiewicz Z, Roederer M & Tanke HJ** (eds) (2004) *Methods in Cell* Biology, Vol. 75. *Cytometry, New Developments.* Elsevier Academic Press, Amsterdam.

★★ 4. **Rasband WS** (1997–2005) *ImageJ.* National Institutes of Health, MD. Available at: http://rsb.info.nih.gov/ij/ – *This is the download site for* IMAGEJ. *This software, written by Wayne Rasband, is described throughout this chapter.*

5. **Wilhelm S, Gröbler B, Gluch M & Heinz H** (2004) *Confocal Laser Scanning Microscopy: Principles.* Carl Zeiss, Jena.

6. **Stelzer E & Grill S** (2000) *Opt. Commun.* **173**, 51–56.

7. **Pawley J** (2000) *BioTechniques*, **28**, 884–888.

8. **Pawley J** (1994) In: *Three-Dimensional Confocal Microscopy: Volume Investigation of Biological Specimens*, pp. 47–94. Edited by J Stevens, L Mills & J Trogadis. Academic Press, San Diego, CA.

9. **Cox GC** (1999) In: *Confocal Microscopy, Methods and Protocols,* pp. 357–371. Edited by SW Paddock. Humana Press, Totowa, NJ.

10. **Lakowicz JR** (1999) *Principles Of Fluorescence Spectroscopy.* Kluwer Academic/Plenum Publishers, New York.

11. **Becker P** (1996) In: *Fluorescence Imaging Spectroscopy and Microscopy,* pp. 1–30. Edited by X Wang and B Hermann. John Wiley & Sons, New York.

12. **Zwier JM, van Rooij GJ, Hofstraat JW & Brakenhoff GJ** (2004) *J. Microsc.* **216**, 15–24.

13. **Sezgin S** (2004) *J. Electronic Imaging,* **13**, 146–165.

14. **Sibarita J-B** (2004) In: *Microscopy Techniques,* pp. 201–244. Edited by J. Rietdorf. Springer, Heidelberg.

★★ 15. **Rasband WS** (1997–2005) *ImageJ documentation.* National Institutes of Health, MD, USA. Available at: http://rsb.info.nih.gov/ij/docs/index.html – *In addition to the download site, full documentation for* IMAGEJ *is maintained at the URL given in (4).*

16. **Stelzer EHK** (1998) *J. Microsc.* **189**, 15–24.

17. **Hirschberg K, Miller CM, Ellenberg J, *et al.*** (1998) *J. Cell Biol.* **143**, 1485–1503.

18. **Murase K, Fujiwara T, Umemura Y, *et al.*** (2004) *Biophys. J.* **86**, 4075–4093.

19. **Dormann D, Libotte T, Weijer CJ & Bretschneider T** (2002) *Cell Motil. Cytoskeleton,* **52**, 221–230.

20. **Miura K** (2005) In: *Advances in Biochemical Engineering/Biotechnology,* pp. 267–310. Edited by J Rietdorf. Springer-Verlag, Heidelberg.

21. **Schütz GJ, Schindler H & Schmidt T** (1997) *Biophys. J.* **73**, 1073–1080.

22. **Anderson CM, Georgiou GN, Morrison IE, Stevenson GV & Cherry RJ** (1992) *J. Cell Sci.* **101**, 415–425.

23. **Tardin C, Cognet L, Bats C, Lounis B & Choquet D** (2003) *EMBO J.* **22**, 4656–4665.

24. **Cheezum MK, Walker WF & Guilford WH** (2001) *Biophys. J.* **81**, 2378–2388.

25. **Ober RJ, Ram S & Ward ES** (2004) *Biophys. J.* **86**, 1185–1200.

26. **Martin DS, Forstner MB & Kas JA** (2002) *Biophys. J.* **83**, 2109–2117.

27. **Thompson RE, Larson DR & Webb WW** (2002) *Biophys. J.* **82**, 2775–2783.

28. **Gelles J, Schnapp BJ & Sheetz MP** (1988) *Nature,* **331**, 450–453.

29. **Geerts H, de Brabander M, Nuydens R, *et al.*** (1987) *Biophys. J.* **52**, 775–782.

30. **Schnapp BJ, Gelles J & Sheetz MP** (1988) *Cell Motil. Cytoskeleton,* **10**, 47–53.

31. **Sheetz MP, Turney S, Qian H & Elson EL** (1989) *Nature,* **340**, 284–288.

32. **Ritchie K & Kusumi A** (2003) *Methods Enzymol.* **360**, 618–634.

33. **Qian H, Sheetz MP & Elson EL** (1991) *Biophys. J.* **60**, 910–921.

34. **Kusumi A, Sako Y & Yamamoto M** (1993) *Biophys. J.* **65**, 2021–2040.

35. **Saxton MJ** (1997) *Biophys. J.* **72**, 1744–1753.

36. **Ghosh RN & Webb WW** (1994) *Biophys. J.* **66**, 1301–1318.

37. **Carlsson AE, Shah AD, Elking D, Karpova TS & Cooper JA** (2002) *Biophys. J.* **82**, 2333–2343.

38. **Berg H** (1993) *Random Walk in Biology.* Princeton University Press, Princeton, NJ.

39. **Saxton MJ & Jacobson K** (1997) *Annu. Rev. Biophys. Biomol. Struct.* **26**, 373–399.

40. **Suh J, Dawson M & Hanes J** (2005) *Adv. Drug Deliv. Rev.* **57**, 63–78.

41. **Witt CM, Raychaudhuri S, Schaefer B, Chakraborty AK & Robey EA** (2005) *PLoS Biol.* **3**, e160.

42. **Cordelieres F** (2004) *Manual Tracking, ImageJ plug-in.* National Institutes of Health, MD, USA. Available at: http://rsb.info.nih.gov/ij/plugins/manual-tracking.html

43. **Meijering E** (2004) *NeuronJ, ImageJ plug-in.* National Institutes of Health, MD, USA. Available at: http://www.imagescience.org/meijering/software/neuronj/
44. **Rasband WS** (2000) *Object Tracker, ImageJ plug-in.* National Institutes of Health, MD, USA. Available at: http://rsb.info.nih.gov/ij/plugins/tracker.html
45. **Kuhn J** (2001) *Multitracker, ImageJ plug-in.* National Institutes of Health, MD, USA. Available at: http://rsb.info.nih.gov/ij/plugins/multitracker.html
46. **Ghauharali RI & Brakenhoff GJ** (2000) *J. Microsc.* **198**, 88–100.
47. **Rietdorf J** (2005) *Bleaching Correction, ImageJ plug-in.* National Institutes of Health, MD, USA. Available at: http://www.embl.de/eamnet/html/bleach_correction.html

6. APPENDIX

6.1. Gaussian fitting method

Sample IGOR PRO procedure functions for fitting a 2D Gaussian curve are shown below. `src3Dwave` is the 2D time series, or 3D matrix (*x*, *y*, *t*). `xleft`, `xright`, `ytop`, `ybottom` are the top left, top right, bottom left, and bottom right *x-y* coordinates, respectively, of the fitting region at the time point `layer`. The first function `gaussfitterInitial` is used to generate guesses of the Gaussian curve width and offset. These values are used for the second fitting of the Gaussian curve, `gaussfitter`, and are also inherited to the next time point, assuming that the shape of the object does not change much.

```
function
gaussfitterInitial(src3Dwave,xleft,xright,ytop,ybottom,layer)
wave src3Dwave
variable xleft,xright,ytop,ybottom,layer
duplicate/o/R=(xleft,xright)(ytop,ybottom)(layer,layer) src3Dwave
  temp2D
Redimension/N=(-1,-1) temp2D //converts 3D matrix to 2D

NVAR/z CurrentChisq
if (NVAR_exists(CurrentChisq)==0)
variable/g CurrentChisq
endif
CurveFit/q Gauss2D temp2D /D
CurrentChisq=V_chisq
wave/z W_coef
variable x0,y0
x0=W_coef[2]
y0=W_coef[4]
variable/g SetGwidth=(W_coef[3]+W_coef[5])/2
variable/g Setz0=W_coef[0]
variable/g SetA=W_coef[1]
printf 'Frame: %d x0: %g y0 %g Chisq:%g \r',layer,x0,y0,V_chisq
  printf '  z0 set to %g Gwidth %g\r',Setz0,SetGwidth
end
```

```
//Gauss width and Z0 are set constant
function gaussfitter(src3Dwave,xleft,xright,ytop,ybottom,layer)
wave src3Dwave
variable xleft,xright,ytop,ybottom,layer
NVAR/z SetGwidth,Setz0    //Global variables
duplicate/o/R=(xleft,xright)(ytop,ybottom)(layer,layer) src3Dwave
  temp2D
Redimension/N=(-1,-1) temp2D //to kill layer

   variable V_FitTol=0.01      //instead of default 0.001 050220
                               //set a tolerance for fractional
                               //decrease of chi-square from
                               //one iteration to the next
variable V_FitError=0
variable V_FitQuitReason
variable V_FitOptions=4   //suppresses curve fit window pop-out
NVAR/z CurrentChisq
K0 = Setz0; K3 = SetGwidth; K5 = SetGwidth; //values from guessing
CurveFit/H='1001010'/n/q Gauss2D temp2D /D
CurrentChisq=V_chisq
wave/z W_coef
variable x0,y0
x0=W_coef[2]
y0=W_coef[4]
if (V_FitError!=0)
printf 'Frame: %d x0: %g y0 %g',layer,x0,y0
      printf 'Error:%d reason %d\r',V_FitError,V_FitQuitReason
endif
return V_FitError
end
```

6.2. Pattern matching method

Sample IMAGEJ macro functions for deriving the 2D normalized cross correlation are as follows. *a* and *b* are 1D arrays containing pixel intensities transformed from the template kernel and a region of interest within the scanning range. `returnsCrossCorMvalue(a,b)` returns the cross-correlation value of these two regions.

```
function returnsCrossCorMvalue(a,b) {
arraysize=a.length;
kbar=findAVEofArray(a);
Mvalue=returnMvalue(a)*returnMvalue(b); //Multiplication of RMS
ccval=0;
for (j=0;j<arraysize;j++) {
```

```
ccval+=(a[j]-kbar)*b[j];
}
ccval/=Mvalue;
return ccval;
}
// finds the average of an array.
function findAVEofArray(a) {
size=a.length;
intsigma=0;
for(i=0;i<size;i++) {
intsigma+=a[i];
}
intsigma/=size;
return intsigma;
}

// root-mean-squared of an array
function returnMvalue(a){
size=a.length;
Isigma=0;
rmsInt=0;
for(i=0;i<size;i++) {
Isigma+=pow(a[i],2);
}
rmsInt=pow(Isigma,0.5);
return rmsInt;
}
```

CHAPTER 4
Transmitted light imaging

Jon D. Lane and Howard Stebbings

1. INTRODUCTION

Since most cells are too small to be seen with the unaided eye, and as the components within cells are even smaller, it is clearly necessary to employ microscopes to study cellular structure and dynamics. Since their invention some 400 years ago, light microscopes have become increasingly versatile and sophisticated. However, the visual detail achievable using light microscopes remains constrained by the resolving power of the instrument, where the resolution obtainable is the closest distance together that two points can be and still be detected as separate. This depends upon the wavelength of the illuminating source – in this case, light in the visible range – and the extent to which light is accepted by the objective lens, denoted by its numerical aperture (NA). Based on these parameters, and provided that high optical quality (NA) lenses and condensers are employed (see *Protocol 1*), resolution approaching a theoretical optimum of ~200 nm can be achieved. Better resolution than this, and indeed a greater magnification, can only be achieved using specialized optical methods such as stimulated emission depletion (1), or nonoptical techniques such as electron microscopy, where the wavelength of the illuminating source is significantly shorter. The latter, while invaluable for studying ultrastructural detail of cells and organelles, has considerable limitations, as it necessarily requires fixed and prepared specimens.

Standard bright-field microscopy is used extensively for histology and pathology applications, where fixed samples can be stained with appropriate dyes. Most living imaging subjects, however, do not provide inherent color or contrast, and so are not well defined by bright-field microscopy. Consequently, procedures that enhance the contrast of transmitted light images of living samples are essential for dynamic studies of cellular processes. Effective microscopy techniques do exactly this – they exploit the physical properties of light to apply contrast to translucent samples and in doing so allow one to *see* (but not necessarily to resolve) small and/or low-contrast objects. For some fields – such as mitosis and apoptosis (see section 2.2.2 and *Protocol 2*) – sufficient information can often be attained from contrast-enhanced transmitted light images using methods such as phase contrast or differential interference contrast to satisfy

Cell Imaging: ***Methods Express*** (D. Stephens, ed.)

certain experimental objectives (see sections 2.2 and 2.3), while less commonly used techniques, such as dark-field or polarized light microscopy, are useful for more specialized applications (see sections 2.4 and 2.5). In many cases, the real value of transmitted light imaging is in its capacity to generate essential morphological information to complement fluorescence imaging of proteins or organelles in dynamic studies of cell function, especially those employing green fluorescent protein (GFP). This combination – especially when enhanced by video imaging – has undoubtedly contributed to the dramatic expansion in the application of live-cell imaging to diverse biological problems in recent years.

Transmitted light microscopy has therefore undergone something of a renaissance during the latter years of the 20th century and the early stages of the 21st, but what can one hope to achieve using this technology and what are the various advantages of one technique over another? This chapter will cover the main transmitted-light microscopy techniques in cell biology, how to set up microscopes in general terms for particular imaging methods, and which techniques are more appropriate for particular applications. In doing so, we will illustrate each approach with published examples of its usage, with emphasis placed on our own research experience. To keep access broad, and due to space constraints, we will explore just a few working examples, and we direct the interested reader to some recent reviews that go into much greater detail about techniques that we will cover (e.g. 2–8).

2. METHODS AND APPROACHES

2.1. The light microscope optical system and bright-field microscopy

In all light microscopes there is a system of glass lenses – the condenser lens that focuses light from the source on to the specimen, the objective lenses that magnify the specimen to different degrees, and the eyepiece lenses that further magnify the primary image. With this arrangement of lenses, one can carry out what is known as bright-field microscopy. This basic form of microscopy is useful if the biological specimens under investigation are pigmented or colored; variously colored vacuoles and green chloroplasts in plant cells are obvious examples, as are the red 'eyespots' in certain protozoa and the pigment granules in vertebrate melanophores (9).

Most cell components are not colored, however, and in these instances contrast needs to be applied to the specimen by staining if bright-field microscopy is to be used. Vital stains such as Janus Green B, which stains mitochondria in living cells, do exist, but in many instances staining can only be carried out following fixation, and often also after embedding and sectioning. These procedures clearly render impossible any studies of cell dynamics. Nevertheless, numerous stains are available that identify a wide range of components within fixed cells with varying degrees of specificity, and while the emphasis on such techniques has waned with the development of more sophisticated approaches, some of these methods are still extremely valuable. For example, stains such as Schiff's reagent that identifies

DNA when applied by the Feulgen method can provide both qualitative and quantitative information. Hence, diploid nuclei stain twice as strongly as haploid nuclei, and microphotometric instrumentation can be used to detect the degree of ploidy in tissues and to compare the genome size (C-value) across a very wide range of invertebrate and vertebrate organisms (10). Meanwhile, the Giemsa stain is an important method for karyotyping (11), and such cytogenetic analysis of chromosome banding has become a standard, and sometimes even automated, way of recognizing not only particular chromosomes, but also inherited defects and abnormalities.

2.2. Phase contrast microscopy

With an increased emphasis on the study of cell dynamics – especially in the fields of motility, division, and cell death – techniques that allow visualization of living cells have become increasingly important. Improved contrast with transparent cells can be gained by closing, to an extent, the condenser diaphragm on a bright-field microscope (see section 2.7), but excessive closure has the deleterious consequence of effectively reducing the NA of the lens systems and thereby the resolution (see *Protocol 1*). To avoid this, all research microscopes can be converted to phase contrast optics. This allows the detection of different components in cells by virtue of slight variations in their refractive indices. The method is straightforward, and specimens mounted in their appropriate media can be observed directly. Invariably, this allows visualization not only of the outline of the cell, but also of many of its organelles and their arrangements (see *Fig. 1*).

Since the phase contrast method allows cells to be observed in their living state, this imaging approach has enhanced diverse fields. Images ranging from flagella beating on protozoa and amoeboid crawling movement to the dramatic cytoplasmic streaming exhibited by plant cells have all been obtained by phase contrast microscopy. More specialized applications have included the study of oocyte lampbrush chromosomes, which appear particularly striking when viewed using phase contrast (12). The ability to study such chromosomes in their native state has facilitated studies of the biophysical properties of different regions of the chromosome (13).

So, what can one hope to see by phase contrast microscopy? Taking the example of the mammalian cell in culture, obvious features are the cell margins and the nucleus – the nuclear envelope is well defined and chromatin domains are apparent – while membrane dynamics at the cell edge (such as ruffling, retraction, and extension of filopodia) are all well described (see *Fig. 1B*). Mitochondria are also clearly resolved (most notably towards the periphery where the cell flattens out) as dynamic, phase-dark, rounded, or elongated organelles (see *Fig. 1B* and *C*). The prominent profile of mitochondria under the phase contrast microscope has allowed the study of their distribution and movement within axons (see, for example, 14). As with most other imaging techniques, whether it is possible to identify particular organelles depends not only on their refractive qualities, but also on whether their shape is sufficiently distinct. Hence, the abundant

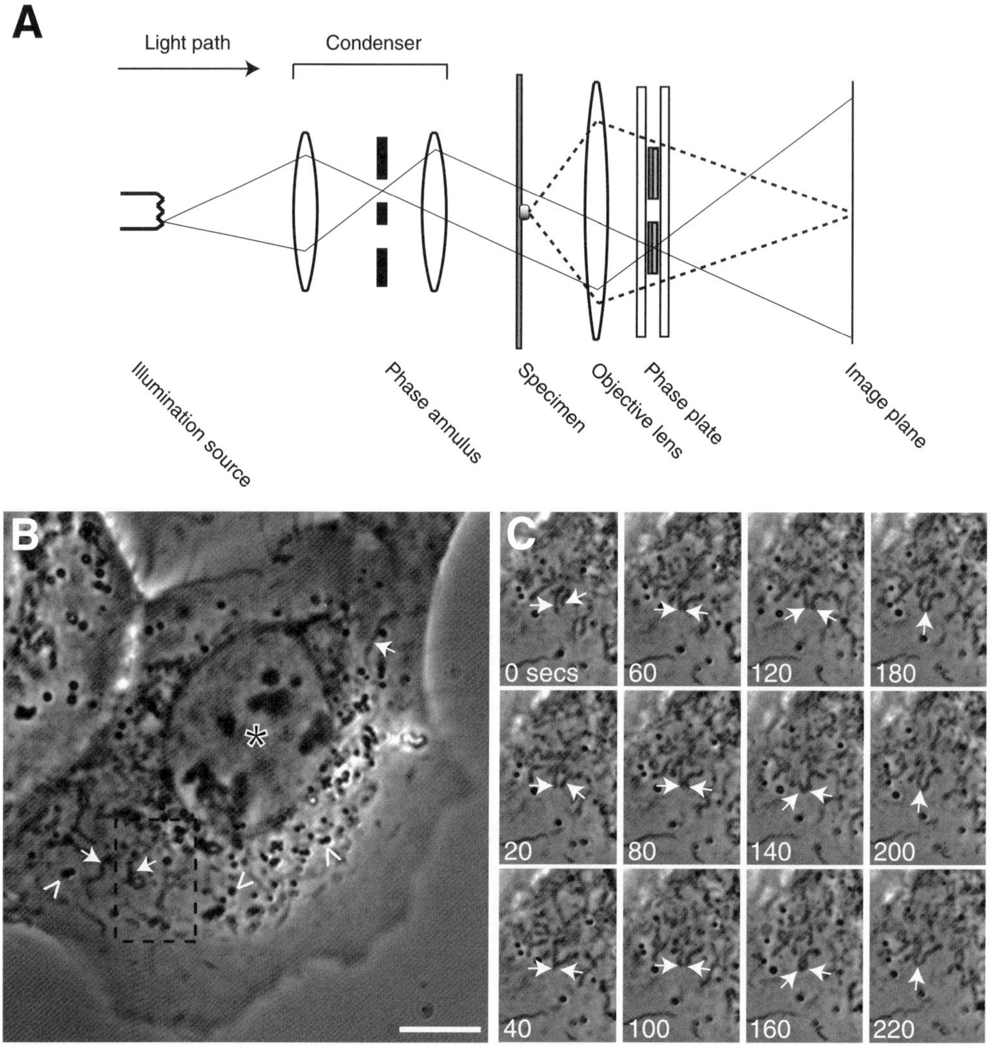

Figure 1. Using phase contrast microscopy to study organelle dynamics in living cells. (*A*) Diagram of the components in a phase contrast microscope. Adapted from (44). (*B*) Phase contrast image of an A431 cell showing the nucleus (*), mitochondria (arrow), and other unidentified organelles (arrowheads). (*C*) Zoomed frames from time-lapse recording of the boxed region shown in (*B*). Complex mitochondrial dynamics, including mitochondrial fusion, are seen (arrows). Bar, 5 μm.

submicrometer spherical organelles distributed throughout the cell are clearly resolved, but remain largely unidentified by phase contrast imaging (see *Fig. 1B*). Some are phase bright, others dark, and they are likely to be comprised of lipid droplets and late endocytic structures (such as lysosomes and multi-vesicular bodies), respectively. Meanwhile, and although not particularly refractive against the plasma membrane, which is closely apposed above and below, the endoplasmic reticulum (ER) can be resolved in the periphery of flattened cells when high-quality phase optics are contrast enhanced by video microscopy – largely on account of its distinctive reticular composition (for a clear demonstration, see 3).

2.2.1. Setting up a microscope for phase contrast

The phase contrast microscopy technique was introduced by the Dutch physicist and Nobel Laureate, Frits Zernike (http://nobelprize.org/physics/laureates/1953/). Phase contrast microscopy is a form of *interferometry*: contrast is derived from refractive elements within the sample, which cause phase shifts in light that are subsequently converted into amplitude changes that are registered as light or dark regions against a uniform reference. Zernike's contribution was to develop a technique for transforming optical phase shifts (undetectable by the eye) into changes in amplitude that can be seen. Although his original approach, which revealed objects to be dark against a bright background (positive/dark phase contrast), has largely been superseded by negative/bright phase contrast, the principles remain the same.

The basic components of a phase contrast microscope are shown in *Fig. 1(A)*. Setting up the phase contrast microscope is straightforward, provided the microscope is equipped with phase lenses (designated by, for example, Ph1 or Ph2) and a corresponding phase ring (annulus) in the condenser. Phase lenses contain a phase plate that works in tandem with the phase annulus to retard direct light (which does not pass through the specimen) by 1/4 of a wavelength. In phase contrast imaging, light that passes through the specimen will be diffracted by organelles, cell margins, etc. and this light is retarded in *phase*. Background light and light that has passed through the specimen interact at the image plane causing wave interference, with objects appearing bright against a uniform background (in negative phase). Having established Köhler illumination (see *Protocol 1*), phase contrast imaging is achieved simply by matching the correct lens type with its corresponding condenser annulus. Optimum alignment is achieved using a phase telescope, which inserts into the eyepiece allowing visualization of the relative positions of the phase plate and annulus.

2.2.2. Working example: using phase contrast to study cellular dynamics during apoptosis

We use phase contrast imaging extensively to complement fluorescence (GFP)-based studies of organelle and cytoskeletal dynamics during apoptosis (see *Protocol 2*). One notable outcome of this approach has been a re-appraisal of the dynamics of the apoptotic execution phase (15) – a good example of how transmitted light imaging when applied to living samples can still yield novel, kinetic information about otherwise well-characterized processes.

Apoptosis is an essential form of programmed cell death, accounting for the majority of developmental and homeostatic cell deaths in multicellular organisms. It is characterized by a series of coordinated changes in behavior, which isolate the dying cell from its neighbors, release it from basal attachments, and culminate in its ordered dismantling into fragments called apoptotic bodies (16). During the apoptotic 'release' phase, cells undergo active surface blebbing – an actin/myosin II-mediated process that is stimulated by caspase cleavage of Rho-activated kinase (17, 18). Meanwhile, neighboring cells squeeze the dying cell out of intact epithelia (19). Finally, apoptotic cells and their fragments are engulfed by macrophages or by nonprofessional phagocytes (albeit with differing efficiencies; see 20). Hence, apoptosis represents a unique and dramatic example of cellular remodeling and cell-to-cell interactions, and lends itself well to studies of cellular dynamics by real-time or time-lapse imaging.

Phase contrast is an essential component of our imaging approach for studying apoptosis, because it allows us to determine not only when a cell enters the execution phase (see 21), but also whether cells are progressing 'normally' through this process. Hence, without the accompanying transmitted light imaging, we would be unable to determine whether changes in organelle dynamics obtained by fluorescence imaging are relevant and how they relate to the progress of cell death. Unfortunately, one of the characteristics of apoptotic cells is that they round up, rendering them very much brighter by phase contrast, so that brightness/contrast or exposure length adjustments must be made 'on the fly'. We overcome this to an extent by using the 'autoscaling' option in our imaging package (METAMORPH; Universal Imaging). Henceforth, details such as organelle positioning gradually become more or less impossible to resolve by phase contrast optics alone.

2.3. Differential interference contrast (DIC) microscopy

One of the disadvantages of phase contrast microscopy is the fact that cells and their components appear surrounded by a white halo caused by interference fringes. This prevents one from observing distinct edges to the specimens. Not only does this restrict the ability to identify and distinguish between organelles within a typical flattened cell, but this limitation also makes phase contrast imaging inappropriate for the study of thicker samples (see section 2.2.2). For example, whilst flat, adherent eukaryotic cells lend themselves relatively well to phase contrast imaging (see *Fig. 1B*), more 'rounded' cells – such as bacteria, yeast, nonadherent eukaryotic cells and higher eukaryotes undergoing mitosis or apoptosis – are less tractable to this approach. For this reason, many researchers favor differential DIC for observing living cells.

DIC was first successfully employed in the 1960s (22). It derives contrast from gradients within the optical path (refractive index and geometric path length), but, in common with other forms of transmitted light microscopy, its resolution is still constrained by the wavelength of light and the NA of the lenses. Importantly, whereas phase contrast microscopy is incapable of reaching the theoretical optimum resolution of 200 nm, because the annular cone of the phase condenser typically reduces NA to ~60% of the objective NA (8), DIC is able to approach this limit of resolution. In addition, the specimen appears in pseudo-3D by DIC, with objects seen as bright projections with dark shadows against a background of intermediate gray (see *Fig. 2B* and *C*). This 3D effect dramatically enhances the

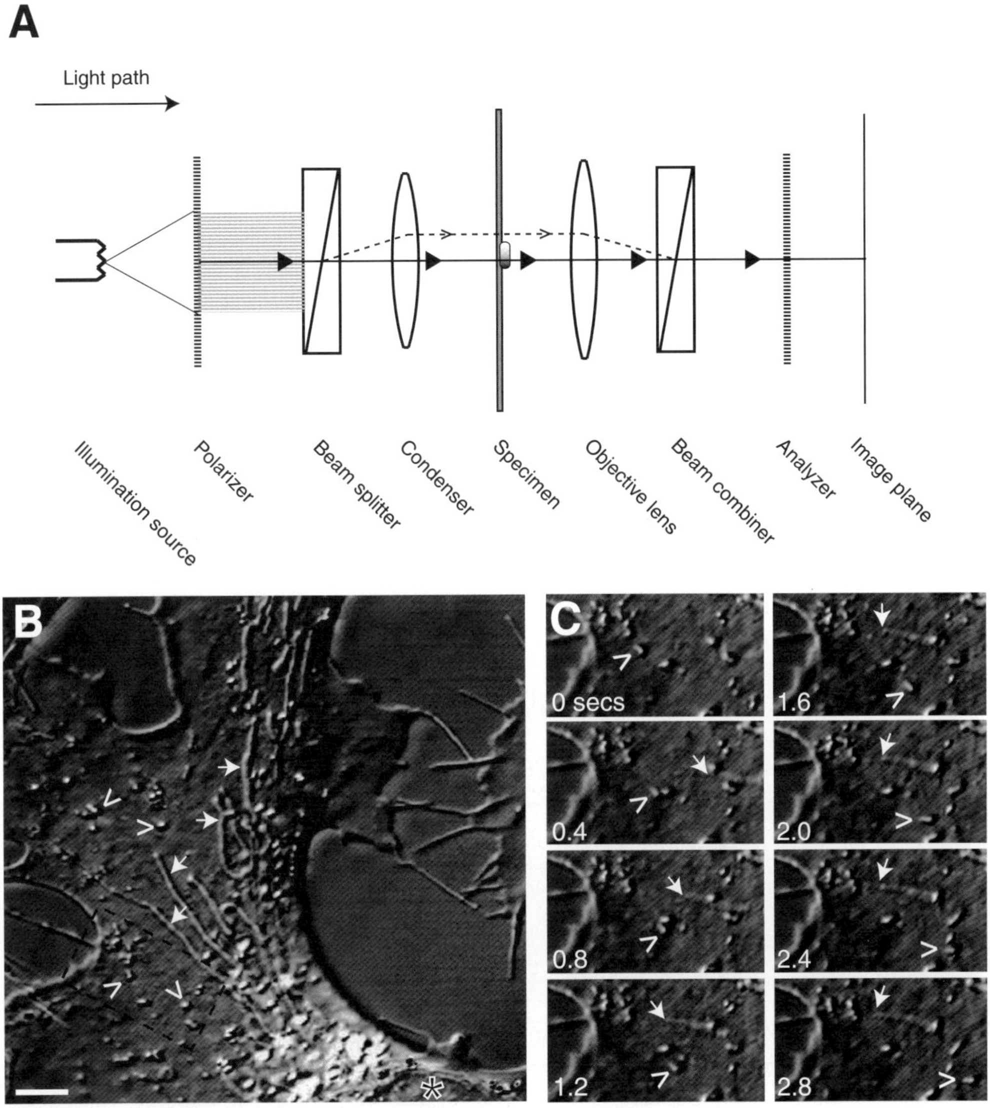

Figure 2. Using DIC microscopy to study organelle dynamics in living cells.
(*A*) Basic components and optical path in a DIC microscope. Adapted from (7). (*B*) DIC image of a *Xenopus* XTC cell showing the edge of the nucleus (*), mitochondria (arrows), and unidentified vesicular structures (arrowheads). The cell margins are clearly distinct. (*C*) Individual frames showing the boxed region in (*B*). Microtubule-based movement of a vesicular structure towards the cell center (arrowheads) and centrifugal extension of a tubular mitochondrion (arrow) are indicated. Frames were collected at video rate (25 frames/s; PAL) directly on to S-VHS. Bar, 2 μm. We thank Dr. Viki Allan (University of Manchester, UK) for permission to use frames from this movie.

visual impact of the image, facilitating object identification – particularly in time-resolved imaging experiments (see *Figs. 2* and *3*) – but it is not a read-out of genuine topology across the sample, because DIC is incapable of distinguishing between refractive index and thickness.

Observing a vertebrate cell in culture by DIC reveals a striking clarity of image, particularly under contrast enhancement afforded by digital imaging techniques (see section 2.3.2). *Fig. 2*(*B* and *C*) shows a still image and subsequent frames from a video of a *Xenopus* XTC cell obtained using DIC optics and minimal contrast enhancement. The cell margins are particularly clear, as is the nuclear envelope. Also prominent are mitochondria and other unidentified spherical vesicular

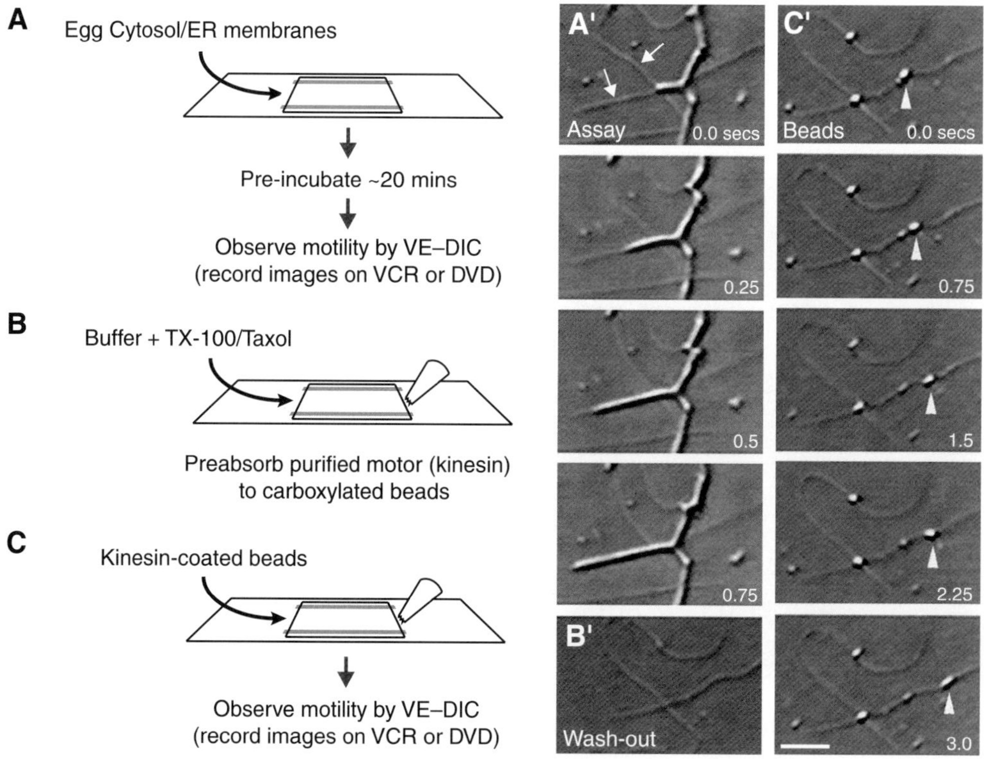

Figure 3. Kinesin bead assay for VE–DIC analysis of microtubule motor directionality.
(*A*)–(*C*) Protocol for retrospective assessment of membrane motor directionality. (*A′*)–(*C′*) Representative frames from a video of membrane dynamics and kinesin-coated bead motility. (*A, A′*) The basic ER motility assay in *Xenopus* egg cytosol (see *Protocol 3*). Microtubules (arrows) polymerize at the coverslip surface and membrane tubules (superimposed) extend by motor activity over time. (*B, B′*) The membrane networks are carefully washed away using buffer containing Triton X-100/Taxol, leaving a field of stabilized microtubules behind. Kinesin-coated beads are then introduced (*C, C′*) and their movement along the microtubule network is monitored. Kinesin-coated beads move exclusively towards the 'plus' ends of microtubules, meaning that the ER tubule extension in *A′* was towards the microtubule 'minus' end. Bar, 0.5 μm. (*A′–C′*) reprinted from *Molecular Biology of the Cell* (28) with permission of The American Society for Cell Biology.

structures that are highly motile and move in a linear, stochastic fashion both towards and away from the cell center. Movement is predominantly microtubule based, although it is not possible to see these 22 nm diameter structures. Under certain conditions (using a well-set-up microscope and in flattened regions of the cell), it is sometimes possible to distinguish the outline of the ER (3), but as with phase contrast imaging of this organelle, relatively poor definition means that this is unlikely to be of any real practical value for studying ER dynamics.

2.3.1. Setting up a microscope for DIC

DIC microscopy utilizes the properties of polarized light to generate contrast from unfixed, unstained specimens and is another example of *interferometry*. Light rays are first polarized and then split into ray pairs. Specimen components in the light path, for example organelles, generate interference by shifting the phase of individual rays within a pair with respect to the other.

The basic components of a Nomarski DIC microscope are shown in *Fig. 2(A)*. A polarizing filter is placed between the light source and condenser to generate plane-polarized light, which passes through a Nomarski-modified Wollaston prism (beam splitter). This separates individual rays of light into ray pairs that are in phase, but are 90° apart and ±45° with respect to the plane of polarized light exiting the polarizer. These divergent light beams are then focused by the condenser to produce two wavefronts (sampling and reference) that are separated spatially by a distance that is less than the resolution limit imposed by the objective and condenser lenses. These parallel wavefronts then pass through the specimen and objective lens, before being recombined by a second Wollaston prism (beam combiner) and analyzer. Wave interference (and hence amplitude differences) due to optical diffractions across the specimen are visualized against a uniform background reference, because spatially separated beam pairs pass through adjacent but distinct points in the sample (see *Fig. 2A*). This has the effect of highlighting edges particularly strongly, one edge of a linear object appearing bright with respect to the gray background, the other dark on account of the 'shear direction' of light produced by the Wollaston prisms (see *Fig. 2B* and *C*). Hence, images are 'shadow-cast' and appear 3D. Note that this biases contrast in favor of objects lying perpendicular to the shear direction (compare tubular mitochondria at different orientations in *Fig. 2B*), meaning that randomly oriented objects in an image field are unevenly contrast enhanced and can 'appear' and 'disappear' should they traverse the field of view in time-resolved experiments.

2.3.2. Video enhancement of the DIC image

Although a basic DIC image is striking and typically more informative than a phase contrast image (compare *Figs. 1B* and *2B*), when video contrast enhancements are applied this technique really comes into its own. Video cameras are sensitive to contrast changes that are undetectable by the native eye, and thus by using contrast and gain functions on the camera, otherwise intractable (small or low-contrast) objects can begin to be seen. In addition, when combined with contrast enhancement afforded by digital image processors, the appearance of the final output image can be dramatically improved. When contrast enhancement and gain are applied carefully to apparently translucent fields with no definition – for example, cytosolic preparations (see section 2.3.3) – objects below the normal resolution of light, such as vesicles (~50 nm), microtubules (22 nm; see *Fig. 3*) and actin bundles (>10 nm) can be detected. The trade-off, however, is that aberrations within the optical path (such as dust), subtle illumination imbalances, and sensitivity variability across the camera detector become particularly apparent. These types of 'fixed pattern' noise can be eliminated by subtracting a stored background image from each and every frame at video rate using a digital image processor such as the Hamamatsu Argus 10. Additional image 'noise' due to voltage fluctuations within the camera can be successfully overcome by frame averaging – typically two frames averaged consecutively are sufficient for most applications.

Our main research expertise with video-enhanced DIC (VE–DIC) has been in the study of microtubule-based membrane dynamics in cell-free motility assays (see section 2.3.3 and *Protocol 3*), but it is also widely used to study chromosome behavior during mitosis in *living* cells. Cells that undergo mitosis with a relatively flat morphology – in particular, newt lung cells or Potoroo kidney PtK1/2 lines – lend themselves particularly well to this approach, and beautiful work has been conducted on the function of the mitotic spindle using VE–DIC (for further details, see 2 and relevant chapters in 5).

2.3.3. Working example: using VE–DIC to study microtubule motor function *in vitro*

Possibly the most valuable contribution that VE–DIC has made to the field of cell biology is in the identification and characterization of microtubule-based motor proteins. These mechano-chemical enzymes derive power from the hydrolysis of ATP to move unidirectionally along microtubules. In doing so, they control the localization of most organelles within animal cells and provide the framework for vesicular–tubular trafficking between membrane compartments (for review, see 23). The study of microtubule motors blossomed following the application of VE–DIC to *in vitro* motility assays of fast axonal transport in extruded axoplasm from giant squid (e.g. *Loligo pealeii*). These preparations, which are essentially pure cytoplasm, retain all cytoskeletal components intact, and when perfused into pre-assembled microscope flow-chambers (see *Protocol 3*), recapitulate microtubule-based motility events *in vitro*. This approach was exploited by Vale and co-workers to characterize microtubule-based organelle translocation, leading to the identification of the anterograde motor, kinesin (24), and later its main retrograde counterpart, cytoplasmic dynein (25).

Since these early studies, VE–DIC has been used to characterize a range of microtubule (and actin)-dependent membrane motility events *in vitro* (e.g. 7). Mostly, the squid axoplasm system has been superseded by extracts prepared from *Xenopus laevis* eggs (e.g. 6, 7). The basic approaches for using VE–DIC to study ER motility in frog egg extracts are outlined in *Protocol 3* (for a more detailed appraisal, see 6 and 7). One obvious advantage of using egg extracts is that they can be prepared in a variety of states that mimic normal cellular processes. For example, interphase (activated), meiotic metaphase (cytostatic factor-arrested) or 'mitotic' metaphase (interphase extracts combined with cytostatic factor extracts) cytosol can be prepared to allow the study of mitotic regulation of motor function (e.g. 26). They can be supplemented with antibodies, recombinant proteins, or drugs, or depleted of key components to facilitate the study of various aspects of motility. For example, we have employed egg extracts supplemented with cytochrome c to drive extracts into apoptosis *in vitro* to investigate apoptotic regulation of cytoplasmic dynein (27), and we have prepared 'micro-extracts' of fertilized eggs and synchronized early embryos to study motor regulation during development (28). In most of these examples, it is the strength of VE–DIC microscopy in allowing visualization of individual microtubules (22 nm, ~1/10 of the resolution of light) and optical distinction between microtubules and vesicles/tubules or latex beads that has proved invaluable. For example, it is possible to determine the directionality of ER tubule extension *retrospectively* by observing the movement of kinesin-coated latex beads on Taxol-stabilized microtubules (see *Fig. 3*) (28).

2.4. Polarizing microscopy

As noted above (section 2.3), DIC imaging requires that a microscope is set up for polarized light optics, but straightforward polarizing microscopy - where two polarizing filters (the polarizer and analyzer) are inserted into the light path - has been particularly useful for investigation of certain biological subjects. As a technique this is valuable, as it generates contrast due to inherent alignments of components within specimens and, as with phase contrast and DIC, it can be performed with living specimens. Polarizing filters allow light with one particular plane of oscillation to pass, so that if two filters lie at right angles to one another, no light passes at all. In practice, when the analyzer is rotated to achieve total extinction with this system, any part of the specimen that possesses distinct alignment alters the plane of oscillation from the polarizer, allowing it to pass through the crossed analyzer, whereupon it appears bright or *birefringent* against a dark background. The best-publicized example of this is striated muscle, which is comprised of myofibrils characterized by banded sarcomeres, the nomenclature of which (isotropic, anisotropic) derives from their appearance under the polarizing microscope (29).

An equally important example that has benefited from study using the polarizing microscope is the mitotic spindle. The spindle is comprised of aggregates of parallel microtubules, which are too small to be resolved using any form of light microscope. However, many decades before this was discovered, it was appreciated that living mitotic spindles could be seen in cells (30). The birefringent properties of spindles - particularly in oocytes of marine invertebrates - have allowed various aspects of spindle function to be studied *in vivo* and the behavior of *in vitro*-reactivated spindles to be monitored.

More recently a new type of polarized light microscope known as the Pol-Scope (Cambridge Research & Instrumentation) has been developed (31). It uses liquid crystal devices instead of the normal crystal compensator and employs digital image processing to measure birefringence in all points of the image at all orientations simultaneously. This circumvents variations seen with traditional polarizing microscopy due to the position of the specimen relative to the orientation of the birefringence axis. The instrument has been used very effectively to investigate the dynamics of filament networks in neuronal growth cones (32) and to analyze meiotic spindles in living human oocytes prior to selection for intracytoplasmic sperm injection (33).

2.4.1. Working example: using polarizing microscopy to study insect ovarian microtubules

For several years, we have been using polarizing microscopy to study microtubule properties in the ovaries of hemipteran insects (see *Protocol 4*). Here, approximately 30 000 parallel microtubules in a bundle up to several millimeters in length occur in so-called nutritive tubes that connect highly synthetic nurse cells to the developing oocytes (34). The microtubule bundles are very strongly birefringent, and since they are so large, they can be manually dissected from the ovaries under polarized light (see *Fig. 4A* and *B*) (35). This has facilitated the biochemical characterization of a discrete but entire microtubule-based intracellular translocation system and has provided microtubule substrates for *in vitro* motility.

2.5. Dark-field microscopy

While excellent for studying aligned specimens – particularly cytoskeletal aggregates – polarizing light does not allow individual components to be discerned with any clarity. This can be achieved using dark-field microscopy. Here, a bright-field microscope is modified using a special condenser lens that diffracts light exiting it wide of the collecting objective lens. In effect, the condenser is designed to form a hollow, rather than a filled, cone of light. Viewing down a microscope set up in such a manner reveals a dark background, but when a specimen is introduced to the stage, some of the light is diffracted back and collected by the objective lens, so that a white image of the specimen appears against a black background. Considerable care must be taken when conducting dark-field microscopy, since dust and other particles also scatter light. Consequently, slides and coverslips must be scrupulously clean and the sample medium may need to be filtered.

Dark-field microscopy is particularly useful for small specimens such as prokaryotes and organelles isolated from eukaryotic cells. Importantly too, the technique allows one to visualize, but obviously not resolve, specimens considerably smaller than the limit of resolution of the light microscope. An excellent example of this was illustrated by experiments investigating the basis of flagellar movement. Using dark-field microscopy, microtubule doublets were

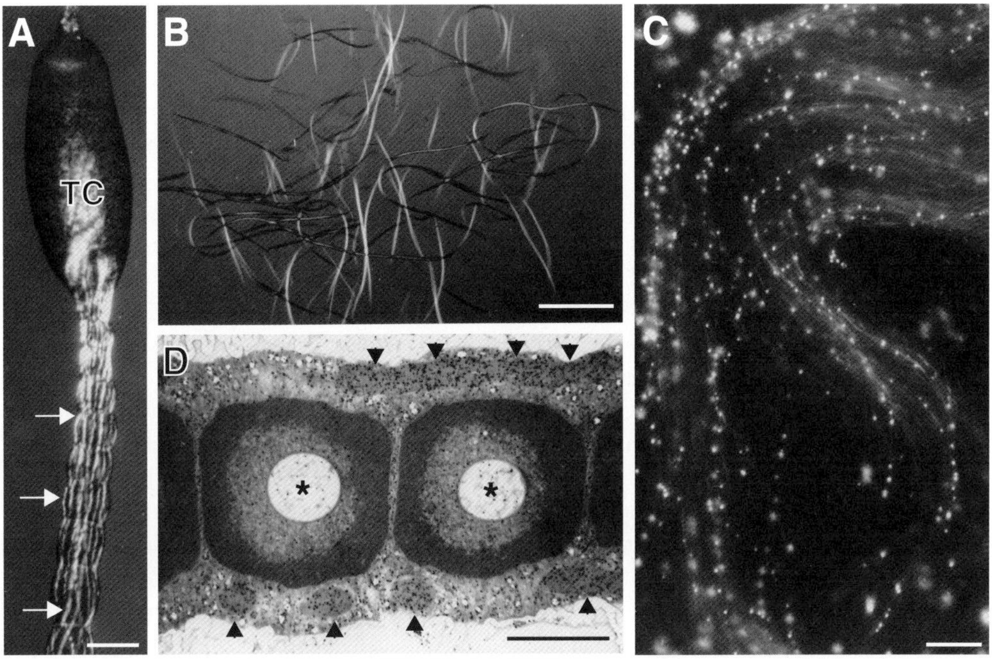

Figure 4. Examples of images of hemipteran insect ovarioles and their components obtained using various transmitted light techniques.
(*A*) An intact ovariole from *Notonecta* seen in polarized light. Under these conditions, transport channels (nutritive tubes; arrows) linking nurse cells in the trophic centre (TC) with the chain of developing oocytes appear strongly birefringent due to the presence of thousands of aligned microtubules. Bar, 250 μm. (*B*) Isolated nutritive tubes from *Notonecta* viewed in polarized light. Bar, 500 μm. (*C*) Dark-field micrograph of an isolated *Notonecta* nutritive tube. This unfixed preparation was treated to disperse the microtubules, whilst retaining the attached mitochondria (bright 'bead-like' objects). Bar, 10 μm. (*D*) Autoradiograph of a *Notonecta* ovariole section after incorporation of [^{3}H]uridine. Label is evident within nutritive tubes (arrows) – the conduits for newly synthesized macromolecules – but sparse in other regions including the oocyte nuclei (*). Bar, 50 μm.

observed sliding from flagellar axonemes that had been demembranated, subjected to partial enzymic digestion, and treated with ATP (36). Other studies have also shown that individual isolated microtubules can be seen using dark-field microscopy (see *Fig. 4C*). This has enabled details of their dynamic instability to be observed (37) and in particular the influence of microtubule-associated proteins on the dynamic properties of microtubules to be assessed (38).

2.6. Autoradiography and the bright-field microscope

Aside from color-based staining, a further way of localizing and monitoring the dynamics of component macromolecules under a bright-field microscope is by autoradiography. This method has been extremely effective when used to study the synthesis of proteins, DNA, and RNA in cells. For autoradiography, appropriate radioactive precursor molecules (usually labeled with tritium) are introduced into cells or tissues. Once these have become incorporated, the material is fixed and prepared for microscopy. Slides are then coated in the dark with photographic emulsion, whereupon silver grains can be seen above the sites of the newly synthesized macromolecules (see *Fig. 4D*). Pulse–chase experiments of this nature reveal not only the sites of synthesis of a component, but also its possible movement.

2.7. Other transmitted light microscopy techniques

There are several other less often utilized microscopy methods that may be useful for certain specialized applications. Oblique or anaxial illumination is a simple technique for enhancing image contrast by partially closing the condenser iris and off-centering the narrowed diaphragm. Although this technique introduces contrast (and simulates DIC-like shadowing), it has the deleterious effect of dramatically reducing image resolution (since resolution is in part governed by condenser NA) and introducing diffraction rings into the field of view.

Hoffman modulation contrast microscopy enhances the contrast within near-transparent objects by transmitting phase gradients within a specimen into levels of gray. Hence, it can be applied to unstained, living material. It was developed by Robert Hoffman in 1975 and consists of a 'modulator' in the objective lens composed of three distinct zones of neutral density: a small zone transmitting only 1% of light (towards the periphery of the back focal plane), a narrow gray zone transmitting 15%, and a clear, transparent zone transmitting 100%. Coupled to slit filters in a modified condenser, and the more recent introduction of paired polarizers that can be crossed to narrow the condenser slit, refracted light from objects within the sample becomes accentuated as light or dark, giving the appearance of a 3D image (for a more detailed discussion, visit the Olympus microscopes website at http://www.olympusmicro.com/primer/techniques/hoffman.html). Variable relief contrast (VAREL) imaging, patented by Zeiss, uses similar principles, but has the advantage that the same modified objective can be used for both VAREL and phase contrast applications.

Rheinberg illumination (first demonstrated in 1896) is a type of 'optical staining'. It uses dark-field illumination coupled with colored filters of an annular nature (colored rings) that collect light refracted by the specimen. Effectively, the color of the background mimics that of the central disc of the filter, with the specimen 'taking on' the color of the outer ring. This technique produces dramatic representative images and can be used to visualize specimens ranging from whole-mount organisms to individual cells and crystals.

Also relevant here is the Richardson contrast microscope system (http://www.richardson-tech.com/p_rtm3.html), which uses a high NA dark-field condenser and finely tuned optical components to achieve a reported resolution of below 125 nm. This stand-alone enhanced contrast/fluorescence live-cell imaging solution reveals living samples in striking clarity and may be useful for samples where phase contrast or DIC fails to provide adequate component definition.

Protocol 1

Correct alignment of a light microscope

For bright-field microscopy of naturally pigmented specimens and stained preparations, and for phase contrast and DIC[a].

Equipment

- Bright-field compound microscope (upright orientation here) with objective lenses of different magnification (10× or lower to 100× oil-immersion) (e.g. Olympus BX-51; Olympus Microscopes)
- Glass slides
- Coverslips (No. 1 or 1.5)

Method

1. Turn the rotating nosepiece to low-power objective (10×) and place the slide on the microscope stage.
2. Switch on the light source and adjust the brightness as necessary by operating the control rheostat. Make sure that the field diaphragm immediately above the light source is fully open and, if present, swing out auxiliary lenses and swing in the diffuser.
3. Focus the objective on to the specimen.
4. Focus the condenser (to set up so-called Köhler illumination and ultimately fill the objective lens with a cone of light) by closing the field diaphragm and bringing this into focus (a pinpoint of light) using the condenser focusing control.
5. Check the centering of the condenser and, if the image of the field diaphragm is not central, adjust using the opposing screw controls on the condenser.
6. Open the field diaphragm to fill the field of view.
7. The substage condenser diaphragm can be partially closed to increase contrast in transparent specimens, although it must be noted that closure reduces the resolving power of the microscope by effectively reducing the NA of the objective[b].
8. Turn the nosepiece to change the objective lens and repeat stages 3–7. When using high powers, if the objective is not in focus, lower the objective towards the slide, carefully viewing from the side to avoid damaging the lens, and focus upwards and away from the slide.
9. With binocular microscopes, the distance between the eyepieces can be adjusted to suit individual users. Also, one of the eyepieces is fixed while the other can be adjusted to suit the user. Focus first with the eye down the fixed eyepiece and then rotate the other eyepiece for compatible viewing[c].

Notes

[a]The technique of Nelsonian or critical illumination can also be used. Here, a focused image of the illuminating source is applied to the plane of the specimen. This has the disadvantage that physical components of the illuminating source (e.g. filaments) produce an uneven illumination field. However, if this is evened out – as is the case when mercury burners are connected via fibre optic cables – critical illumination can be used, for example, to improve VE-DIC imaging (see 6).

[b]The resolving power of a microscope (or the minimum resolved distance) depends on the wavelength of illumination (λ; in this case, white light) and the extent to which light is accepted by the objective lens (determined by its NA). The NA of a lens is shown alongside its magnification. Resolution (r) is governed by the equation $r = 1.2\lambda/(NA_{obj} + NA_{cond})$, where the value 1.2 is a

theoretical adjustment to compensate for physical inefficiencies within the optics. The best resolving power – the lowest minimum resolved distance – is achieved using the largest NA possible (the maximum NA available commercially now exceeds 1.4; e.g. Olympus PLAPON 60XO, NA = 1.42; PLAPON 60XOTIRFM (specialized for total internal reflection fluorescence), NA = 1.45; Zeiss α Plan-FLUAR 100×, NA = 1.45), and shortest wavelength.
[c]For useful reading, see (39–41).

Protocol 2

Using phase contrast microscopy to monitor apoptosis by time lapse

Equipment and Reagents

- An inverted microscope equipped for phase contrast and epifluorescence[a] (e.g. Olympus IX71)
- Automated shutters for transmitted light and epifluorescence and control unit (e.g. from Prior)
- Suitable camera and image acquisition software (e.g. CoolSNAP HQ CCD camera (Roper Scientific) and METAMORPH software)
- Environmental chamber and accessories for CO_2 enrichment (e.g. from Solent Scientific)[b]
- Incubator (5% CO_2, humidified)
- Appropriate cell lines
- Glass-bottomed imaging dishes (e.g. from MatTek)
- Appropriate inducers of apoptosis (e.g. cytotoxic drugs or UV)
- Fluorescent Annexin V (various suppliers)

Method

1. Grow adherent cells (e.g. HeLa, Cos-7, A431, NIH-3T3) to ~75% confluency on glass-bottomed imaging dishes.
2. Pre-warm the microscope chamber to 37°C.
3. Induce apoptosis by the addition of drug or by UV irradiation.
4. Add fluorescent Annexin V (0.1%) to the imaging medium.
5. Transfer cells to the microscope chamber.
6. Identify fields of view with good image definition using phase contrast.
7. Set up the image acquisition software to collect phase contrast images every 1–3 min and fluorescence images less frequently (this reduces phototoxic damage to the cells).
8. If not using automated Z-stepping/autofocus, be prepared to adjust the focus manually when cells enter the apoptotic execution phase and round up.
9. When apoptotic cells label with Annexin V (as a consequence of plasma membrane phosphatidylserine flipping; 42), the apoptotic process has more or less run its course.

Notes

[a]We have equipped our microscope with an automated *x*-*y* stage (Prior) and Piezo Z-stepper (Physik Instrumente), enabling us to monitor several fields in a single experiment and to respond to focus changes automatically as apoptotic cells round up. We also use halogen lamp illumination for both transmitted light and for epifluorescence, as this is less photodamaging to fragile apoptotic cells.
[b]CO_2-independent medium can be used if CO_2-enrichment equipment is unavailable.

Protocol 3

Using VE–DIC to study microtubule-based membrane motility *in vitro*

Equipment and Reagents

- A microscope system set up for DIC (various suppliers)
- Suitable camera (e.g. Hamamatsu Newvicon)
- Image processor (e.g. Hamamatsu Argus 10)
- Monitor and VHS video recorder or recordable DVD
- *Xenopus* egg extract[a]
- Acetate sucrose buffer (100 mM potassium acetate; 3 mM magnesium acetate; 5 mM EGTA; 10 mM HEPES; 150 mM sucrose; pH 7.4)
- Energy mix (7.5 mM creatine phosphate; 1 mM MgATP; 0.1 mM EGTA; pH 7.7)
- Ultracentrifuge
- Standard glass slides and coverslips (No. 1; 18 × 18 mm)[b]
- Sellotape
- Vacuum grease (e.g. Apiezon M)

Method

(a) *Preparation of* Xenopus *egg cytosol and membrane fractions*

1. Dilute crude *Xenopus* egg extracts[a] with two volumes of acetate sucrose buffer containing energy mix.
2. Centrifuge the diluted cytoplasm for 30 min at 117 000 ***g*** at 4°C (we use a Beckman TL-100 bench-top ultracentrifuge and TL100 rotor).
3. Collect the clear cytosolic layer above the sticky, yellow ribosome/glycogen pellet.
4. Carefully dislodge the membrane fraction at the interface between cytosol and pellet by resuspending in ~12 µl acetate sucrose buffer.
5. Recombine cytosol and membranes[c] (at a ratio of 10–20:1) by pipetting up and down.
6. Fill the flow-chamber (see below) with cytosol/membrane mix by gentle aspiration using a pipetteman (ensuring that no air bubbles are introduced).

(b) *Preparing flow-chambers for imaging motility by VE–DIC*

1. Assemble microscope flow-chambers from standard glass slides and coverslips[b].
2. Apply thin strips of Sellotape, about 2 cm apart, in tramlines along the edges of the central third of the slide.
3. Seal the flow-chamber laterally by applying thin strips of vacuum grease at the internal edge of the Sellotape.
4. Set the coverslip carefully on top to produce an open-ended flow-chamber of ~10 µl volume.
5. Introduce *Xenopus* egg cytosol/membranes by careful pipetting[d].
6. Incubate the flow-chamber, inverted, in a humidified box for 10–30 min to allow spontaneous microtubule polymerization at the coverslip surface.

(c) *VE–DIC microscopy to study microtubule-based membrane motility*

1. Oil the objective lens and condenser.
2. Introduce the flow-chamber and establish Köhler (or critical) illumination[e].
3. Offset the adjustable Wollaston before diverting light to the camera path (to prevent camera damage due to bright illumination).
4. Allow light to enter the camera and adjust the Wollaston to increase light intensity gradually to optimum (on some camera control boxes, a indicator of optimum brightness is present).
5. Using the monitor, identify a region free from large objects, and defocus a little away from the plane of the coverslip.
6. Apply background subtraction.
7. Refocus and record images on to VHS tape or DVD.

Notes

[a]For the preparation of *Xenopus* egg extracts, see (6) and (43).
[b]Problems with both image quality and extract behavior (e.g. capacity to polymerize microtubules) can arise through dirt or grease on the coverslip surface. Some coverslip batches also appear 'pitted', and whilst this does not appear to affect extract behavior, it deteriorates image quality somewhat.
[c]Other sources of membrane including rat liver ER/Golgi can also be used.
[d]Open-ended flow-chambers allow easy exchange of cytosol by resting a small triangle of filter paper at one end (tear the point to provide more rapid capillary suction) and exchange fluid in a pipette tip at the other (see *Fig. 3*).
[e]Since extracts usually do not provide adequate contrast by eye, it may be necessary to focus on the edge of the vacuum grease to establish the focal plane.

Protocol 4

Using polarizing light microscopy to study aligned microtubules arrays

Aligned cytoskeletal aggregates can be seen in living cells and tissues using polarized light optics[a]. Hemipteran insect ovarioles are used here as an example.

Equipment and Reagents

- Compound research microscope equipped with strain-free lenses (10× to 40×), with facilities to insert polarizing filters into the light path, and with a rotating stage (various suppliers)
- Glass microscope slides
- Glass coverslips
- Dissecting microscope (various suppliers)
- Hemipteran insect - *Notonecta* (aquatic backswimmer) is suggested as it is large (~1.5 cm) and occurs widely
- Insect saline solution (154 mM NaCl; 5.6 mM KCl; 1.7 mM $CaCl_2.2H_20$; 2.4 mM $NaHCO_3$; 13.8 mM glucose)
- Watchmakers' forceps

Method

1. Dissect the ovaries from female *Notonecta* under saline using the dissecting microscope and watchmakers' forceps.
2. Separate the ovaries into constituent ovarioles[b].
3. Using forceps, place an ovariole into a drop of saline on a microscope slide.
4. Cover carefully with a coverslip, raised to avoid excessive squashing.
5. Observe using bright-field microscopy (see *Protocol 1*).
6. Insert a polarizing filter into the holder below the condenser and a second (the analyzer) above the objective lens.
7. Rotate one of the polarizing filters to achieve total extinction, whereupon birefringent (white or dark depending on the rotational position of the stage) nutritive tubes[c] can be seen[d] connecting the nurse cells at the anterior of the ovariole to the chain of developing oocytes.

Notes

[a]A similar protocol can be used to investigate a variety of aligned cytoskeletal aggregates including mitotic spindles and the filamentous systems in striated muscle.

[b]There are two conspicuous ovaries, each consisting of seven ovarioles (see *Fig. 4A*). These run the length of the abdomen and into the thorax of the insect.

[c]Each nutritive tube is approximately 30 μm in diameter and up to several millimeters in length, and is comprised of ~30 000 parallel microtubules - the largest bundle so far described (see section 2.4). Their strong birefringence has meant that they can be dissected in polarized light (see *Fig. 4B*) both for biochemical analysis and to act as substrates in microtubule motility assays (32).

[d]Birefringence can be measured by an extension of this method (not discussed here).

4. TROUBLESHOOTING

As with all microscopy-based experiments, the clarity of the images – and thus the quality of any data generated – depends to a very large extent upon the quality of the equipment and the accuracy with which it has been set up. Microscope components (lenses, etc.) are fragile and have to be kept clean, but provided they are well cared for – during both day-to-day use and whenever stored – there is no reason why they should not function well for many years. Setting up a microscope for many of the protocols outlined above can be technically demanding; however, major microscope equipment suppliers will normally be willing to assist in aligning your microscope for a particular technique, the details of which will obviously vary among equipment manufacturers.

Acknowledgements

We are grateful to Viki Allan (Manchester, UK) for permission to use images for *Fig. 3*. We thank Virginie Betin, David Moss, and Ruth Rollason for critical comments on the manuscript. Work in the Lane laboratory is supported by a Research Career Development Fellowship from the Wellcome Trust (No. 067358) and by a Wellcome Trust Project Grant (No. 074208). H.S. is an Honorary Fellow of the Peninsula Medical School.

5. REFERENCES

1. **Klar TA, Jakobs S, Dyba M, Egner A & Hell SW** (2000) *Proc. Natl. Acad. Sci. U. S. A.* **97**, 8206–8210.
★★ 2. **Rieder CL & Khodjakov A** (2003) *Science*, **300**, 91–96. – *Excellent, brief overview of light microscopy for cell biology applications.*
3. **Stephens DJ & Allan VJ** (2003) *Science*, **300**, 82–86.
4. **Sluder G & Wolf DH** (eds.) (2003) *Digital Microscopy: a Second Edition of Video Microscopy.* Academic Press, San Diego.
5. **Rieder CL** (ed.) (1999) *Mitosis and Meiosis.* Academic Press, San Diego.
6. **Allan VJ** (1993) *Methods Cell Biol.* **39**, 203–226.
★★ 7. **Allan VJ** (1998) *Methods Enzymol.* **298**, 339–353. – *In-depth coverage of VE-DIC microscopy.*
8. **Salmon ED & Tran P** (2003) *Methods Cell Biol.* **72**, 289–318.
9. **Nascimento AA, Roland JT & Gelfand VI** (2003) *Annu. Rev. Cell Dev. Biol.* **19**, 469–491.
10. **Macgregor HC** (1993) *An Introduction to Animal Cytogenetics.* Chapman & Hall, London.
11. **Verma R & Babu A** (1989) *Human Chromosomes: Manual of Basic Techniques.* Pergamon Press, Oxford and New York.
12. **Macgregor HC** (1987) *J. Cell Sci.* **88**, 7–9.
13. **Morgan GT** (2002) *Chromosome Res.* **10**, 177–200.
14. **Morris RL & Hollenbeck PJ** (1995) *J. Cell Biol.* **131**, 1315–1326.
15. **Lane JD, Allan VJ & Woodman PG** (2005) *J. Cell Sci.* **118**, 4059–4071.
16. **Mills JC, Stone NL & Pittman RN** (1999) *J. Cell Biol.* **146**, 703–708.
17. **Coleman ML, Sahai EA, Yeo M, Bosch M, Dewar A & Olson MF** (2001) *Nat. Cell Biol.* **3**, 339–345.
18. **Sebbagh M, Renvoize C, Hamelin J, Riche N, Bertoglio J & Breard J** (2001) *Nat. Cell Biol.* **3**, 346–352.

19. **Rosenblatt J, Raff MC & Cramer LP** (2001) *Curr. Biol.* **11**, 1847–1857.
★★★ 20. **Parnaik R, Raff MC & Scholes J** (2000) *Curr. Biol.* **10**, 857–860. – *Seminal paper describing the DIC microscopy set-up.*
21. **Lane JD, Lucocq J, Pryde J, Barr FA, Woodman PG, Allan VJ & Lowe M** (2002) *J. Cell Biol.* **156**, 495–509.
★ 22. **Allen RD, David GB & Nomarski G** (1969) *Z. Wiss. Mikrosk.* **69**, 193–221. – *One of several key papers utilizing VE-DIC to characterize microtubule motors.*
★ 23. **Lane J & Allan V** (1998) *Biochim. Biophys. Acta,* **1376**, 27–55. – *One of several key papers utilizing VE-DIC to characterize microtubule motors.*
24. **Vale RD, Reese TS & Sheetz MP** (1985) *Cell,* **42**, 39–50.
25. **Vale RD, Schnapp BJ, Mitchison T, Steuer E, Reese TS & Sheetz MP** (1985) *Cell,* **43**, 623–632.
26. **Allan VJ & Vale RD** (1991) *J. Cell Biol.* **113**, 347–359.
27. **Lane JD, Vergnolle MA, Woodman PG & Allan VJ** (2001) *J. Cell Biol.* **153**, 1415–1426.
28. **Lane JD & Allan VJ** (1999) *Mol. Biol. Cell,* **10**, 1909–1922.
29. **Huxley H & Hanson J** (1954) *Nature,* **173**, 973–976.
30. **Inoue S** (1953) *Chromosoma,* **5**, 487–500.
31. **Oldenbourg R** (1996) *Nature,* **381**, 811–812.
32. **Katoh K, Hammar K, Smith PJ & Oldenbourg R** (1999) *Mol. Biol. Cell,* **10**, 197–210.
33. **Rienzi L, Ubaldi F, Iacobelli M, Minasi MG, Romano S & Greco E** (2005) *Reprod. Biomed. Online,* **10**, 192–198.
34. **Macgregor HC & Stebbings H** (1970) *J. Cell Sci.* **6**, 431–449.
35. **Stebbings H** (1986) *Int. Rev. Cytol.* **101**, 101–123.
36. **Summers KE & Gibbons IR** (1971) *Proc. Natl. Acad. Sci. U. S. A.* **68**, 3092–3096.
37. **Horio T & Hotani H** (1986) *Nature,* **321**, 605–607.
38. **Permana S, Hisanaga S, Nagatomo Y, Iida J, Hotani H & Itoh TJ** (2005) *Cell Struct. Funct.* **29**, 147–157.
39. **Bradbury S & Bracegirdle B** (1998) *Introduction to Light Microscopy.* BIOS Scientific Publishers, Oxford.
40. **Oldfield R** (1994) *Light Microscopy: an Illustrated Guide.* Wolfe Publications, London.
41. **Keller HE** (2003) *Methods Cell Biol.* **72**, 45–55.
42. **Martin SJ, Finucane DM, Amarante-Mendes GP, O'Brien GA & Green DR** (1996) *J. Biol. Chem.* **271**, 28753–6.
43. **Murray AW** (1991) *Methods Cell Biol.* **36**, 581–605.
44. **Ruzin S** (1999) *Plant Microtechnique and Microscopy.* Oxford University Press, New York.

CHAPTER 5

Spectral imaging techniques for fluorescence microscopy

Timo Zimmermann

1. INTRODUCTION

The visualization of biological structures is an essential tool of modern biological research. In cellular and molecular biology, fluorescence labeling techniques and epifluorescence microscopy provide highly detailed images of specific structures against a minimal background. New developments in the design of fluorescent labels, in microscope technology, and in image processing continue to improve our capacity to analyze complex biological samples.

New fluorophores have been introduced (1, 2) and an increasing number of spectrally different fluorescent proteins have become available (3–5). On the microscope side, multi-channel fluorescence imaging uses the availability of distinct markers to visualize different aspects of the same specimen. During cellular processes such as cell division (6) or secretion (7), several components can thus be analyzed together and their interactions can be understood. The information from the sample is collected in increasing numbers of detection channels, and quantitative analysis is required for a good understanding of the data. Quantitative analysis depends on the accuracy of the data. In multi-channel fluorescence imaging there are some inherent problems with this. The signals of the fluorescent labels may not be completely contained in just one specific detection channel. A clear identification of the specific signals, a prerequisite for quantitative analysis, may therefore not be possible. Although this is not the case for all multi-channel fluorescence imaging, it is an important consideration for mixtures of strong and weak signals of the different labels or when the choice of available dyes is limited, as it is in the case of fluorescent proteins. For an accurate understanding and quantitation of such signals, techniques that reliably identify and separate the multi-channel data are needed.

The problems facing multi-channel fluorescence imaging are similar to those encountered in multi-band satellite imaging. Recently, methods developed to separate information in multi-band satellite data have successfully been applied

Cell Imaging: ***Methods Express*** (D. Stephens, ed.)

to the analysis of multi-channel microscope images (8, 9). The technique used, called linear unmixing, separates the fluorophore signals in the sample and allows unambiguous representations of the labels and therefore quantitative analysis of the signals.

It is the aim of this chapter to explain potential problems in multi-channel fluorescence imaging and to describe methods for the correction of spectral image data. Relevant parameters for these methods and possible limitations are discussed and application examples for these techniques are given. Protocols for the acquisition of spectral image data and for the processing of the data by linear unmixing are also provided.

1.1. Definition of the overlap of fluorophore signals (crosstalk)

When excited with light of appropriate wavelengths, a fluorescent molecule will re-emit some of its acquired energy as fluorescence light. The distribution of emission wavelengths around a maximum is determined by the chemical properties of the fluorophore and represents its characteristic emission spectrum. Emission spectra are asymmetrically distributed around their maximum. The fluorescence emission generally shows a steep increase from the shorter wavelengths to the emission peak. The decrease from the peak to longer wavelengths is much less steep and can cover up to 100 nm of spectral range behind the peak emission. It contains a significant amount of the total emission of a fluorophore.

Because of the wide emission spectra of fluorophores, spectral overlap of different fluorochromes is almost unavoidable and is thus an inherent problem for the imaging of multiple fluorescent labels (see *Fig. 1*). In the presence of only a single type of fluorophore in the sample, all of the available emission of the fluorophore can be detected with a long-pass filter. If, however, additional fluorophores are present, a band-pass filter around the emission peak is required to reduce the crosstalk by constraining the detected spectral range.

The problem of crosstalk appears not to be as critical for observations through the eyepiece. Fluorophores can often be distinguished by eye due to their different colors. A green fluorescent protein (GFP) signal can thus easily be distinguished from an additional orange Cy3 label or from broadly distributed, yellowish autofluorescence. Most detection devices that are used for fluorescence imaging, however, are not capable of color distinction. The spectral information (color) of the signal is lost and only the intensity information is acquired. In this monochromatic detection, it is no longer possible to distinguish between different colors, as in a black and white photograph. Under these conditions, a specific fluorescent label cannot be clearly associated with an observed structure. For co-localizing fluorescence signals, analysis becomes impossible.

If such fundamental problems for multiple fluorescent labels exist, why has multi-channel fluorescence imaging been used successfully in biology for the last few decades? The problem of crosstalk between fluorophores can easily be overcome for many standard fluorescence imaging applications. The contribution of a specific fluorophore to a detection channel is determined by two separate factors: the spectral characteristics of the emission filter (detection range) and the

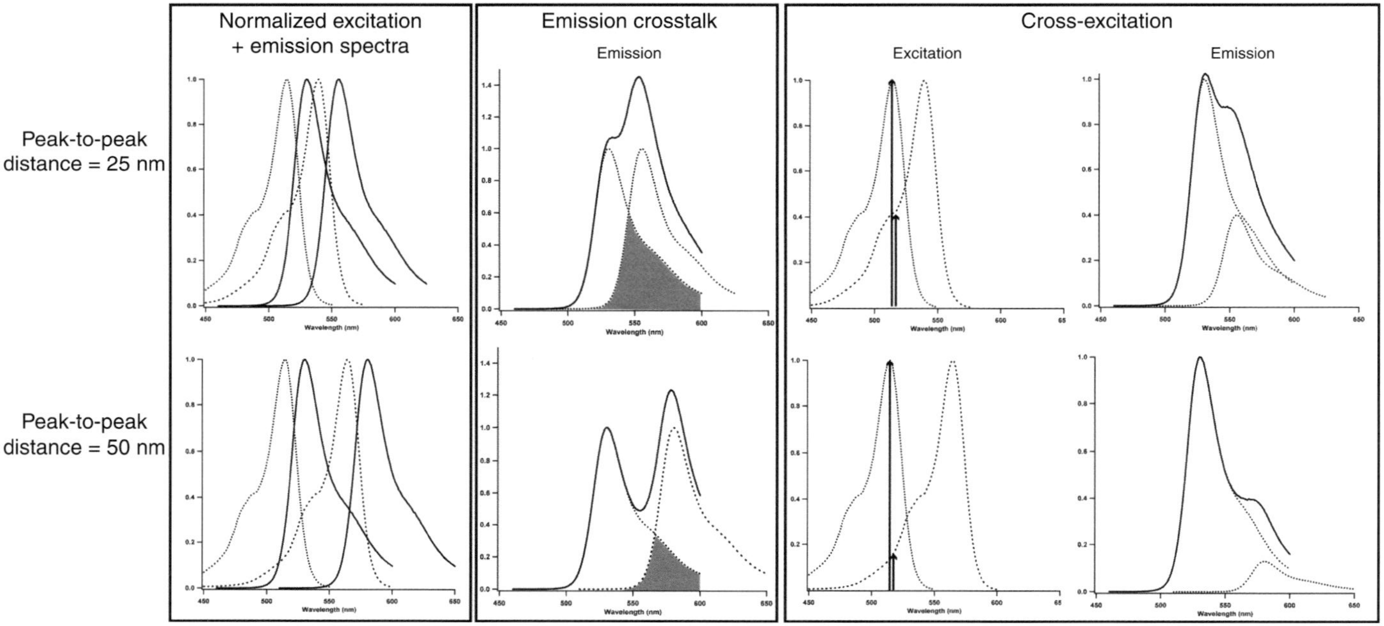

Figure 1. Emission crosstalk and cross-excitation.
The two factors contributing to the spectral overlap are shown for two fluorophore combinations with different degrees of overlap (upper row, peak-to-peak distance 25 nm; lower row, 50 nm). Emission crosstalk, the overlap of the emission spectra (shaded gray) causes a mixed signal that cannot easily be separated for the single contributions when both fluorophores are excited at the same time. Cross-excitation, optimal excitation of the shorter-wavelength fluorophore often also excites fluorophores with longer-wavelength properties, so that a mixed emission signal is detectable that depends on how efficiently the second fluorophore is excited. A comparison of the upper and lower rows shows that the problems increase with decreasing distance between the spectra.

efficiency of the fluorophore in absorbing the wavelengths of the excitation light used. This efficiency is determined by the excitation spectrum of the fluorophore. Therefore, even a fluorophore with significant overlap in its emission with another fluorescent label may not contribute much signal as crosstalk if the excitation wavelengths are chosen carefully. Since the excitation efficiency of a fluorophore decreases significantly for the longer wavelengths after the excitation peak, fluorophores with longer-wavelength properties can be excited at their excitation maximum without co-exciting shorter-wavelength labels.

For standard fluorescence imaging applications, two complementary approaches can therefore be used to separate multiple fluorescent labels reliably:

1. The use of an emission band-pass filter around the emission maximum separates the fluorescence emission of a fluorophore from additional signals of co-excited longer-wavelength fluorophores.
2. Fluorophores of shorter wavelength are not co-excited by light best suited for a specific longer-wavelength fluorophore. As they are not excited, they do not contribute to the signal by crosstalk.

Standard fluorescence imaging combines these two steps by the appropriate selection of excitation filter, beam splitter and emission filter to detect specifically one label with one filter combination. In this way, standard labels for fluorescence imaging can be reliably distinguished (e.g. DAPI, FITC, TRITC, Cy5 or spectrally similar labels). This well-established approach does, however, have clear drawbacks and limitations:

1. Channels are acquired sequentially. At any time, only one excitation can be used in combination with one band-pass filter to avoid crosstalk between fluorophores.
2. Only approximately 50% of the available signal of a fluorophore is detected through a band-pass emission filter. Signal from longer wavelengths makes up the rest, but is not detected, as it overlaps longer-wavelength fluorophores.
3. The method does not work efficiently with highly overlapping emission spectra or for fluorophores with unusual Stokes shifts (spectral distance between excitation and emission) For example, band-pass-based multi-color fluorescence *in situ* hybridization methods for karyotyping use little of the available signal due to the filter restrictions (10).

The shortcomings of the method result in limitations in speed (sequential image acquisition), sensitivity (amount of detected signal), and labeling (choice of fluorophores). These limitations are very significant for advanced imaging applications, especially in the field of *in vivo* imaging:

1. In sequential image acquisition, more time is required for the recording of a single time point, so that temporal resolution is impaired for fast time-lapse imaging. Additionally, there may be a mismatch between the fluorescence channels if fast movements occur during the acquisition of the channels.
2. When imaging very weak or sensitive samples, loss of detection sensitivity by stringent band-pass filters is a problem.

3. The choice of spectral variants of fluorescent proteins is still very limited. Those most suited for *in vivo* labeling have considerable spectral overlap that cannot be separated by the use of specific filter sets (see *Fig. 1*). In addition, the spectra of many of the available *in vivo* dyes are not suitable for standard filter sets.

Methods to correct reliably for crosstalk between fluorophores instead of avoiding it by sequential imaging can overcome these limitations.

2. METHODS AND APPROACHES

2.1. Spectral imaging

The problems with spectral overlap in multi-channel fluorescence imaging are similar to those in the field of satellite imaging. The multi-band images acquired by satellites contain spectrally different aspects of the same scene. The information in the image bands is not clearly separated, but is instead distributed over several image channels. Specific objects in the scene do, however, show a characteristic distribution over the image bands, a 'signature'. In the field of remote sensing, methods have been developed to extract information about specific objects by using the known signatures of these objects and calculating their contributions to the total signal of the multi-band data set (11).

The spectral signature of objects in multi-band satellite images is similar to the representation of the emission spectrum of a fluorophore in multi-channel fluorescence imaging. Similar approaches may therefore also be applicable in analysis of the microscopic data. Recently, processing methods used in remote sensing have also been applied to fluorescence imaging (8, 9, 12).

Three analysis methods commonly used in satellite imaging have also been tested for multi-channel microscopy (8): (i) supervised classification analysis; (ii) primary component analysis; and (iii) linear unmixing. The first two methods are based on classification. This approach has been used for some time in spectral karyotyping using multi-color fluorescent *in situ* hybridization (13) where the labeled chromosomes have only one characteristic signature. Classification techniques, however, do not work for the quantitative analysis of samples that may have co-localized fluorophore signals. This is frequently the case for immunolabelling or fluorescent protein labeling of tissues or cells. It has been shown that the third method, linear unmixing, is best suited for analyzing mixed contributions to a pixel, as in the case of co-localizing labels (8, 9).

2.2. Linear unmixing

Linear unmixing is based on the assumption that the total detected signal (*S*) for every channel (λ) can be expressed as a linear combination of the contributing fluorophores (*Fluo*X):

$$S(\lambda) = A_1 \times Fluo1(\lambda) + A_2 \times Fluo2(\lambda) + A_3 \times Fluo3(\lambda)\ldots \qquad \text{(Equation 1)}$$

where A_x represents the amount of contribution by a specific fluorophore.

More generally this can be expressed as:

$$S(\lambda) = \sum A_i \times R_i(\lambda) \qquad \text{(Equation 2)}$$

or

$$S = A \times R \qquad \text{(Equation 3)}$$

where R represents the reference emission spectra of the fluorophores (12).

If the reference spectra (R) are known, the contributions (A) of the fluorophores in the sample can be calculated for the detected signal (S). This is done by calculating contribution values that most closely match the detected signals in the channels. A least-squares fitting approach minimizes the square difference between the calculated and measured values with the following set of differential equations:

$$\frac{\partial \sum_j \{S(\lambda_j) - \sum_i A_i R_i(\lambda_j)\}^2}{\partial A_i} = 0 \qquad \text{(Equation 4)}$$

where j represents the number of detection channels and i the number of fluorophores.

The linear equations are usually solved with the singular value decomposition (SVD) method (9, 14), so that after calculation of the weighing matrix (A), clear representations of the separated fluorophores can be created (see *Fig. 2*, also available in color section, and *Fig. 3A* and *B* in color section).

Two basic requirements have to be fulfilled for the linear unmixing of spectral data (see *Table 1*):

1. The number of spectral detection channels must be at least equal to the number of fluorophores in the sample, otherwise multiple solutions are possible and spectral separation cannot be performed.
2. All fluorophores contributing to the signal have to be considered for the unmixing calculation. If this is not done, the spectral separation procedure will give false results! Considering more spectra than present in the signal does not

Figure 2. Spectral imaging and linear unmixing of fluorescence signals (see page xix for color version). Contributions of the fluorescent proteins cyan fluorescent protein (CFP), green fluorescent protein (GFP) and yellow fluorescent protein (YFP) to eight successive spectral channels are shown. The distribution of the emission signal to the channels is a direct representation of the fluorophore emission spectrum and thus constitutes its spectral signature. Using these spectral signatures as reference, even combined and mixed signals can be clearly separated by linear unmixing into the fluorophores that contribute to the total signal.

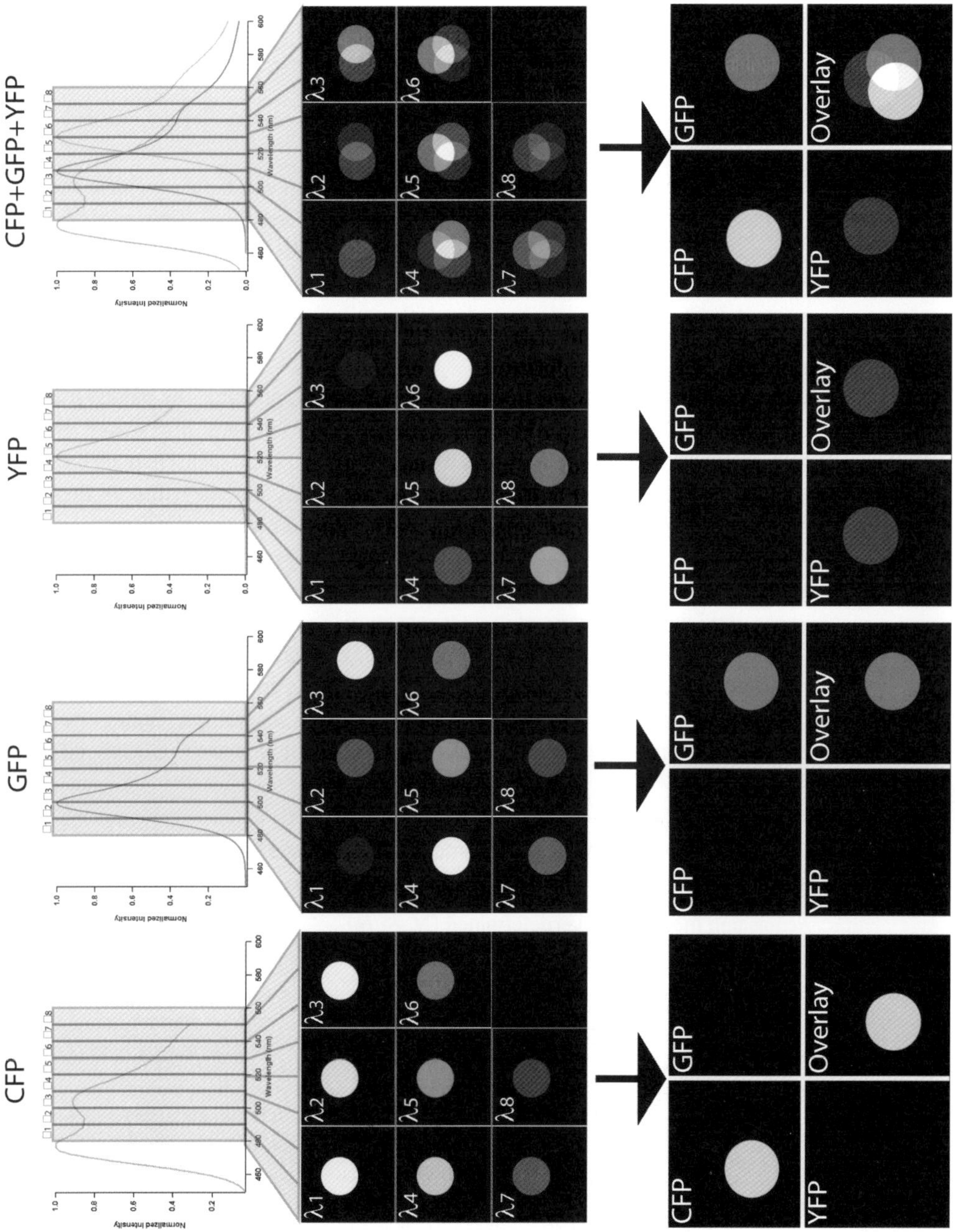
CFP
GFP
YFP
CFP+GFP+YFP
Normalized Intensity
Wavelength (nm)
λ1
λ2
λ3
λ4
λ5
λ6
λ7
λ8
CFP
GFP
YFP
Overlay

Table 1. Sampling requirements for linear unmixing

Detection channels	$< N_{fluorophores}$	$= N_{fluorophores}$	$> N_{fluorophores}$
Solution	No unique solution (underdetermined)	Unique solution (determined)	Unique solution (overdetermined)
Image output	False result	Correct result	Correct result

$N_{fluorophores}$, number of fluorescent labels present in the sample.

affect the unmixing calculation. No contribution will be assigned to them. If background is present in the signal (e.g. autofluorescence), these contributions also have to be considered. This can be done either by defining them spectrally for the unmixing calculation (in the case of structural autofluorescence) or by performing a background subtraction or flat-field correction (in the case of microscope or detection device contributions).

Linear unmixing is performed on measured microscope data. The values used for the unmixing calculations therefore contain inaccuracies introduced by the measurement procedure. The use of a fitting approach (least-squares fit, see above) provides valid results, even with such measurement errors. If there are more detection channels than fluorophores (an overdetermined system, see *Table 2*), an estimate of the quality of the fit is possible by calculating the mismatch of the calculated signal with the original data. Factors influencing the quality of linear unmixing are discussed in a later section.

It is an advantage of measured fluorescent signals that not all cases possible for equation systems in linear algebra have to be considered. Since the signal of a fluorophore is either present or absent in the combined signal, negative contributions are not possible and accordingly do not have to be considered (non-

Table 2. Overview of different commercial microscope set-ups for spectral imaging

Acquisition mode	Overdetermined ($N_c >> N_f$)	Determined ($N_c \approx N_f$)
Parallel	Zeiss LSM 510 Meta[a] Nikon Digital Eclipse C1 si	Leica SP/SP2/SP2-AOBS[b] Olympus FluoView 1000[b] Beam splitter set-ups Confocal microscopes
Sequential	Zeiss LSM 510 Meta[a] Leica SP/SP2/SP2-AOBS[b] Olympus FluoView 1000[b] Liquid crystal tunable filter detection SpectraCube (FTS)	Wide-field microscopes with filter wheels/filter cubes Optical Insights Spectral-DV

[a]The Zeiss Meta detector can read out eight of its 32 channels at the same time. To read out all 32 channels, four sequential acquisitions have to be made.
[b]Sequential: λ-series into a single detector. Parallel: multiple detectors (from two to four).
N_c, number of detection channels; N_f, number of fluorophores

negativity constraint). Using constrained unmixing algorithms, the calculated fluorophore contributions are made to sum to unity to avoid leftover values. This facilitates the comparison between separated fluorophores and keeps the unmixed data in close relation to the measured values.

In addition to the described approach for linear unmixing, alternative approaches for the spectral separation of fluorescence data contained in a few spectral channels have also been implemented (15). By plotting the intensities of a pixel in different spectral channels to separate axes, scatter plots similar to those used in cytofluorometry can be generated. The characteristic signatures of the pure fluorophores are found by cluster analysis and by fitting lines along which the pixels with only one fluorophore are located. Spectral separation is achieved by orthogonalizing these lines into separate channels ('stretching' them on to different axes of the plot). This method does not require *a priori* information about the fluorophore spectra, as the main distributions are found by line fitting. However, it only works reliably if significant amounts of the fluorophores are present without other signals contributing to the same pixel.

2.3. Methods for spectral imaging and linear unmixing

2.3.1. Emission unmixing

Several solutions for spectral imaging have been implemented on wide-field and confocal microscopes. Some of these set-ups require specific instrumentation, but the method is also generally applicable. Any multi-channel fluorescence image can be considered as a series of spectral images. For standard wide-field fluorescence imaging, spectral separation has previously been shown to be improved by determining and correcting for the contributions of individual fluorophores to the different filter sets (16–18). Fourier transform spectroscopy (FTS) allows detailed spectral information to be obtained at every position in a sample by coupling an interferometer to a wide-field microscope (9, 19). This method was initially used for classification-based analysis in spectral karyotyping (13), but was subsequently also applied to the study of microscope samples with co-localizing labels. Using linear unmixing by SVD for the spectral separation of the FTS image data, up to seven fluorophores were imaged together in fixed tissue preparations (9). The same linear unmixing procedure was then also applied to spectral data acquired with two-photon confocal microscopy and a liquid crystal tunable filter to separate up to four fluorescent proteins (8). In the described methods, spectral data are taken sequentially as a series of images (λ-stack). Sequential imaging is time-consuming and a single λ-stack can require up to several minutes. Fluorophore-specific photobleaching rates are another problem of sequential image acquisition. In live-cell imaging, the localization of the fluorescent proteins may rapidly change during the data acquisition and the photobleaching caused by repeated acquisitions may affect the observed cells. Sequential approaches are therefore not favorable for such measurements. The problems of sequential acquisition can, however, be overcome by acquiring the spectral image series into parallel detection channels. This has recently been implemented on commercial confocal microscopes. The first system implementing spectral imaging in combination with linear unmixing was the LSM 510 Meta (Carl Zeiss) in which a grating is used to disperse the signal spectrally on to a multi-detector array (12, 14). Flexible parallel acquisition of spectral data sets has also been realized for prism-based spectral confocal microscopes from Leica (TCS SP, SP2, SP2 AOBS) (15, 20). Bio-Rad (the microscopy arm of which is now owned by Carl Zeiss) has provided a filter-based spectral confocal solution (Radiance 2100 Rainbow) for spectral imaging. Recently, Olympus introduced its variant of a confocal microscope with spectral detection capabilities (FluoView 1000). In this approach, a movable grating for spectral dispersion in combination with a slit of variable width allows the fast and flexible adjustment of the spectral detection range separately for several detectors. The most recent addition to the range of spectral confocal microscopes comes from Nikon (Digital Eclipse C1 si). Here, a multi-detector similar to that of the Zeiss LSM 510 Meta is used. Differences to the existing systems are that all 32 channels can be read out in just one acquisition step, that the spectral resolution can be varied down to 2.5 nm, and that a diffraction efficiency enhancement system decreases losses at the spectral grating.

For all spectral confocal solutions except Bio-Rad, only one confocal pinhole is in the spectral detection light path. This may cause a wavelength-dependent change in the optical slice thickness for the spectral channels. In practice this is not so much of a consideration as many confocals are built with just one pinhole for all detectors; similarly, dyes for which spectral separation is required are, by definition, in a similar spectral range. The Zeiss LSM 510 Meta offers both solutions – separate pinholes for the standard detectors and a single pinhole for the spectral detector.

Spectral imaging solutions for wide-field microscopes are also available. Optical Insights developed a variant of its multi-channel beam splitters that functions as a spectral slit scanner (Spectral-DV) and can be used to analyze samples spectrally line by line. Olympus BioSystems offers an excitation-based solution for spectral imaging (see below).

In addition to the hardware-based confocal solutions for spectral imaging, it has become apparent that linear unmixing of multi-fluorescence data can be realized on a wide variety of microscope systems. As the number of fluorophores in a typical biological sample is often limited to two or three, only a few channels ($N_{channels} \geq N_{fluorophores}$) are required for spectral imaging and most devices capable of multi-channel imaging can be adapted for use in spectral imaging and unmixing. *Table 2* provides an overview of spectral microscope systems and their capabilities.

2.3.2. Excitation unmixing

The possibilities for linear unmixing can be significantly extended by including the second characteristic property of fluorophores – the excitation spectra. As shown for the emission wavelengths, linear unmixing can also be applied based on the excitation spectra of fluorophores. Instead of keeping the excitation constant and collecting the fluorescence emission into spectral channels (see *Fig. 3C* in color section), the sample can be excited sequentially at different wavelengths and the emitted fluorescence can be collected with just one detector (see *Fig. 3D* in color section) (21, 22). As the total emission is collected in just one channel, the signal intensity of the acquired data is very high and the noise contribution low, an advantage for processing by linear unmixing. The acquired data can be analyzed with the same linear unmixing calculations used for emission-based unmixing (see above). Excitation-based spectral unmixing is ideal for microscope systems with excitation light sources that allow fast switching of the excitation light.

Commercial solutions using excitation unmixing exist for wide-field as well as for two-photon confocal imaging. Unlike emission unmixing, parallel detection is not possible, so that excitation unmixing is unavoidably based on sequential data acquisition.

Combining excitation and emission unmixing is a possibility to increase the number of spectral channels available on a microscope system significantly (21, 23). As the maximal number of separable fluorophores is determined by the number of detection channels, this increases the range of possible applications. Spectral data created by combining excitation and emission channels can be processed with the same methods as data created exclusively by one approach.

2.4. Spectral unmixing by subtraction

In some cases of spectral overlap between fluorophores, a more straightforward subtraction approach, as used in many formulas for sensitized emission detection (such as in fluorescence resonance energy transfer experiments; see below) can be used instead of linear unmixing. In the subtraction method, the contribution of the first fluorophore (*A*) into the detection channel for the second fluorophore (*B*) is determined and expressed as a normalized value, R_A:

$$R_A = \frac{A_2}{A_1} \qquad \text{(Equation 5)}$$

To get the second fluorophore without the contribution of the first, a simple subtraction is performed using the information from the first channel:

$$B = Ch_2 - (R_A \times Ch_1) \qquad \text{(Equation 6)}$$

This approach has been used for imaging fluorescent proteins (24) and can be used for more than two fluorophores (see below).

The technique is limited by the requirement of at least one spectral channel to contain just one fluorophore without any other contribution:

$$R_x(\lambda) - \Sigma\, _iR_i(\lambda) = 0\ ,\ x \in \{1...i\} \qquad \text{(Equation 7)}$$

This channel can be used for the subtraction. The method is not applicable in cases of mixed contributions in all channels! In cases of multiple fluorophores, the condition has to be met at the beginning of the calculation and again after each subtraction of a fluorophore. Thus, the subtraction method is limited to a subset of the cases that can be solved by linear unmixing. This means that not all fluorophore combinations can be used and that more restrictive filter settings have to be used that do not sample the entire signal available.

A second limitation of the subtraction method is that while bleed-through is corrected, the additional signal of a fluorophore that is contained in other channels is not utilized further. In linear unmixing, the total signal of all channels is distributed on to the fluorophores according to their contributions. The unmixed signals can therefore be significantly brighter than their representation in the single channels of the raw data. With the subtraction method, this is not possible. Both the subtraction technique and the linear unmixing method are equally susceptible to noise artifacts. In summary, comparing linear unmixing and subtraction for spectral analysis, linear unmixing performs better because it is more flexible and uses more of the available information; it is thus the best approach for processing data with overlapping fluorophores.

2.5. Applications

Almost any fluorescence microscope can be used to acquire spectral image data. For processing by linear unmixing, tools are provided in the operating software of spectral confocal microscopes (see *Table 2*). Solutions are also available for wide-field systems (see *Table 2*). In addition, it is possible to implement linear unmixing in any image processing package and several plug-ins for IMAGEJ are available for download (see *Appendix*, section 6.2). A numerical 'recipe' for simple linear unmixing that can be used in any software is provided in the *Appendix* (section 6.1). With the tools available, spectral imaging and linear unmixing can be applied to many forms of biological imaging.

2.5.1. Time-lapse imaging

Processes in living samples can be highly dynamic. This makes it hard to image two or more fluorescent labels in exactly the same state, as is required for *in vivo* co-localization or ratiometric measurements. We still lack spectral variants of fluorescent proteins that can be excited simultaneously and imaged without too much overlap in their emissions; this means that it is necessary to clean up the signals acquired. The problem of bleed-through with GFP variants can be partially overcome by sequential image acquisition (25). However, this cannot be done when the need to resolve fast processes, e.g. the movement of membrane transport carriers (26), precludes the use of time-consuming sequential image acquisition. In this case, spectral imaging offers the possibility of exciting two spectrally similar fluorophores (e.g. enhanced GFP and enhanced yellow fluorescent protein (YFP)) with just one wavelength and detecting them simultaneously without significant losses in the fluorescence emissions. In contrast to the established methods of using defined restrictive band-pass filters to separate the emission signals of distinct fluorophores, spectral imaging collects almost all of the emitted fluorescence. This is an advantage when working with the weak signals of living samples. Additionally, the possibility of exciting two fluorophores efficiently with the same wavelength helps reduce the total amount of illumination applied to living samples. Therefore, spectral imaging, even if it involves only two fluorophores, should be considered the method of choice when working with living samples.

Another recent *in vivo* application of spectral imaging has been the detection of multiple components in the fluorescence signals of single cells and the subsequent analysis by blind spectral decomposition (23). A combination of excitation and emission properties was used to resolve more than ten spectral components and to track their changes over time. No spatial resolution (images) of the cellular signals was involved in this application, but the method can be extended for microscopic imaging, as described in section 2.3.2 for combinations of excitation and emission unmixing.

2.5.2. Fluorescence resonance energy transfer (FRET)

Another application of spectral imaging is FRET microscopy, which is an important tool for imaging the dynamics of cellular processes such as protein–protein interactions (27). An efficient FRET signal requires significant overlap between the emission spectrum of the donor and the excitation spectrum of the acceptor fluorophore. This requirement almost inevitably also leads to a significant overlap of the emission spectra of the donor and the acceptor. This complicates the determination of FRET efficiencies (27–29). Since spectral imaging separates even highly overlapping donor and acceptor emissions, FRET imaging with already established donor–acceptor pairs can be facilitated (14) and can even be applied to new donor–acceptor combinations with increased FRET efficiencies due to the spectral overlap of the donor emission and acceptor excitation (20).

FRET in living samples is often observed with methods that detect and quantify the sensitized emission of the acceptor fluorophore. In the most common methods, the crosstalk between measurement channels is corrected by the subtraction of image channels (27) (see above). The subtraction step can easily be replaced by linear unmixing. Since linear unmixing also covers cases of strongly overlapping fluorophores, the method can be extended, and less restrictive filter settings that collect more of the signal can be used (contributions of both fluorophores to both channels are allowed). By possibly correcting for additional signal contributions and by providing proper representations of the total fluorescent signals, quantitative FRET analysis can be improved (29). By combining spectral and polarization imaging of FRET signals, all the required information for donor, acceptor, and FRET pair concentrations can be acquired with just one exposure instead of two or three as is normally required (30). Spectral imaging and linear unmixing is thus a potent addition to the existing FRET detection and analysis techniques.

3. RECOMMENDED PROTOCOLS

The following protocols are intended as a guideline for simple spectral acquisitions on different microscope system configurations. The section focuses on the general requirements for a good spectral data set. It covers the linear unmixing process in less detail, as this is usually performed within the microscope operating software and thus varies among manufacturers.

Protocol 1

Acquiring a spectral data set with a multi-detector array (confocal)

Equipment

- Spectral confocal microscopes with a single multi-channel detection unit (Zeiss LSM 510 Meta, Nikon Digital Eclipse C1 si)
- System specification example:
  LSM 510 Meta confocal laser scanning microscope (Carl Zeiss)
  Laser lines: 405, 458, 488, 514, 543, and 633 nm
  Detectors: two photomultiplier tubes (filter-based) and one Meta detector (32 channels, spectral range 350–700 nm)

Method

1. Activate the spectral detection mode (Zeiss: Lambda mode)
2. Select the required number of spectral channels to cover the spectra of interest. For most applications, a detection range of approx. 100 nm may be enough. This can be a speed or illumination consideration, as using more than eight spectral channels requires additional acquisition passes on the Zeiss LSM 510 Meta. Also, sampling channels with no relevant spectral information simply increase the amount of data, but not information.
3. Set the spectral width of the detection channels. Narrower detection channels provide more spectral resolution, but wider channels sample weak signals more efficiently. The spectral width of the detection channels varies from 11 to 44 nm for the Zeiss LSM 510 Meta.
4. If possible, do not use spectral channels that contain reflections of the excitation laser light. The problem can be encountered near the coverslip when imaging weak samples with a high level of excitation or in the case of refractive index mismatches (e.g. oil-immersion/aqueous medium). Reflection problems can be improved by using a beam splitter that is matched to the excitation instead of a neutral beam splitter, but this will affect the representation of the spectral information (see *Protocol 4*, Notes). Using an objective and immersion medium that matches the refractive properties of the sample also helps.
5. When setting up your acquisition, set the excitation intensity and detection gain so that your structures of interest are not saturated in the image. Exceeding the detection range by saturation means that your data at this position of the image cannot be reliably unmixed. If strong signals outside your structure of interest are present, it is acceptable to saturate them to get a good image of your structure. However, this information will then not be accurate in the unmixed image.
6. Acquire the spectral images.
7. For the acquisition of reference spectra, proceed according to *Protocol 4*.

Protocol 2

Acquiring a spectral data set with multiple detectors (confocal)

Equipment

- Spectral confocal microscopes with flexible spectral detectors (Leica SP or SP2 series, Olympus FV 1000)[a]
- System specification example:
  TCS SP2 AOBS laser scanning confocal microscope (Leica Microsystems)
  Laser lines: 405, 458, 476, 488, 514, 543, and 633 nm
  Detectors: four photomultiplier tubes with adjustable, nonoverlapping detection ranges between 400 and 800 nm

Method

1. Set the detection channels so that most of the signal of interest is spectrally covered. Use at least as many detection channels as there are different fluorophores. The positioning of the channels should be chosen according to the considerations of maximal spectral separation (see section 4.1.4).
2. If using several excitation laser lines, avoid reflections of the excitations by positioning the detection channels accordingly. Reflection can be decreased by using a multi-dichroic beam splitter adjusted to the excitations used. Matching the refractive index of the objective and immersion medium to the sample also helps.
3. When setting up your acquisition, set the excitation intensity and detection gains so that your structures of interest are not saturated in the image. Exceeding the detection range by saturation means that your data at this position of the image cannot be reliably unmixed. If strong signals outside your structure of interest are present, it is acceptable to saturate them to get a good image of your structure. However, this information will then not be accurate in the unmixed image. The gains of the spectral detectors can be set separately so that they are optimal for each channel. The references then have to be taken with exactly the same gain settings.
4. Acquire the spectral images.
5. For the acquisition of reference spectra, proceed according to *Protocol 4*.

Notes

[a]Multi-detector array spectral confocals can emulate the use of spectral channels of variable width (e.g. Meta-channels). They can then be used similarly to the separate spectral detectors described here, except for the gain settings in simultaneous image acquisition.

Protocol 3

Acquiring a spectral data set with wide-field microscopes

Equipment

- Wide-field fluorescence microscopes with filter cubes or filter wheels and/or monochromators
- System specification example:

  Inverted Zeiss Axiovert 100 microscope (Carl Zeiss) with a 100 W mercury lamp for fluorescence illumination

  Spectral control: separate excitation and emission filter wheels plus control unit (Ludl Electronic Products)

  Detection: Photometrics CoolSNAP HQ CCD camera (Roper Scientific)

  Operated under METAMORPH software (Universal Imaging)

Method

1. Choose instrument-dependent filter settings that give good signals for the fluorophores of interest[a].
2. Acquire the spectral images as a multi-channel measurement.
3. For the acquisition of reference spectra, proceed according to *Protocol 4*.

Notes

[a]Depending on the system, both excitation and emission (filter cubes) or excitation and/or emission settings (separate excitation and emission filter wheels, monochromator) may vary between the channels.

Protocol 4

Acquiring reference spectra

Equipment

- As for *Protocols 1–3*

Method

1. If possible, use separate samples for the reference spectra (for example, living cells transfected with one fluorescent marker; fixed cells labeled with one fluorophore).
2. Use the identical spectral acquisition settings that were used for the spectral data set to be unmixed[a].
3. Make sure that the reference signal is not saturated, but is clearly visible. If the intensity differs significantly between the sample to be analyzed and the references, the excitation intensity (confocals) or the exposure time of the camera (wide-field systems) can be adjusted to avoid saturation, as this will not change the spectral distribution and representation of the reference emission spectrum.

Notes

[a]The selected beam splitters (e.g. double or triple dichroics) in the light path strongly affect the representation of the emission spectra in the measurement by introducing dips in the signal where the beam splitter is sensitive for excitation light. Only in the case of neutral beam splitters (e.g. 80/20) is the emission unaffected by the beam splitter characteristics. The references used for unmixing *must* be taken with the same beam splitter. References from a spectral library with a different beam splitter will give false results.

4. OPTIMIZATION AND TROUBLESHOOTING

The quality of the unmixed result depends directly on the acquisition settings and the quality of the spectral data set. A thorough understanding of the parameters influencing spectral imaging and linear unmixing is thus required to obtain good results and to be able to evaluate the advantages and disadvantages of the available acquisition possibilities.

4.1. Parameters affecting spectral imaging and linear unmixing

Spectral imaging methods have already been used in satellite imaging and remote sensing for some time and factors influencing the quality of the result have been identified and analyzed (11). The technique is predominantly limited by factors such as image background and detector noise, but it has also been shown that appropriate selection of the number and spectral range of the detection channels in relation to the fluorophores to be distinguished also plays an important role (21,

31). In the imaging of living specimens, these considerations become important, as the signals of interest are usually weak, and bleaching and photodamage by overintense irradiation has to be avoided.

4.1.1. Reference spectra

The use of accurate reference spectra is essential for correct unmixing results. The best way to acquire spectra for unmixing is by using reference samples with only one fluorophore. This avoids the dangers of accidental contributions of other fluorophores to the measured spectrum. Any distortion of the reference spectra by other signals would result in assignment errors that would make a quantitative analysis of the images impossible. If co-localization of signals can be excluded with certainty in the regions used for reference, it is also possible to use samples containing multiple labels to obtain the reference spectra.

Spectral analysis of microscope data differs from that of satellite images in several aspects. One of them is that in microscopy more *a priori* knowledge exists about the sample, as the dyes used for staining are known. This can be used as an initial step to improve the results. In the processing of spectral data, only the reference spectra of fluorophores present in the sample should be used. The inclusion of additional spectral references could degrade the resulting image because noise and other aberrations in the spectral data result in a higher unmixing error under these conditions. In advanced spectral unmixing procedures, the exclusion of absent reference spectra can also be done by an optimization step (e.g. iterative endmember ejection, see below), but using knowledge of the fluorophores present also helps here, as the substantial noise in microscope images makes the automated decision-making less reliable.

4.1.2. Background correction

It is an essential prerequisite of linear unmixing analysis that any signal not coming from the fluorophores to be analyzed has to be removed by background correction (9). As is also the case for ratiometric measurements (32), intensity-dependent artifacts in the process can be caused by the background contribution. Background can often be corrected by the subtraction of a constant or a reference image. In the case of a spectrally homogeneous background, the background can be treated as a further fluorophore and can be separated from the fluorophore signals by linear unmixing.

4.1.3. Noise

When applied to microscope images, linear unmixing is a pixel-based method. It is thus susceptible to the errors introduced into the images by the Poisson noise of the fluorescence light and by detector readout noise. These noise contributions become relevant at the low light levels present in live-specimen imaging, where the illumination intensity and the exposure time usually have to be kept to a minimum to preserve the sample. In the absence of detector readout noise, the unmixing efficiency is independent of the number of detection channels used, as has been shown by computer simulations (21). In the presence of detector readout noise, the errors in the unmixed image increase with the number of detection channels (21). If the detector readout noise is equal for all detection channels, the detector signal-to-noise ratio, (dSNR, no other source of noise included) decreases according to $1/n$ (n = number of detection channels) as each detector receives smaller amounts of the total signal. The decrease in dSNR is only partially compensated by the averaging effect of increased numbers of detection channels. The averaging improvement corresponds to $n^{-1/2}$ (33). Combining the effects of the decrease of dSNR by $1/n$ with increasing channel numbers and the improvement by averaging of $n^{-1/2}$ leaves an overall decrease in the dSNR of $n^{-1/2}$. Sampling the fluorescence signals into fewer, wide detection channels should therefore result in an improvement of the unmixed data compared with using more but narrower channels. This effect is relevant for measurements that contain high detector readout noise and/or low fluorescence signals. As there are multiple sources of noise in a microscope image, the total SNR for all noise types in a measurement is affected by the number of detection channels proportional to the contribution of detection noise to the total noise.

Detection noise is signal independent, so additional detectors increase the total noise in a measurement. This is not the case for another significant source of noise of weak fluorescence signals. The Poisson-distributed noise of the emission light (photon shot noise) is *not* affected by the number of detection channels. The standard deviation of Poisson noise depends on the strength of the signal. For signal-dependent noise, fewer signals per channel means less noise in that channel. Although the photon shot noise SNR ($N/\sqrt{N}$ for N photons) is lower for weaker signals, the total noise of the measurement improves according to $1/\sqrt{N}$ by the averaging of n channels. Results obtained with few or many channels are thus similar for the signal-dependent photon shot noise.

4.1.4. Detector channel characteristics and fluorophore overlap

Acquiring spectral data with few or many channels differs in another aspect. If working with significantly more channels than fluorophores, the spectral positioning of the channels does not affect the measurement significantly. If working with approximately as many channels as fluorophores, then filter settings must be chosen carefully for optimal results (21, 31). The best arrangement of the detection channels can be determined according to the spectra of the fluorophores. The optimal solution corresponds to maximal separation. This can be found by calculating the endmember separability using the SVD method (see section 2.2) for diverse filter set-ups (21, 22). It can also be found using a 'figure of merit' calculation as a measure of the efficiency in the use of emitted photons (31). In contrast to potentially more complex mathematical functions, the relative simplicity of fluorophore spectra (increase, peak, decrease) helps intuitive targeting of the optimal placement of detection channels. For most fluorophore combinations, the optimal border between two channels is located at the intersection of the falling slope with the rising slope of the second fluorophore spectrum.

Even with optimized settings of the detection channels, the unmixing efficiency is directly affected by the degree of spectral overlap between fluorophores. Errors in the unmixed data increase with increasing overlap of the fluorophores. Beyond optimal placement of the detection channels (see above), an improvement in this case can be achieved by sequential acquisition with partially overlapping detection channels (spectral oversampling) (10). In the case of significantly overdetermined set-ups (more channels than fluorophores, nonoverlapping channels), the separation of the fluorophores is solely influenced by the amount of spectral overlap between the used fluorophores. If all optimization steps for the best-suited filter settings have been taken, the next step for improving the unmixed images of overlapping fluorophores would be to improve the SNR of the data by more intense illumination, longer exposures, etc. As the noise in the data causes the unmixing errors, the SNR is the factor to improve. As in all imaging experiments, a compromise between illumination, bleaching, and viability has to be found for an optimal result.

4.1.5. Number of detection channels required

Since both determined (channels = fluorophores) as well as overdetermined (channels > fluorophores) systems can be used for spectral imaging and linear unmixing, where are the differences in performance (21, 31)?

Readout noise increases with the number of detection channels used (see above). In cases of significant readout noise, the quality of the unmixed result is affected. On the other hand, an estimate of the error of the fitted unmixing solution is possible for overdetermined but not for determined systems by comparing the fit with the raw data from each detection channel.

The positioning of the detection channels is highly relevant for determined set-ups (see above), whereas significantly overdetermined systems are less affected by the filter settings. This allows the following conclusions:

1. The best results for unmixing are obtained with few and wide channels, since this gives higher signal levels per channel and possible readout noise problems are minimized. If the gains of the channels can be set independently, the separation of the signals can be enhanced additionally. However, fine-tuning of the settings is required, as only optimized settings will provide an improvement in the result. On the other hand, choosing the wrong settings will give inferior results.
2. Overdetermined systems may not give better results compared with optimized few-channel solutions, but they can be used without fine-tuning of filter settings.

It is often possible to operate detection systems with many spectral channels in modes that work similarly to systems with fewer detection channels, so that a maximum degree of flexibility is given. For example, the Zeiss LSM 510 Meta can reduce its spectral resolution from 11 to 44 nm per channel by detector binning. In addition, even more flexible filter settings are possible by the defining 'Meta' channels, which can be handled like normal filter-based detection channels and allow the setting of different detection sensitivities for each channel.

4.1.6. Dynamic range for signal detection

It is a prerequisite for spectral unmixing that the signal to be analyzed is not saturated in any of the detection channels used. This can cause problems for the settings of detection sensitivity when strong and weak signals have to be detected in the same image. Here, detection set-ups with separate sensitivity controls for the detectors have an advantage as they can flexibly adjust the dynamic range to cover strong and weak signals optimally. However, the spectral references have to be taken under the same conditions, as different gain settings distort the spectra. Set-ups with just one gain control for the spectral detection unit are at a disadvantage here, although they are easier to handle and understand as the displayed signal directly corresponds to the undistorted fluorophore spectra. In addition, using modes that emulate the flexibility of filter-based channels allows separate control of the gains for different channels.

4.2. Optimization of linear unmixing calculations

In addition to optimizing the conditions of detection of the spectral data, there are also possibilities for improving the linear unmixing step (see *Fig. 4* in color section). As mentioned before, the inclusion of increasing numbers of reference spectra decreases the quality of the unmixed result. A noisy spectral data set containing two fluorophore signals will give an inferior result when unmixed for three fluorophores instead of two. An initial optimization step is therefore to limit the unmixing to only the fluorophores present in the sample. It would, however, be hard to unmix a spectrally complex sample like this, as the sample may contain, for example, five different fluorophores. Here, an optimization step in the unmixing algorithm can help. As most of the time not all spectra present in the sample are present in a single pixel, the unmixing result can be improved significantly by excluding signals not present in a pixel from the calculation. An initial step is to use constrained unmixing using the knowledge that a fluorophore can be absent but cannot give a negative signal. Constrained linear unmixing disregards negative contributions of fluorophores and makes the sum of the remaining calculated contributions equal to the intensity of the unmixed signal. This improves the result significantly, but making this calculation with all the reference spectra in the sample can still give a significant unmixing error. A second step is therefore to redo the unmixing calculation with only the reference spectra that gave no negative contribution in the first calculation. This process of iterative endmember ejection can be repeated until only the fluorophores actually contributing to the pixel are used for the calculation. Application of this procedure to every pixel results in significantly less error during unmixing (see *Fig. 4* in color section).

Acknowledgements

I thank Jens Rietdorf, Stefan Terjung, and Rainer Pepperkok for helpful discussions on the subject of this chapter.

5. REFERENCES

1. **Zhang J, Campbell RE, Ting AY & Tsien RY** (2002) *Nat. Rev. Mol. Cell Biol.* **3**, 906–918.
2. **Chan WC, Maxwell DJ, Gao X, Bailey RE, Han M & Nie S** (2002) *Curr. Opin. Biotechnol.* **13**, 40–46.
3. **Lippincott-Schwartz J & Patterson GH** (2003) *Science*, **300**, 87–91.
4. **Miyawaki A, Sawano A & Kogure T** (2003) *Nat Cell Biol.* **Suppl.**, S1–S7.
5. **Hu CD & Kerppola TK** (2003) *Nat. Biotechnol.* **21**, 539–545.
6. **Gerlich D, Beaudouin J, Gebhard M, Ellenberg J & Eils R** (2001) *Nat. Cell Biol.* **3**, 852–855.
7. **Stephens DJ, Lin-Marq N, Pagano A, Pepperkok R & Paccaud JP** (2000) *J. Cell Sci.* **113**, 2177–2185.

★★★ 8. **Lansford R, Bearman G & Fraser SE (2001)** *J. Biomed Opt.* **6**, 311–318. – *First publication demonstrating the power of spectral imaging and linear unmixing for light microscopy, with a clear and concise explanation of the concept that was later implemented in the LSM 510 Meta.*

★★★ 9. **Tsurui H, Nishimura H, Hattori S, Hirose S, Okumura K & Shirai T** (2000) *J. Histochem. Cytochem.* **48**, 653–662. – *First demonstration in fluorescence imaging of the now commonly used approach for linear unmixing. Good explanation of the mathematical steps required.*

10. **Garini Y, Gil A, Bar-Am I, Cabib D & Katzir N** (1999) *Cytometry*, **35**, 214–226.
11. **Landgrebe D** (2000) *IEEE Sig. Proc. Mag.* **19**, 17–28.

★★ 12. **Dickinson ME, Bearman G, Tille S, Lansford R & Fraser SE** (2001) *BioTechniques*, **31**, 1272, 1274–1276, 1278. – *Good technical description of the first commercial confocal microscope with linear unmixing functionality, clearly demonstrating the Meta concept and its uses.*

13. **Schrock E, du Manoir S, Veldman T, *et al.*** (1996) *Science*, **273**, 494–497.
14. **Hiraoka Y, Shimi T & Haraguchi T** (2002) *Cell Struct. Funct.* **27**, 367–374.
15. **Olschewski F** (2002) GIT *Imaging & Microscopy*, **2**, 22–24.
16. **Castleman KR** (1993) *Bioimaging*, **1**, 159–165.
17. **Castleman KR** (1994) *Bioimaging*, **2**, 160–162.
18. **Kato N, Pontier D & Lam E** (2002) *Plant Physiol.* **129**, 931–942.
19. **Malik Z, Cabib D, Buckwald RA, Talmi A, Garini Y & Lipson SG (1996)** *J. Microsc.* **182**, 133–140.
20. **Zimmermann T, Rietdorf J, Girod A, Georget V & Pepperkok R** (2002) *FEBS Lett.* **531**, 245–249.

★★★ 21. **Zimmermann T, Rietdorf J & Pepperkok R (2003)** *FEBS Lett.* **246**, 87–92. – *Covers a wide range of spectral microscope set-ups and discusses factors affecting the quality of linear unmixing.*

22. **Zimmermann T** (2005) In: *Advances in Biochemical Engineering/Biotechnology*, Vol. 95. *Microscopy Techniques*, pp. 245–266. Edited by J Rietdorf. Springer-Verlag, Berlin/Heidelberg.
23. **Shirakawa H &Miyazaki S** (2004) *Biophys. J.* **86**, 1739–1752.
24. **Zimmermann T & Siegert F** (1998) *BioTechniques*, **24**, 458–461.
25. **Ellenberg J, Lippincott-Schwartz J & Presley JF** (1999) *Trends Cell Biol.* **9**, 52–56.
26. **Shima DT, Scales SJ, Kreis TE & Pepperkok R** (1999) *Curr. Biol.* **9**, 821–824.
27. **Wouters FS, Verveer PJ & Bastiaens PI** (2001) *Trends Cell Biol.* **11**, 203–211.
28. **Gordon GW, Berry G, Liang XH, Levine B & Herman B** (1998) *Biophys. J.* **74**, 2702–2713.
29. **Neher RA & Neher E** (2004) *Microsc. Res. Tech.* **64**, 185–195.
30. **Mattheyses AL, Hoppe AD & Axelrod D** (2004) *Biophys. J.* **87**, 2787–2797.

★★★ 31. **Neher R & Neher E** (2004) *J. Microsc.* **213**, 46–62. – *Provides a quantifiable basis for optimizing spectral detection parameters.*
32. **Bolsover SR, Silver RA & Whitaker M.** (1993) In: *Electronic Light Microscopy*, pp. 181–210. Edited by D Shotton. Wiley-Liss, New York.
33. **Sheppard CJR, Gu M & Roy M** (1992) *J. Microsc.* **168**, 209–218.
34. **Power C** (2003) Spectral Reassignment in LaserSharp2000. Bio-Rad technical note (15).

6. APPENDIX

6.1. Numerical recipe for linear unmixing

To facilitate an understanding of the principle of linear unmixing, the solution for a two channel/two fluorophore situation is explained here.

$Ch_{x,y}$ represent the signals in detection channels x and y, and A_x, B_x and A_y, B_y the normalized contributions of *FluoA* or *FluoB* to channels x and y as they are known from the spectral signatures of the fluorescent proteins.

Deduction:

$$Ch_x = A_x FluoA + B_x FluoB$$
$$Ch_y = A_y FluoA + B_y FluoB$$

$$Q = \frac{Ch_x}{Ch_y} \Rightarrow Ch_x = QChy \Rightarrow$$

$$A_x FluoA + B_x FluoB = Q(A_y FluoA + B_y FluoB)$$
$$A_x FluoA + B_x FluoB = QA_y FluoA + QB_y FluoB$$
$$A_x FluoA + QA_y FluoA = QB_y FluoB - B_x FluoB$$
$$FluoA(A_x - QA_y) = FluoB(QB_y - B_x)$$

$$\frac{FluoA}{FluoB} = \frac{QB_y - B_x}{A_x - QA_y}$$

Application:

In order to determine the fluorescence emitted by each of two individual fluorophores (*FluoA*, *FluoB*) in co-localization or FRET experiments, only four equations have to be applied for every image pixel (i):

$$Q(i) = \frac{Ch_x(i)}{Ch_y(i)} \qquad \text{(Equation A1)}$$

$$R(i) = \frac{FluoA(i)}{FluoB(i)} = \frac{B_y Q(i) - B_x}{A_x - A_y Q(i)} \qquad \text{(Equation A2)}$$

The total signal $S(i) = \sum_{k=1}^{n} Ch_k(i)$ as the sum of all channels can then be divided into the contributions of *FluoA* and *FluoB* by:

$$FluoA(i) = \frac{S(i)}{1+\frac{1}{R(i)}} \qquad \text{(Equation A3)}$$

$$FluoB(i) = \frac{S(i)}{1+R(i)} \qquad \text{(Equation A4)}$$

or computationally more simply as:

$$FluoB(i) = \frac{S(i)}{1+R(i)} \qquad \text{(Equation A3')}$$

$$FluoA(i) = S(i) - FluoB(i) \qquad \text{(Equation A4')}$$

In the above example, the computational steps follow the same path as the shown deduction. The two-equation system can, however, be solved in several ways. An even faster spectral reassignment computation can be performed with just two equations, solving directly for *FluoA* and *FluoB* (34):

$$FluoA(i) = \frac{B_x Ch_y(i) - B_y Ch_x(i)}{B_x A_y - B_y A_x} \qquad \text{(Equation A5)}$$

$$FluoB(i) = \frac{A_y Ch_x(i) - A_x Ch_y(i)}{B_x A_y - B_y A_x} \qquad \text{(Equation A6)}$$

6.2. Links for linear unmixing tools

For the free and platform-independent image processing software IMAGEJ (http://rsb.info.nih.gov/ij/), linear unmixing tools are available in the form of plug-ins and macros at http://rsb.info.nih.gov/ij/plugins/spectral-unmixing.html and http://www.embl.de/eamnet/html/linear_unmixing.html, respectively.

CHAPTER 6

Measurement of protein motion by photobleaching

John F. Presley

1. INTRODUCTION

Proteins often move from one location to another within the cell. The use of *in vivo* fluorescent tagging techniques (e.g. green fluorescent protein (GFP)) has made direct observation a viable approach for studying this transport. In the most favorable case, the protein is transported in large carriers or even organelles, which can be directly imaged (1, 2). When such individually visible carriers do not exist, detection and measurement of protein flux is more difficult. Rates of protein synthesis and degradation are usually slow compared with transport processes, resulting in a cellular distribution that is at steady state prior to the beginning of the experiment. A consequence is that, although individual protein molecules may be extremely mobile, the overall distribution of protein in the cell does not change with time. Thus, pharmacological or temperature blocks must be used prior to observing and quantitating motion of the protein (e.g. 3). However, situations in which pharmacological perturbation from steady state is impossible or undesirable are common. The most notable example of such a situation is furnished by the diffusive movement of individual integral membrane proteins in a membrane (4).

Photobleaching and photoactivation provide tools for detecting movement in these situations by effectively 'tagging' or 'untagging' molecules that reside in particular regions of a cell at a particular time. Both techniques have a long history in cell biology. FRAP (fluorescence recovery after photobleaching), also known as FPR (fluorescence photobleaching and recovery), was first used in the 1970s to measure the motions of integral membrane proteins in the plasma membrane of cells (5). Photoactivation was first used to 'tag' and track portions of microtubules in studies of mitosis (6). The recent ready availability of confocal microscopes and the widespread use of GFP tagging has created an increased interest in these photobleaching techniques among cell biologists.

Cell Imaging: ***Methods Express*** (D. Stephens, ed.)

1.1. Physics of photobleaching

Fluorescent proteins used as tracers in live-cell experiments can be derivatized *in vitro* with fluorophores and then introduced into cells by techniques such as microinjection or endocytosis. More recently, the use of GFP (7), which eliminates the need for invasive introduction of the fluorescent protein into the cell, has allowed fluorescent tagging of an increased range of intracellular proteins. Chemical fluorophores (GFP included) are driven into an excited state upon absorbing a photon of light of appropriate wavelength. While in the excited state, the molecule is extremely reactive, particularly with molecular oxygen. Typically, the chemical reaction will result in irreversible loss of fluorescence. As a side effect, potentially cytotoxic free radicals are also generated, although this side effect is minimized with GFP (the radicals probably tend to react with the interior of the GFP before reaching the cellular environment). As the fluorophore remains in the excited state for only 1–10 ns (8, 9), the potential for photobleaching after a single excitation is very low. However, as the fluorophore may be excited many times a second, the cumulative time in the excited state can be much greater, leading to a significant cumulative risk of photobleaching.

Usually, as in fluorescence observations in fixed cells, photobleaching is considered a nuisance and steps are taken to minimize it, including limiting exposure to excitation light and the use of antifade reagents to scavenge molecular oxygen. In a photobleaching experiment, by contrast, the goal is to maximize photobleaching within a defined region of the cell.

1.2. Motivation for photobleaching experiments

Many polypeptides in a cell can diffuse freely through a 3D medium such as the cytosol or within the 2D surface of a membrane. Such a freely diffusing polypeptide can be part of a small complex, the diffusion of which may be rapid, or of a much larger complex, which may diffuse more slowly. Diffusion of a polypeptide in a lipid bilayer can also be affected by the fluidity or presence of microdomains in the lipid bilayer. Thus, photobleaching experiments aimed at measuring diffusion of a protein (often expressed in terms of a diffusion coefficient (*D*) with units of distance squared per time and an immobile fraction; 5) are often used to probe the protein microenvironment. Such experiments are of great historical importance, e.g. in helping to establish the fluid mosaic model of the plasma membrane (10). More recently, photobleaching experiments have shown the existence of barriers to diffusion on the plasma membrane (reviewed in 11), rapid unrestrained diffusion of proteins on internal membranes of cells (12), and have been used to argue for or against the existence of lipid rafts in cellular membranes (reviewed in 13). These more biophysical uses of photobleaching are extensively covered in the literature. Photobleaching techniques have also been used to measure the flux of protein between one intracellular organelle and another (3, 14). The more biochemistry-oriented cell biologist will often be concerned with another feature of the microenvironment of a protein – the presence of other interacting polypeptide chains.

1.3. Scope

At one time, specialized 'photobleaching machines' were required for FRAP experiments (5). As such systems were usually not commercially available, FRAP was restricted to specialists and as a result was used only by a small biophysically oriented subset of the cell biology community. More recently the situation has changed. The wide availability of confocal microscopes and GFP-tagged proteins has made FRAP experiments almost trivial to perform.

Unfortunately, ready availability of equipment and reagents has not led to a similar wide dissemination of expertise in design and interpretation of such experiments or in awareness of the common pitfalls of FRAP experiments. We describe here methods for analysis of two simple but common situations: determination of the diffusive mobility of a protein in membrane or cytosol (the classical FRAP application), or the exchange of a protein between a large organelle such as the Golgi apparatus or the nucleus and the rest of the cell (see *Fig. 1* for an illustration of some major categories of photobleaching experiments). Mathematical sophistication is not expected or needed by the reader beyond an understanding of simple first-order kinetics and some common sense in understanding when the simple assumptions made here fail for a particular system. The two simple applications described go surprisingly far, but more complex situations (such as a protein shuttling between multiple compartments) cannot be discussed in detail due to space limitations. These more complex applications will likely require the additional assistance of a mathematically sophisticated expert.

2. METHODS AND APPROACHES

2.1. Diffusion of a fluorescently tagged protein within a membrane

Photobleaching experiments (referred to in the literature as FRAP (15) or FPR (5)) were classically first used to measure mobility of proteins in the plasma membrane or in model membranes. Typically the protein was labeled by the binding of fluorescently tagged F_{ab} fragments, hence the original restriction of diffusion measurements to the exposed plasma membrane of cells and to model membranes. An area of known geometry in a 2D membrane would be photobleached and a recovery curve obtained from the same area. Typically, two parameters would be determined – a diffusion coefficient (D) and an immobile fraction. The diffusion coefficient would be determined from an equation involving the recovery curve, and the immobile fraction would be the degree to which the fluorescence in the bleached area failed to reach its original value. A detailed discussion of these methods including derivation of equations for a variety of different geometries is given in (5). Protocols will subsequently be given for two simple bleaching geometries easily achievable with confocal microscopes – a uniform circular bleach (5) and a long rectangular strip (16). An alternate approach to measuring diffusion coefficients that does not require a well-defined

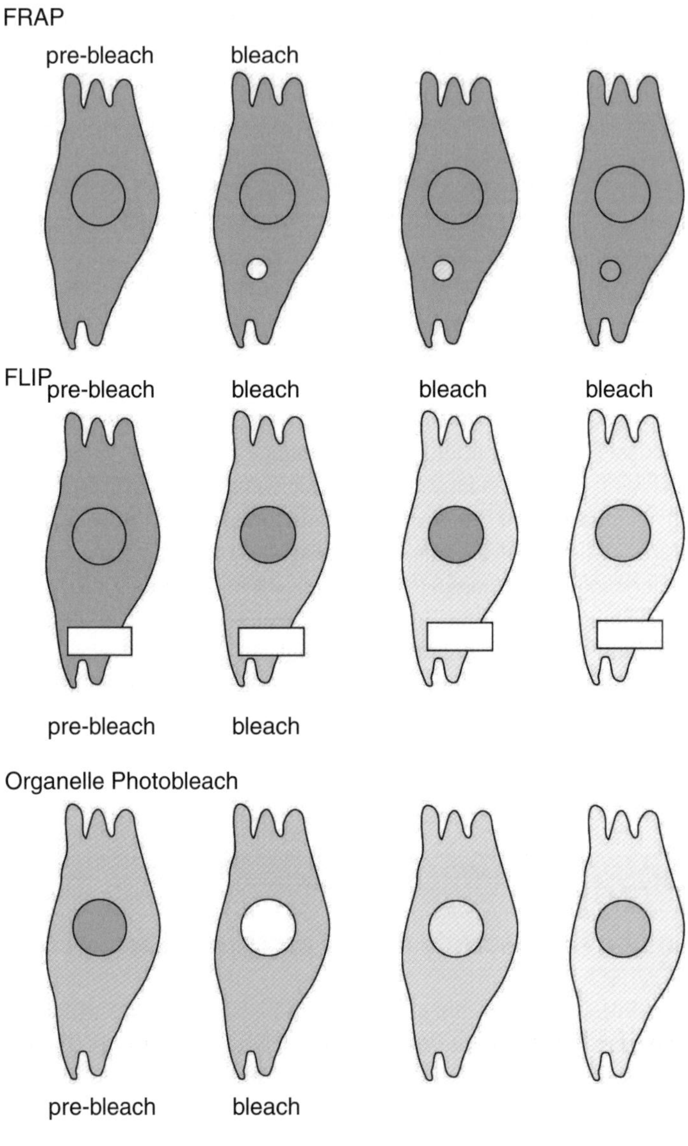

Figure 1. Major categories of photobleaching experiments described in the text (see page xxii for color version).

geometry for the bleached area is given in (17), but involves nontrivial computer-based analysis.

Determination of *D* and the immobile fraction on intracellular membranes is not conceptually different from such measurements on the cell surface, the primary difficulty being the introduction of the fluorescent protein. The advent of GFP tagging has largely removed this technical barrier. When the organelles in question can be taken as effectively 2D, which is often the case in tissue culture cells, procedures identical to those used for measuring the diffusion coefficient of cell-surface proteins will be effective (e.g. 12, 16, 18). In some cases, however, the 3D organization of the organelle must be taken into account in the experimental design. In work by Ellenberg and coworkers in which the goal was to determine the mobility of nuclear envelope proteins (16), this was done by making a photobleached region on the bottom of the nuclear envelope small relative to the total size of the organelle so that only the local neighborhood (largely 2D) would contribute to the bleach recovery.

2.2. Determination of diffusion coefficients for cytoplasmic proteins

Cytoplasmic proteins can be treated as effectively 2D in flat tissue culture cells and an effective *D* calculated (which will be smaller than the true *D* in 3D by a factor of 2/3). Cytoplasmic proteins present the additional technical problem that they can diffuse extremely fast, up to two orders of magnitude faster than membrane proteins (19). Thus, the bleach and post-bleach image acquisition must also be very fast. With many confocal systems, effective diffusion coefficients can be calculated by FRAP for cytoplasmic polypeptides in large complexes (which tend to diffuse slowly in cytoplasm), but not smaller polypeptides, as recovery into the bleached region is essentially complete even after the first post-bleach image is taken. When *D* for rapidly diffusing cytoplasmic proteins must be obtained, the most favorable conditions are created when a powerful laser is used, permitting a rapid bleaching. With many confocal microscope systems, post-bleach image acquisition is much faster if only a small region of interest is scanned. Even when these optimizations are made, measurements can be difficult. Thus, Elsner *et al.* (20) could measure a *D* of 0.5 μ^2/s for the complete coat protein I (COPI) complex (800 000 Da) and a *D* of 16 μ^2/s for the free subunit εCOP–GFP (60 000 Da) by fluctuation-correlation spectroscopy. However, these experiments primarily detected diffusion of the complete complex by FRAP, as complete recovery of the much smaller free subunit had occurred by the time of capture of the first post-bleach image.

2.3. Use of the diffusion coefficient to detect entry of a protein into a complex

Many protein–protein interactions have been found by traditional biochemical techniques such as co-immunoprecipitation or in genetic screens (21). Newer techniques including yeast two-hybrid screens (22) and mass spectroscopy-based approaches (23) have streamlined the process of finding protein–protein interactions and increased the pace of their discovery. Many of these protein–protein interactions are spatially and temporally regulated in cells. Light microscopy-based methods for further characterizing protein–protein interactions in living cells are therefore highly desirable. Energy transfer-based single-cell methods, i.e. FRET (24; see also Chapter 12 by Varma and Mayor) and bioluminescence resonance energy transfer (25), are most commonly used. However, these methods are technically difficult and do not work for most protein pairs (discussed in 26).

Photobleaching techniques can be easily implemented and only require that a single protein is fluorescently tagged. Photobleaching-based techniques can measure diffusion coefficients (5), residence time in an organelle (3, 27), and immobile fractions (5). In the simplest case, complexing of the protein of interest to a polypeptide chain on a large structure or with a very large complex can immobilize it. If candidate binding partners have been identified using other techniques, a photobleaching assay for an immobile fraction in combination with RNA interference ablation of the potential interaction partners (see 28) could be implemented without great difficulty.

Complete immobilization of a protein by an interaction partner is not needed in order to detect the entry of a protein into a complex by photobleaching. Small soluble polypeptides will have diffusion coefficients very different from integral membrane polypeptides. The value of *D* for a small cytoplasmic protein such as a Ras family GTPase can be in the range of 20–40 μ^2/s (19) compared with 0.4 μ^2/s for the most mobile integral membrane proteins in a lipid bilayer (29) or 0.05 μ^2/s for a typical cell-surface protein (30). Thus, there can potentially be a 50-fold difference, or more, in mobility between a membrane and a cytoplasmic protein. Binding of a cytoplasmic protein to membrane would result in a corresponding reduction in *D*, effectively rendering the protein immobile on the timescale required for measuring cytoplasmic diffusion coefficients. Entry of an integral membrane protein into a freely diffusing complex of other integral membrane proteins could be more difficult to detect. The value of *D* will scale with the cube root of the complex size, assuming other factors are constant. Thus a 100 000 Da protein would have to enter an 800 000 Da complex or oligomerize to form an octamer to achieve a twofold difference in *D*. Large cytoplasmic complexes may diffuse slowly due to interactions with the cytoplasmic matrix (19, 31), so the association of a small cytoplasmic protein with such a complex may be easier to detect

2.4. Fluorescence loss in photobleaching (FLIP)

FLIP is a variation on FRAP in which a region of the cell is bleached not just once but at regular time intervals (see *Fig. 1*, also available in color section) (12). The result is that all fluorescence in the cell that can access the bleached area is eventually bleached. FLIP is most often used as a sensitive means of mapping an immobile fraction of the fluorescent protein (16), showing in a sensitive way that such an immobile fraction does not exist (12), or for testing whether a pool of fluorescent protein is confined to a particular region of the cell (14). In principle, diffusion coefficients can also be determined by FLIP (17), but this has not been frequently done.

2.5. Photoactivation

Photoactivation is a property of some fluorophores. Photoactivatable fluorophores typically can be altered by short-wavelength (usually UV) light such that they become excitable with longer-wavelength light. The first photoactivatable reagents widely known to cell biologists were 'caged' fluoresceins, which were nonfluorescent but could be 'uncaged' by brief exposure to UV light (6). The Mitchison laboratory labeled tubulin with caged fluoresceins and microinjected it into cells in order to observe poleward flux of microtubules in kinetochores (6). A stripe crossing the microtubules was activated and the movement of this stripe could be followed over time over a dark background. This emphasizes one of the main uses of photoactivatable fluorophores, i.e. to follow a small object or small numbers of objects originating from a particular location.

Recently, photoactivatable GFPs have become available, usually activated by short visible wavelengths in the range of 400–420 nm (32). Photoactivatable GFPs are not as bright as conventional GFPs, even after photoactivation (32), but have some advantages. In long-duration FRAP experiments (as in a study of the very slow turnover of a protein between two organelles), the introduction of new protein into the system by protein synthesis will be an issue. This new protein will not appear randomly, but most likely will first be detected in the cytoplasm or the endoplasmic reticulum (ER). If this new protein contains a GFP, it will be fluorescent and could affect the experimental results. The usual solution to this problem is treatment of the cells with a protein synthesis inhibitor such as cycloheximide. The pharmacological inhibition of protein synthesis can itself have significant effects on the cell, especially if the experiment is of long duration. Photoactivatable GFPs neatly avoid this problem as any new GFP-tagged protein synthesized will be nonfluorescent and will remain so unless intentionally photoactivated. We have not made extensive use of photoactivatable GFPs.

2.6. Exchange of a protein between a large organelle and the cytoplasm

Cytoplasmic proteins can be recruited to organelles by binding to membrane lipids, by high- or low-affinity interactions with proteins on the surface of the organelle, or by more complex mechanisms (e.g. nuclear import). Binding of a cytoplasmic fluorescent protein to a large organelle such as the Golgi apparatus or the nucleus can easily be detected without the use of photobleaching techniques. Whether the protein is attached indefinitely to the organelle or is rapidly exchanging between the organelle and the cytoplasm will not be apparent on visual inspection, but can readily be determined by FRAP. Measurement of residence time of a poorly characterized protein on a particular membrane is quite feasible, but may require additional information to interpret. Pharmacological or RNA interference approaches combined with photobleaching can be a powerful tool for obtaining this additional information if potential interaction partners can be found.

2.7. Qualitative analysis of exchange between an organelle and the rest of the cell

When a protein can be fluorescently labeled, the stability with which the protein associates with a large, easily visible organelle can be estimated by photobleaching. The half-time for recovery (defined as the time at which the recovery curve crosses the midpoint between the initial post-bleach value and a plateau value) gives a rough timescale for exchange between organelle and cytoplasm. An immobile fraction can also be measured. Half-time can be determined empirically for any monotone increasing function that rises to a plateau, even if a mathematical model of the system has not been constructed. Half-lives cannot be determined if there is no recovery after the photobleaching or if the fluorescence oscillates over the timescale of the experiment (e.g. for cell-cycle proteins).

Frequently, the exchange between the organelle and the cytoplasm can be approximated as a first-order kinetic process involving only two compartments (see *Fig. 2*, also available in color section). First-order kinetics arise when the set of cytoplasmic molecules and the set of organelle-associated molecules can be treated as two well-mixed pools with molecules drawn randomly from each pool with characteristic probabilities per unit time. Given two pools, organelle and cytoplasm, and the a priori knowledge that exchange between these compartments follows first-order kinetics, the exchange can then be represented

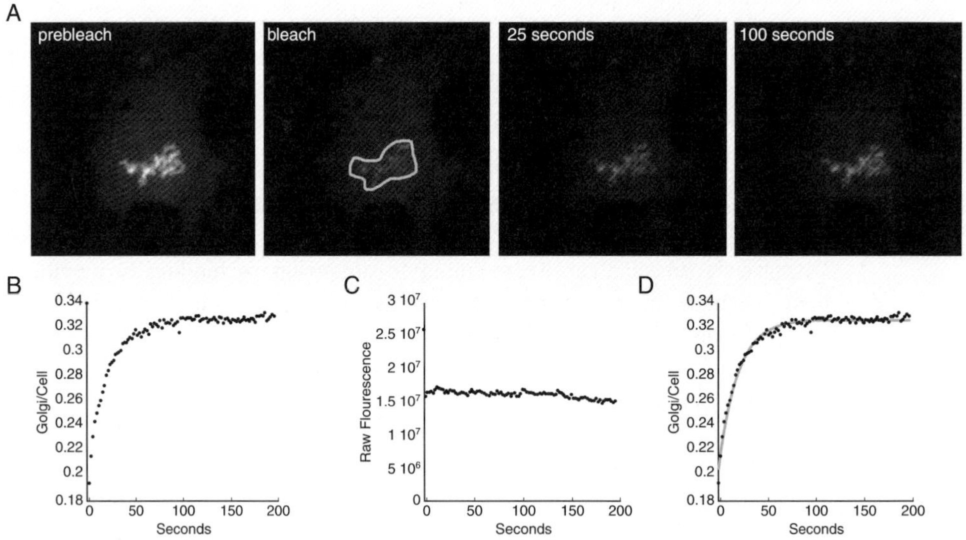

Figure 2. Example of an organelle photobleaching experiment (see page xxiii for color version). The experiment was performed using a Zeiss LSM 510 confocal microscope with a 25× 0.8 NA oil-immersion objective and a fully open pinhole. (*A*) Photobleaching of Arf1–GFP localized on the Golgi apparatus of a normal rat kidney cell. Note that a substantial pool of ADP-ribosylation factor 1 (Arf1)–GFP can be detected on the Golgi apparatus even after the bleaching as this protein cycles between the cytosol and the Golgi on a timescale that is comparable to the total bleaching time. (*B*) Recovery curve for the Golgi area shown as a fraction of total cell fluorescence. (*C*) Total cell GFP fluorescence as a function of time. The drop in fluorescence between the first and second frame is due to photobleaching of the Golgi-associated pool. Total cell fluorescence remains roughly constant for the remainder of the experiment, even though redistribution between the Golgi and the cytosol occurs. This is an important test of whether all fluorescence in the cell is accounted for. (*D*) Curve fitted to date from (*B*) (as described in *Protocol 5*).

by a pair of linear differential equations involving a rate constant k_{in} for transport from the cytoplasm to the organelle and another k_{out} for transport from the organelle to the cytoplasm:

$$\frac{dO}{dt} = k_{in}C - k_{out}O \qquad \text{(Equation 1)}$$

$$\frac{dC}{dt} = k_{out}O - k_{in}C \qquad \text{(Equation 2)}$$

In these equations, O represents the organelle-associated protein, and C represents the cytoplasmic protein. Both are dependent variables and therefore functions of time.

The solution to these equations can be expressed as the sum of exponentials:

$$O = A_1\,[1 - e^{-rt}] + A_2\,e^{-rt} \qquad \text{(Equation 3)}$$

where the parameter r represents the sum of the rate constants:

$$r = k_{in} + k_{out} \qquad \text{(Equation 4)}$$

O is the organelle-associated fluorescence as a function of time and can be measured experimentally. The parameter r can be obtained by using a curve-fitting algorithm as supplied in most graphing packages (the Levenberg–Marquard method is most usual; this is the algorithm used in SIGMAPLOT (Systat Software) and in KALEIDAGRAPH (Synergy Software)). However, what is fitted is the sum of the two rate constants when what is really desired are k_{in} and k_{out}. Additional information is needed – the fraction of fluorescence associated with the organelle and with the cytoplasm in an equilibrium state (prior to photobleaching or after complete recovery to a plateau). Once O_∞ (the organelle-associated fluorescence at equilibrium) and C_∞ (cytoplasmic fluorescence at equilibrium) are known, k_{in} and k_{out} can be determined as follows:

$$k_{in} = \frac{r}{1 + \frac{C}{O}} \qquad \text{(Equation 5)}$$

$$k_{out} = \frac{r}{1 + \frac{O}{C}} \qquad \text{(Equation 6)}$$

Residence times (which are intuitively interpretable) can then be easily calculated. Residence times in the organelle or in the cytoplasm represent the mean time a molecule of the protein of interest spends in the compartment. In this simple two-compartment case, the residence time in the organelle is $1/k_{out}$ while the residence time in the cytoplasm is $1/k_{in}$ (see *Protocol 4*).

The preceding discussion assumes that there is not a large immobile fraction. An immobile fraction can be determined by comparing O/C before the bleach to O/C after complete recovery. If there is no immobile fraction, they should be the same. The applicability of a first-order, two-compartment kinetic approach is conditional on several factors. The cytoplasmic protein should diffuse quickly so that the mixing time over the width of the cell is fast compared with the exchange time between the organelle and cytoplasm. This is usually a valid assumption because of fast diffusion of cytosolic proteins, but may fail in the case where the protein diffuses unusually slowly. It is common that there are multiple subcompartments either within the cytoplasmic or the organelle compartment, each containing a measurable portion of the total protein (see *Fig. 3B*). In this case, there will likely be an initial rise after photobleaching that is faster than predicted by first-order kinetics for a two-compartment system, followed by a slower-than-expected rise to a plateau. From inspection of *Fig. 3(B)*, it should be apparent that the introduction of several subcompartments into systems of the type we have just discussed may produce only subtle changes in the recovery curves produced if equilibrium between the subcompartments can be achieved on timescales roughly similar to the overall recovery. If flux into or out of a large subcompartment is sufficiently faster or slower than flux into or out of other compartments, it will have an easily detectable effect on the shape of the recovery curve (see *Fig. 3C* for the effect of a fourfold slowing of trafficking into and out of a subcompartment constituting 50% of the overall compartment at equilibrium).

It should be noted that if the system described in *Fig. 3(B)* is treated as a two-compartment system (as it likely would be in a real study), the internal structure within the two major compartments would be missed, but the calculated residence times within these visible compartments would be roughly correct. The rate constants obtained would, however, not specify the rates of any single well-defined physical process. The two-compartment system governed by first-order kinetics that we have just discussed is often only an idealization, although a useful one.

Another caveat is that it is likely that binding of a protein to a receptor on an organelle will be saturable, and thus, strictly speaking, that first-order kinetics will not apply. Within a particular cell, however, the concentration of protein and receptor will be constant over the course of the experiment if it is not too long and the methods of analysis described in the protocols are valid within that particular cell. This assumes that photobleaching and photoactivation are not changing the number of protein molecules within a cell, but that they are simply rendered visible or invisible.

2.8. Photobleaching of fragmented organelles

Analysis of cycling dynamics has to this point concentrated on the relatively simple case where an organelle concentrated in a single location in the cell (e.g. nucleus or Golgi apparatus) exchanges a protein with the remainder of the cell. Some organelles (Golgi in nocodazole-treated cells, mitochondria, endosomes) consist of multiple smaller structures dispersed across the cell. If exchange is between smaller but individually visible structures and a dispersed compartment (ER, plasma membrane or cytoplasm), slight modifications of the methods above can be used.

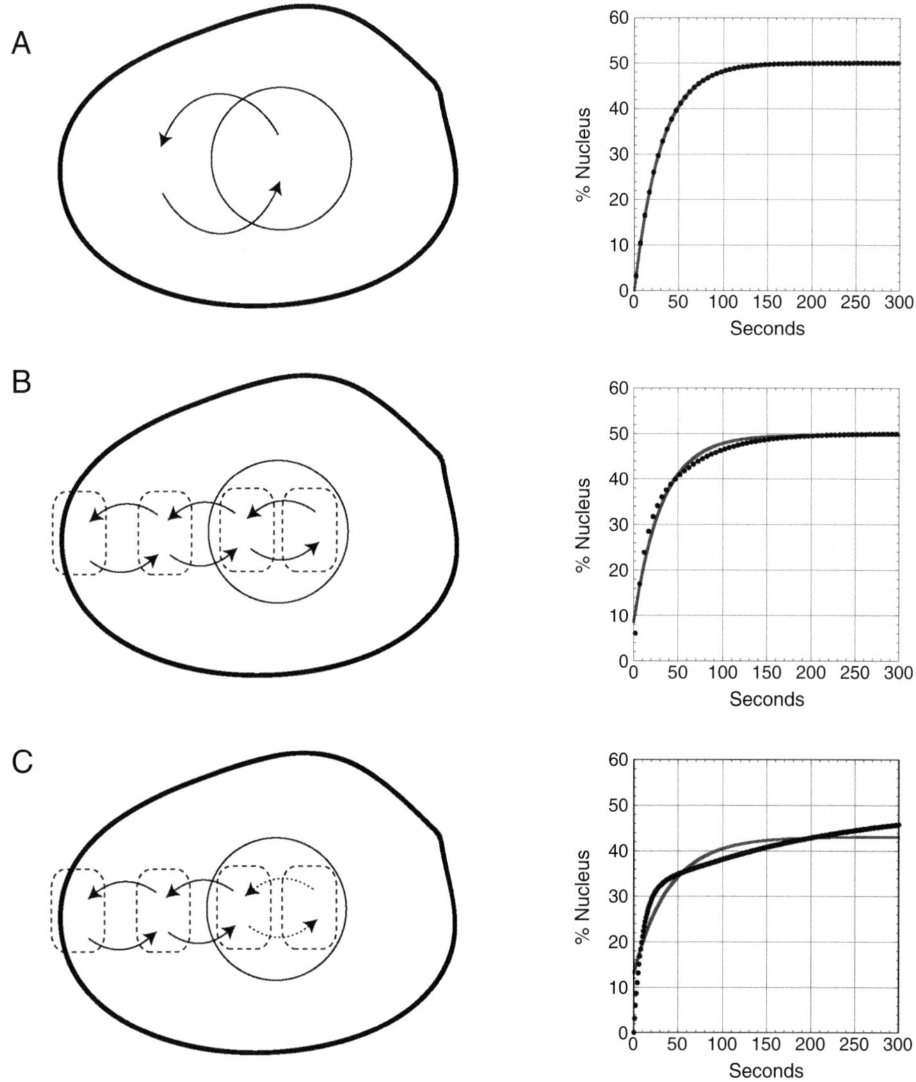

Figure 3. Determination of rate constants from simulated data.
(*A*) A model for exchange of a protein between nucleus and cytoplasm is shown on the left. In this model there are only two compartments and exchange between them involves pure first-order kinetics. A simulated recovery curve for the nucleus after photobleaching with curve fit is shown on the right (k_{in} = 1/60 s^{-1}, k_{out} = 1/60 s^{-1}). (*B*) A more complex model yielding similar observations. A model in which the nucleus and cytoplasm are each divided into two compartments is shown on the left. However, all kinetics are still first order. Compartments are connected as shown by arrows with all rate constants equal to 1/30 s^{-1} so that all compartments are of equal size at equilibrium and therefore the concept of a rate-limiting step cannot be used. A simulated recovery curve after photobleaching of the nucleus is shown on the right. The curve fit is as described in *Protocol 5* and incorrectly assumes the model in (*A*). Note that deviations of the data from the fit are not large. (*C*) Same model as (*B*) but with exchange between the two nuclear pools (left; dashed arrows) occurring with a slower rate constant of 1/240 s^{-1}. Other rate constants remain as 1/30 s^{-1} as in (*B*). The result is that molecules within the nucleus are partitioned into a fast pool and a slow pool of roughly equal size. Note that now the two components are clearly visible in the simulated data and that the fit is obviously poor.

The easiest method to adopt is to bleach an area of the cell containing <20% of the total organelle-associated fluorescence and take a post-bleach time course as described above. The limited area of the bleaching will allow the simplifying assumption that the cytosolic pool is unaffected by exchange with the bleached organelle, eliminating the need to account explicitly for exchange of cytosol with both bleached and unbleached organelles. After the sequence is acquired, the next step is background correction of the entire sequence to remove to the greatest extent possible the contribution made by the background to this recovery, and then obtaining a recovery curve over the bleached region of interest. The hope is that by averaging over many structures it will be possible to obtain a recovery curve of sufficient quality to identify a rise to a plateau and to measure half-times. Because the fluorescence to be quantitated will be dispersed in multiple, possibly dim, structures, which will be superimposed over a diffuse fluorescent background, quantitation will be considerably more difficult and probably less accurate than for the previously discussed experimental designs.

The first major difficulty is that the fluorescent background given by GFP-tagged protein in the dispersed compartment will vary in different regions of the image. The approach suggested in the previous section (to determine the fluorescent background manually for each structure by measuring fluorescence in a nearby region of interest) is potentially the most accurate, but is likely to be prohibitively time-consuming as there may be hundreds of structures to correct. If the structures move over the timescale of interest, manually background correcting each one is simply not possible for a large number of structures.

It is more convenient and almost as accurate to determine a background over the entire image using image-processing software and then subtract the background from the image. A median filter using a neighborhood width of about twice the diameter of the largest structure of interest is our preferred method (see 33 for rationale). An appropriate median filter function is present in some commercial software including 'Background correct' in the METAMORPH package (Molecular Devices). Other methods include replacing a pixel by a mean or a Gaussian blur of its neighborhood. The 'rolling ball' background correction in the freely available NIH IMAGE package seems to undercorrect backgrounds compared with other methods, but is better than no correction at all. In general, all of these methods will give very good results with a smoothly varying background such as given by a soluble cytoplasmic protein and the worst results with a highly nonsmooth distribution of fluorescence such as sometimes given by ER-localized proteins. It can be helpful in this case to blur the image slightly prior to applying the background correction to smooth out ER fine structure.

Once the background correction is applied, a second step of thresholding can be applied to isolate the structures. There are routines for applying a threshold to an image series in all major image-processing packages. A curve showing recovery of fluorescence in the photobleached region can be obtained by quantitating over the bleached region on the processed images. If it is possible to compare this curve with a recovery curve measured from images showing background only, this should be done. The background will consist largely of fluorescent protein, which will diffuse in over a time ranging from seconds (for cytosolic proteins) to minutes for integral membrane proteins. The approximate rate of this process should be known. If the two recovery curves (organelle and background) are similar, it is likely that either the background correction was not very successful (this should be apparent by eye) or that the rate of exchange between organelle and dispersed compartment is fast, with equilibration occurring on the same timescale as diffusion of the protein in the dispersed compartment.

If first-order kinetics are assumed, only a small fraction of either the dispersed or nondispersed compartment (<20% of either) has been bleached, the recovery is not diffusion limited, and the fraction of total cellular fluorescence in the compartment of interest can be determined with reasonable accuracy, it is not difficult to obtain approximate rate constants for entry and exit from the organelle. The first step is to fit the recovery curve to Equation 3. In this case, it is assumed that fluorescence outside the bleached region does not significantly change over the course of the experiment. This leads to a simplification. Unlike the preceding case, $k_{out} = r$ and so can be fitted directly. k_{in} can be determined by the ratio of the organelle to cytoplasm (or other dispersed compartment) as follows:

$$k_{in} = (O/C)k_{out} \qquad \text{(Equation 7)}$$

The fluorescence in the dispersed compartment should be determined by first determining the total cell fluorescence (accounting for background outside the cell) for the pre-bleached image and then subtracting organelle-associated fluorescence determined for the same image. It is likely that organelle-associated fluorescence will be underestimated if the organelle consists of many dim subcompartments. This measurement error will affect k_{in} but not k_{out}. There will be a small error (in the order of 10%) in determining k_{out} since the assumption that fluorescence associated with the dispersed compartment and with the unbleached organelle will remain constant over the course of the experiment is only approximately true. However, useful information can still be obtained if it is possible to obtain a recovery curve.

2.9. Analogous problems

Flux of a cytoplasmic protein into or out of an organelle can occur for reasons other than a ligand–receptor relationship. A classical example is nuclear import/export in which a large protein is imported into the nucleus through a pore in a process requiring energy and an import signal on the protein. Proteins can also be exported from the nucleus in a similarly signal-mediated manner. Thus, a protein containing an import and an export signal can cycle between nucleus and cytoplasm. While mechanistically nuclear import differs greatly from the binding of a peripheral membrane protein, the difference is not important for light microscopic analysis. The protein in the cell can easily be partitioned between a cytoplasmic and nuclear pool, while the nucleus or the cytoplasm can be bleached selectively and recovery monitored. If the kinetics of the particular process are approximately first order and neither the nucleus nor cytoplasm contains large subcompartments, a rate constant into and a rate constant out of the nucleus can be determined using exactly the methods used to characterize binding and unbinding of a peripheral membrane protein.

Another case similarly analyzed is the cycling of a GFP-tagged integral membrane protein between two membrane-bound compartments via vesicular intermediates. If one compartment is compact and easily isolated in an image (e.g. Golgi) and the other dispersed (e.g. ER), then rate constants into and out of the compact compartment can be calculated precisely as above (18). The lesson here is that whether a microscopic approach can be applied depends more often on the geometry of the cell than the underlying nature of the physical process.

2.10. More complex kinetic models

Kinetic analyses of more complex multi-compartment situations are possible. When all of the individual compartments can be distinguished visibly and kinetics of transport between them is strictly first order, approaches similar to those we have discussed (but slightly more complex) will work. An alternative approach is to simulate numerically the trafficking process on a computer and fit all parameters simultaneously. Hirschberg and co-workers described kinetic modeling of the trafficking of the cargo protein vesicular stomatitis virus G protein from the ER through the Golgi to the cell surface (3). It is also possible to determine rate constants one at a time, an approach used by Maxfield and colleagues in studies of the endocytic pathway (33–35). These approaches are beyond the scope of this chapter, but an overview is given in (36).

3. RECOMMENDED PROTOCOLS

All photobleaching experiments discussed here require maintaining tissue culture cells on a microscope stage in a satisfactory condition for a sufficient period of time to conduct the experiment. Normally this requires maintaining the cells in the appropriate medium and at normal pH and temperature.

Protocol 1

Time-lapse imaging of living cells on a microscope stage

Equipment and Reagents

- Computer-controlled inverted microscope
- Coverslip-bottomed dish (MatTek) or chamber (e.g. LabTek chamber slide; Nunc) suitable for tissue culture cells
- Means of heating stage to 37°C (Nevtek air stream incubator or alternative)
- Thermometer (mercury or digital)

Method

1. Plate cells into coverslip-bottomed dishes or LabTek chambers[a] using usual laboratory methods[b].

2. Transfect cells with the expression plasmid for the GFP construct. We use FuGENE 6 (Roche Molecular Biochemicals) according to manufacturer's instructions.
3. After 24–48 h, change the medium by adding new medium with 20% serum and 50 mm HEPES if the microscope does not possess a CO_2-controlled chamber. Fill the dish completely. Put petroleum jelly on the lid so as to form an airtight seal and replace on the dish.
4. Pre-heat the microscope stage to 37°C. When CO_2 control is not needed, we use the NevTek air stream incubator[c]. A thermometer is placed on the microscope stage to verify the temperature. It is important to give the temperature time to stabilize before moving the cells to the microscope stage both to avoid focus problems due to expansion of metal components of the stage during the course of the experiment and to avoid accidental heat shock to the cells.
5. Place the dish with cells in the stage holder[d].
6. Acquire a time series using microscope software[e]. On the Zeiss LSM 510, this involves opening the time series window, setting the number of images to acquire and the interval between images and pressing 'Start time series'.

Notes

[a]Most high-resolution objectives have insufficient working distance to look through the bottom of a plastic dish and are highly optimized for thin coverglasses. For short-duration experiments it is possible to plate cells on sterile coverslips on the bottom of a normal plastic Petri dish or six-well plate and then mount them on a media-containing hole in a rubber gasket attached to a slide with petroleum jelly. More permanent containers containing a coverslip bottom are, however, preferable.

[b]Many tissue culture cells (e.g. COS, HeLa) can be plated directly on glass. Others will require specific substrates.

[c]Any device that provides a stream of hot air will suffice. The temperature can be adjusted by varying the distance between the device and the microscope stage. Hairdryers produce strong vibrations but can be used if the microscope is on an air table and the hairdryer is on a support separate from the air table.

[d]We prefer to use 3.5 and 6 cm coverslip-bottomed dishes as they are low cost compared with LabTek chambers. However, coverslip-bottomed dishes may not fit on the slide holders on some microscopes. In this case, a holder can be ordered or machined, but for small numbers of experiments LabTek chamber slides, which are coverslip-bottomed, have the same shape as slides and can therefore be used on any inverted microscope. These may be easier to use.

[e]When the desired interval between images is very long (>1 min), it is probably best to take the images manually rather than using an automated time series procedure as it will be very difficult to compensate for focus drift otherwise. In this case, one is strongly advised to follow the same convention of sequentially numbering the images as would have been produced by the software. This will prevent confusion at later points in the analysis.

In general, the location and shape of the photobleached region and the time between images in the post-bleach sequence will differ between different kinds of photobleaching experiments, but the experimental design is identical. Protocol 2 assumes the use of the Zeiss LSM 510 confocal microscope (see Note a). Steps should roughly correspond for newer confocal microscopes from other manufacturers. Photobleaching experiments with older confocal microscope designs will be more difficult. Suggested modifications will be described in the notes.

Protocol 2

Photobleaching and data acquisition

Equipment and Reagents

- Confocal microscope (the procedures described assume the use of the Zeiss LSM 510[b]; some modifications will be required for other confocal microscopes)
- 488 nm Argon laser with 20–80 mW power
- 28× or 40× Oil-immersion objective with numerical aperture (NA)>0.7

Method

1. Select a 25× or 40× objective (oil-immersion if possible)[c].
2. Set up cells on the microscope stage at an appropriate temperature as described in *Protocol 1*.
3. Find a cell[d]. Open the pinhole to maximum. Focus the cell to maximize brightness. Optimize the image so that the background between cells is nearly zero (a few blue dots are seen using the 'Range indicator' palette). This optimization should not have to be repeated over the course of the session. Now adjust the 'Detector gain' so that the brightest point in the cell is as bright as possible without saturation. Saturated pixels are red when using the 'Range indicator' palette.
4. Zoom in on the cell. The entire cell should be in the field of view. It is helpful to use a standard zoom for all photobleaching experiments as the number of scans needed for photobleaching depends on the zoom.
5. Take an image with 'Single scan'. We typically use line averaging of 4, but this can be adjusted as per usual practice. With the Zeiss LSM 510 confocal microscope, there can be a slight difference in alignment between 'Fast XY'[e] and 'Single scan'.
6. Go to the 'Edit bleach' window. Select the box indicating 'Acquire bleach after scan'. For the scan number, put '1' (one image taken before scan). For GFP, set the 488 line of the argon laser to 100% in the bleach window. Draw the bleached region on the cell as desired. If the bleached region is elongated, it is best if it is oriented up/down or left/right[f].
7. In the bleach window, set 'Number of iterations' to a sufficient number of scans to completely bleach out the GFP but no more. This may require trial bleaches, which can be done either on live cells or on unimportant fixed samples. If the number of iterations required to bleach is unknown, start with 50 and then increase or decrease as required. The number of iterations required to bleach will also depend strongly on the zoom factor and thus it may be useful to optimize for a standard zoom[g]. Different objectives pass different amounts of light, so will require separate optimization.
8. Open the time series window. Select the number of images and spacing between images required to cover the expected timescale of the experiment. In general, for fast processes (e.g. diffusion of a protein in the ER) images should be acquired as quickly as possible. For slower processes, adjust the interval between image so that no more than 200 images are acquired over the total timescale of the process to prevent additional photobleaching during the image-acquisition phase.
9. Take the image again to check the focus. Readjust the focus if required. This step is especially important if the preceding steps took a substantial length of time.
10. Press 'StartB' in the time series window. Acquisition of a single pre-bleach image, photobleaching, and a subsequent time series will occur automatically. Focus may need to be adjusted manually during acquisition of the time series. This requires some practice.
11. Save the time series to a file for later analysis.

Notes

[a]The Zeiss LSM 510 confocal microscope, as well as high-end microscopes from other manufacturers, generally have acousto-optical tunable filters (AOTFs) or other ultrafast shuttering mechanisms, which permit the bleaching of arbitrary regions of interest. Confocal microscopes that lack ultrafast shuttering can be used in photobleaching experiments as well, but will be limited to bleaching square or rectangular areas. If no special photobleaching features are provided, this can be done by going to high or even maximum zoom for the bleaching, scanning continuously with the laser at full power (no neutral density filters) for sufficient time to bleach, and reversing the zoom to its previous value. As this technique leads to bleaching of a fixed-size square located at a fixed location in the center of the field, special care must be taken to position the target cell properly. A useful trick is to place an overhead transparency over the computer screen, bleach a fixed sample, and then mark the bleached area on the transparency. The transparency can be labeled with the zoom and objective used, and kept as a reference.

[b]Our Zeiss LSM 510 confocal microscope is equipped with a 405 nm diode laser, a 40 mW argon laser with 488 and 514 nm lines, helium neon lasers with lines at 543 and 633 nm, two photomultiplier tubes and the META detector. In practice, we rarely use any other laser line than the 488 nm line for photobleaching. Objectives most frequently used for photobleaching are the Plan-Apochromat 63×/1.4 NA oil and the Plan-Neofluar 25×/0.8 NA immersion corrected objective (normally used with oil).

[c]In our system, the combination of 25×/0.8 NA immersion corrected objective (using oil) with a fully open pinhole gives a confocal slice thickness greater than the thickness of typical tissue culture cells. This has two virtues: (i) changes in focus caused by stage heating, vibrations, or settling of the dish are minimized due to the depth of field of the objective; and (ii) fluorescence is effectively collected from the entire depth of the cell. However, the 25× objective would not be the best choice if maximum resolution is desired. We have found the Zeiss 40× 1.2 NA Fluor (which is optimized for passing large amounts of light) is also an excellent objective for this type of experiment, and can yield better resolution when needed. Use of either a 63× or a 100× 1.4 NA objective will give maximum resolution, but fluorescence cannot be simultaneously acquired from all points in the cell. Additionally, it will be difficult to maintain focus. If all fluorescence in and outside of an organelle must be accounted for, it is possible to *z*-section the cell completely prior to conducting the photobleaching and then acquire a typical image sequence afterwards from a single plane. The ratio of organelle to total cell fluorescence can be determined by summing over the *z*-series.

[d]It is often useful to find a pair of cells, only one of which will be photobleached. The unbleached cell will likely not change significantly over the course of the experiment and can be used as a reference for checking and adjusting focus.

[e]'Fast XY' on the Zeiss system is a rapid scan for purposes of focusing, scanning the coverslips, etc., while 'Single scan' is slower and allows averaging for acquisition of high-quality images. Most other confocals have similar modes. On the Zeiss confocal microscope, there is typically an offset between the images captured with the two modes. Thus, it is important that any regions of interest to be bleached be defined on an image taken with 'Single scan', as the bleach and time series will use this mode. This caution may not apply to other confocal microscope systems.

[f]This is because the scanning mirrors will always scan the minimum-sized rectangle that completely bounds the bleached region. The Zeiss LSM 510 shutters the laser using an AOTF while scanning over regions not to be bleached, which is why bleached regions of arbitrary shape can be defined. However, the time spent per pixel in nonbleached regions in the scanned rectangle is equal to time spent per pixel in bleached regions.

[g]Bleaching efficiency will be proportional to energy per unit area × time. Increasing the number of iterations over the bleached area will increase bleaching time, while the energy per unit area is proportional to the square of the zoom factor. It may be easiest to optimize for a single zoom, which will be used for all experiments. However, if the zoom is changed, the number of iterations can be adjusted as follows: NewIterations = OldIterations * $(\text{OldZoom}/\text{NewZoom})^2$.

Protocol 3

Simple method for determination of diffusion coefficients

Method

1. Photobleach a circular region within a cell using *Protocol 2.* The bleach should be in a region of relatively uniform fluorescence, small relative to the size of the cell, and not near the edge of the cell.
2. Obtain a recovery curve from the bleached area and a control curve from elsewhere in the same cell. This can be done using 'Quantitate ROI' in the Zeiss LSM 510 software or using other software such as METAMORPH, IMAGEJ or NIH IMAGE[a]. The recovery curve should show a smooth rise gradually approaching a plateau, while the control curve should show little change (possibly a gradual decline due to photobleaching). Verify that the time series is sufficiently long that the recovery curve effectively reaches a plateau.
3. Produce a normalized recovery curve by dividing the recovery curve by the control curve. Spreadsheet software (e.g. EXCEL) will work. However, there are advantages to using a specialized plotting package such as SIGMAPLOT or KALEIDAGRAPH. This will correct for possible photobleaching during the acquisition of the time series.
4. Determine the immobile fraction. The mobile fraction is the final value (at plateau) in the time series divided by the initial value in the bleached region. The immobile fraction is (1.0 – mobile fraction).
5. Determine the half-time for recovery by inspection of the graph. This is the time at which the graph crosses the average of the immediate pre-bleach value and the plateau value.
6. Determine the diffusion coefficient, *D*[b]:

$$D=\left[\frac{w^2}{4t_{1/2}}\right]0.88 \qquad \text{(Equation 8)}^{c}$$

where w is the radius of the bleach, and $t_{1/2}$ is the half-time for recovery (the time at which the curve crosses the average between the immediate post-bleach value and the final value).

Notes

[a]NIH IMAGE automatically rescales 12- or 16-bit images to 8 bit. Since individual images in a time series may be scaled differently, incorrect recovery curves may be obtained. Use NIH IMAGE *only* for quantitation of 8-bit images. More recent image processing packages (IMAGEJ or METAMORPH) will handle 12- or 16-bit images correctly.

[b]When using a confocal microscope restricted to photobleaching of rectangles, a thin rectangle (2–4 μM in width) totally crossing the cell can be used rather than a circular bleach. The immobile fraction can be used as described above. *D* can be found by fitting the recovery points to the following equation:

$$I(\infty)\left[1-\sqrt{\frac{w^2}{w^2+4\pi Dt}}\right] \qquad \text{(Equation 9)}$$

[c]This equation will give a correct answer if there is a single diffusing component. Its main disadvantage over a curve fit[b] approach is that a half-time can be obtained even if there is a mixture of diffusing components each with its own *D*. In similar situations, curve fits tend to be obviously poor and thus give warning of problems.

Protocol 4

Model-independent estimation of exchange between a compact organelle and the cell

Method

1. Photobleach the organelle of interest and acquire an image sequence using *Protocol 2*. The organelle should be relatively compact (e.g. nucleus or Golgi apparatus) and should be photobleached over its whole area. Complete removal of fluorescence in the bleached area is desirable if it can easily be achieved, but is not required. The image sequence should be sufficiently long that a plateau value is reached.
2. Acquire a recovery curve using 'Mean of ROI' in the Zeiss LSM 510 software or similar functions in other software. For this protocol, it is immaterial whether mean pixel intensity or sum of pixel intensity is acquired. The rise after photobleaching to a plateau value should be visible in the graph. If not, repeat step 1 with a longer-duration time series.
3. Determine the value halfway between the first post-bleach value in the recovery curve and the plateau value.
4. The time at which the curve crosses the mid-range value is the half-time.

Protocol 5

Determination of rate constants for exchange between a compact organelle and the cell given first-order kinetics of exchange between the two compartments

Method

1. Photobleach the organelle of interest and acquire an image sequence using *Protocol 2*. The organelle should be relatively compact (e.g. nucleus or Golgi apparatus) and should be photobleached over its whole area. Complete removal of fluorescence in the bleached area is not required. The image sequence should be sufficiently long that a plateau value is reached.
2. Quantitate the following regions over the entire time series using a function (such as 'Quantitate ROI' in the Zeiss LSM 510 software) that gives a mean of ROI (region of interest) for: (i) the entire organelle (*AveOrganelle*); (ii) the entire cell (*AveCell*); (iii) a background region outside the cell (*Background*); and (iv) a cytoplasmic region near the organelle (*OrganelleBackground*). Additionally, record the area of each region in consistent units (e.g. all areas in units of pixels).
3. Background correct as follows for each timepoint[a]:

$$AveCell = AveCell - Background \quad \text{(Equation 10)}$$

$$AveOrganelle = AveOrganelle - OrganelleBackground \quad \text{(Equation 11)}$$

4. Compute the total fluorescence in the cell (*TotalCell*) and organelle (*TotalOrganelle*) as follows:

$$TotalCell = AveCell \times \text{number of pixels in cell} \qquad \text{(Equation 12)}$$

$$TotalOrganelle = AveOrganelle \times \text{number of pixels in organelle} \qquad \text{(Equation 13)}$$

5. The fraction of the fluorescence that is organelle-associated (*FractOrganelle*) can then be calculated for each time point:

$$FractOrganelle = \frac{TotalOrganelle}{TotalCell} \qquad \text{(Equation 14)}$$

6. *FractOrganelle* should reach a plateau value after the recovery. If this is less than the pre-bleach value and there are no focus or other problems with the experiment, there is likely an immobile fraction, *I*:

$$I = 1 - \frac{FractOrganelle_{\text{post-bleach}}}{FractOrganelle_{\text{pre-bleach}}} \qquad \text{(Equation 15)}$$

7. Fit *FractOrganelle*(*t*) to the following equation[b,c]:

$$A_1 * [1 - e^{-rt}] + A_2 * e^{-rt} \qquad \text{(Equation 16)}$$

The pre-bleach image should not be included in the fit. Time 0 should be defined as time of acquisition of the first post-bleach image. The A_2 term accounts for incomplete photobleaching of the organelle and may be disregarded if there is no organelle-associated fluorescence on the organelle after the first post-bleach image. A_1 should fit to a value close to the plateau value estimated by eye and A_2 should fit a value close to the organelle fluorescence in the first post-bleach image. If this is not true, something is wrong with the curve fit or sequence.

8. The value of *r* fitted in Equation (16) is the sum of the rate constants k_{in} and k_{out} (Equation 4). To determine the individual rate constants, compute *FractCytoplasm* as (1 – *FractOrganelle*). Compute *FractOrganelle*/*FractCytoplasm* from the pre-bleach image. Compute as follows[d]:

$$k_{in} = r/(1 + FractCytoplasm/FractOrganelle) \qquad \text{(Equation 17)}$$

$$k_{out} = r/(1 + FractOrganelle/FractCytoplasm) \qquad \text{(Equation 18)}$$

Notes

[a]This is most easily done using spreadsheet software or graphing programs. Most graphing programs can perform least squares curve fits to arbitrary equations (usually using the Levenberg–Marquard method). As such a curve fit will be required at a later stage, it is most convenient to do all steps in the same software. We use the Macintosh program KALEIDAGRAPH. Other options are SIGMAPLOT and MATLAB.

[b]It is best to fit only the first 70% of the recovery, with the plateau value A_1 determined by the fit. This should be compared with the plateau value determined by eye. If the assumptions previously described hold, the two values should be close. This is an important consistency check.

[c]In KALEIDAGRAPH, go to 'General curve fit' and enter the equation as follows:
m1 * (1 – exp(– m2 * m0)) + m3 * exp(– m2 * m0). Initial guesses must be entered for some values. It is best to use the same initial values for all sequences if possible.

[d]If there is a significant immobile fraction, this will not be quite accurate.

4. TROUBLESHOOTING

4.1. Focus problems

The most common problems with analysis of FRAP experiments result from failure to maintain focus. The loss of some experiments to focus problems is inevitable so it is important to recognize recovery curves seriously perturbed by unstable focus (see *Fig. 4A* for an example). In general, focus is most unstable during the initial period when heating of the stage begins, and also within a few minutes of sample loading. Pre-heating the stage for an hour or more and giving the sample time to settle (at least 15 min) prior to beginning the experiment will reduce focus problems. In general, objectives of higher power and higher NA have less depth of field and are more sensitive to focus drift. The 63× and 100× 1.4 NA objectives are quite difficult to keep in focus, while a 25× 0.8 NA objective does not present great problems. A wide-open pinhole also reduces sensitivity to focus.

Manually correcting focus during time series acquisition is needed with the higher-powered objectives but requires practice. In general, the faster the images are acquired in the time series, the easier it will be to maintain focus. For very long time series, it may be easier to take a relatively small number of images manually rather than using an automated time series function.

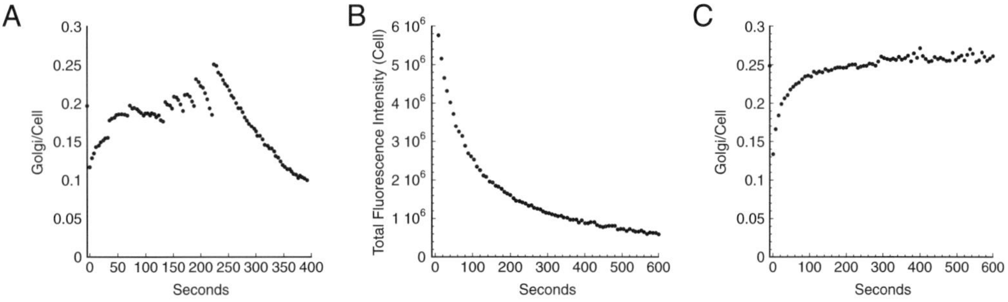

Figure 4. Problems with photobleaching experiments.
(A) Golgi-localized Arf1–GFP was photobleached and recovery measured as described in *Fig. 2*. The graph shown corresponds to *Fig. 2(b)*. However, the sequence commenced immediately after switching on an air stream incubator with resulting focus instability. Several characteristic focus problems are visible in this sequence. There is an abrupt jump in Golgi intensity at around 50 s due to a sudden change in focus. The value for Golgi intensity at 250 s (after the operator attempted to recover focus) is much higher than the pre-bleach value, indicating that the initial pre-bleach value was likely already out of focus. Either of the preceding problems renders a sequence unusable as they affect the critical early portions of the sequence. After 250 s, there is a decline in Golgi intensity due to a gradual uncorrected shift in focal plane. While this is the most dramatic defect in the sequence, it would in fact have no effect on analysis of this experiment because it occurs after recovery to a plateau has occurred. (*B, C*) Effects of photobleaching during acquisition of the recovery sequence. (*B*) Total fluorescence in a cell after photobleaching. The cell was imaged as described in *Fig. 2* but with overly high laser intensity (AOTF at 50%). Note the smooth and roughly exponential decay of total cellular fluorescence with time. (*C*) Golgi fluorescence normalized to total cell fluorescence. Data is from the same sequence as (*B*). Despite the severe photobleaching that occurred during acquisition of the recovery sequence, this is usable data.

4.2. Photobleaching during sequence acquisition

Photobleaching during sequence acquisition can be minimized by reducing the illumination per image (by reducing either laser intensity or amount of averaging) or the total number of images. Quantitative FRAP experiments involve averaging over many pixels, so noisy images involving high gain and low illumination do not present problems. Photobleaching is proportional to energy per unit area, which can also be reduced by reducing zoom (spreading the same energy over a larger area). Photobleaching during sequence acquisition often does not compromise data analysis (unless very severe) as normalization to total cell fluorescence will compensate (see *Fig. 4B* and *C* for an example).

4.3. Difficult curve fits

Curve fits can fail either because of bad data (usually due to focus problems or cell movement) or bad models. If the fits are 'bad' in a consistently similar manner, it is likely the model that is at fault. If not much is known about the underlying process, it may be safest not to overinterpret the data.

4.4. Other problems

Cells can move significantly during a long-duration experiment, which is likely to affect quantitation done using fixed ROIs. It may be necessary to redraw manually all regions of interest on every image, which can be a time-consuming process. In this case, it suffices to use images taken at regular intervals such that at least five to six images cover the main recovery phase.

Phototoxicity is usually not an issue with GFPs except in long-duration experiments if 488 nm or longer wavelength light is used. Short-wavelength light (e.g. 405 nm) can kill cells rapidly at high intensities and thus photobleached areas should be kept small and the minimum light intensity required for adequate photobleaching should be used. In long-duration experiments (many hours), if cells are observed going into mitosis, check that they proceed to cytokinesis. Some nonGFP dyes (e.g. NBD) produce significant quantities of free radicals on photobleaching, which can cause significant damage. Keeping dye concentrations low is helpful.

The intracellular distribution of GFP constructs should always be compared with the endogenous protein by immunofluorescence if possible. If not, it should give a distribution consistent with known biochemical data. It is possible for uncharacterized constructs to misfold or grossly mislocalize. Even correctly targeted and folded GFP constructs can be overexpressed. A specific example is the εCOP–GFP construct, which incorporates into the COPI complex. A cell contains a finite number of COPI complexes (an estimated 400 000–800 000 in Chinese hamster ovary cells; 27) and these can be rapidly saturated in a transient transfection. The remainder of the uncomplexed εCOP–GFP simply accumulates in the cytoplasm. If a problem like this is suspected, a weaker promoter can be used.

An alternative is to construct stable cell lines so that clones with a suitable expression level can be chosen. This proved to be a satisfactory solution in the case of εCOP–GFP (27).

We have found that with some commercial GFP expression vectors containing strong ribosome-binding sites immediately adjacent to the GFP (i.e. the Clontech N1 family of vectors), internal initiation of translation can occur in some constructs leading to the generation of free GFP. This can be detected by Western blotting using commercial antibodies against GFP. Point mutations in the ribosome-binding site solve the problem.

5. REFERENCES

1. **Presley JF, Cole NB, Schroer TA, Hirschberg K, Zaal KJ & Lippincott-Schwartz J** (1997) *Nature,* **389**, 81–85.
2. **Harder Z, Zunino R & McBride H** (2004) *Curr. Biol.* **14**, 340–345.
3. **Hirschberg K, Miller CM, Ellenberg J, *et al.*** (1998) *J. Cell Biol.* **143**, 1485–1503.
4. **Frye LD & Edidin M** (1970) *J. Cell Sci.* **7**, 319–335.
5. ★ **Axelrod D, Koppel DE, Schlessinger J, Elson E & Webb WW** (1976) *Biophys. J.* **16**, 1055–1069. – *The first application of photobleaching to measure the dynamics of membrane proteins.*
6. **Mitchison TJ** (1989) *J. Cell Biol.* **109**, 637–652.
7. **Chalfie M, Tu Y, Euskirchen G, Ward WW & Prasher DC** (1994) *Science,* **263**, 802–805.
8. **Brismar H, Trepte O & Ulfhake B** (1995) *J. Histochem. Cytochem.* **43**, 699–707.
9. **Pepperkok R, Squire A, Geley S & Bastiaens PI** (1999) *Curr. Biol.* **9**, 269–272.
10. **Singer SL & Nicolson GL** (1972) *Science,* **175**, 720–731.
11. **Edidin M** (1992) *Trends Cell Biol.* **2**, 376–380.
12. **Cole NB, Smith CL, Sciaky N, Terasaki M, Edidin M & Lippincott-Schwartz J** (1996) *Science,* **273**, 797–801.
13. **Edidin M** (2003) *Annu. Rev. Biophys. Biomol. Struct.* **32**, 257–283.
14. **Tao T, Lan J, Presley JF, Sweezey NB & Kaplan F** (2004) *Am. J. Respir. Cell Mol. Biol.* **30**, 350–359.
15. **Jacobson K, Derzko Z, Wu ES, Hou Y & Poste G** (1976) *J. Supramol. Struct.* **5**, 565–576.
16. **Ellenberg J, Siggia ED, Moreira JE, *et al.*** (1997) *J. Cell Biol.* **138**, 1193–1206.
17. **Siggia ED, Lippincott-Schwartz J & Bekiranov S** (2000) *Biophys. J.* **79**, 1761–1770.
18. **Zaal KJ, Smith CL, Polishchuk RS, *et al.*** (1999) *Cell,* **99**, 589–601.
19. **Luby-Phelps K, Taylor DL & Lanni F** (1986) *J. Cell Biol.* **102**, 2015–2022.
20. **Elsner M, Hashimoto H, Simpson JC, Cassel D, Nilsson T & Weiss M** (2003) *EMBO Rep.* **4**, 1000–1005.
21. **Novick P, Field C & Schekman R** (1980) *Cell,* **21**, 205–215.
22. **Fields S & Song O** (1989) *Nature,* **340**, 245–246.
23. **Venable JD, Dong M-Q, Wohlschlegel J, Dillin A & Yates JR** (2004) *Nat. Methods,* **1**, 39–45.
24. **Jares-Erijman EA & Jovin TM** (2003) *Nat. Biotechnol.* **21**, 1387–1395.
25. **Xu Y, Piston DW & Johnson CH** (1999) *Proc. Natl. Acad. Sci. U. S. A.* **96**, 151–156. – *A description of bioluminescence-based resonance energy transfer.*
26. ★★ **Presley JF** (2005) *Biochim. Biophys. Acta,* **1744**, 259–272. – *A recent overview of imaging techniques with bias towards the field of membrane trafficking but with general application to diverse subject areas.*
27. **Presley JF, Ward TH, Pfeifer AC, Siggia ED, Phair RD & Lippincott-Schwartz J** (2002) *Nature,* **417**, 187–193.
28. **Elbashir SM, Harborth J, Weber K & Tuschl T** (2002) *Methods,* **26**, 199–213.
29. **Wier ML & Edidin M** (1988) *Science,* **242**, 412–414.

30. **Edidin M** (1994) In: *Mobility and Proximity in Biological Membranes*, pp. 109–135. Edited by S Damjanovish, M Edidin, J Szollosi & L Tron. CRC Press, Boca Raton, FL.
31. **Weiss M, Hashimoto H & Nilsson T** (2003) *Biophys. J.* **84**, 4043–4052.
32. **Patterson GH & Lippincott-Schwartz J** (2002) *Science*, **297**, 1873–1877.
33. **Dunn KW, McGraw TE & Maxfield FR** (1989) *J. Cell Biol.* **109**, 3303–3314.
34. **Dunn KW & Maxfield FR** (1992) *J. Cell Biol.* **117**, 301–310.
35. **Presley JF, Mayor S, Dunn KW, Johnson LS, McGraw TE & Maxfield FR** (1993) *J. Cell Biol.* **122**, 1231–1241.
★★ 36. **Phair RD & Misteli T** (2001) *Nat. Rev. Mol. Cell Biol.* **2**, 898–907. – *An excellent review of kinetic modeling approaches applied to live-cell imaging.*

CHAPTER 7

Imaging calcium and calcium-binding proteins

Burcu Hasdemir, Robert Burgoyne and Alexei Tepikin

1. INTRODUCTION

In this chapter, we will discuss the imaging of calcium signals in the cytosol, mitochondria, and the lumen of endoplasmic reticulum (ER). We will describe the protocols for combined imaging of calcium responses in different organelles and for simultaneous calcium imaging and calcium uncaging experiments. We will also consider the imaging of calcium extrusion using dextran-bound fluorescent indicators. Finally, we will discuss practical aspects of monitoring the activation and translocation of calcium-binding proteins (CaBPs).

2. METHODS AND APPROACHES

2.1. Imaging calcium in the cytosol

Ratiometric fluorescent indicators fura-2 and indo-1, developed by Tsien and colleagues (1), were the instruments of choice for calcium imaging during the pre-confocal era. The acetoxymethyl ester (AM) forms of these indicators allow simple loading of the probes. AM forms are membrane permeable and they become de-esterified in the cytosol with release of charged, membrane-impermeable, and calcium-sensitive indicators. The probes can also be loaded into the cytosol by pressure injection or by perfusion via a patch pipette in the whole-cell configuration. The advantage of these popular indicators is that upon calcium binding the excitation (for fura-2) or emission (for indo-1) spectra of the probes change (shift along the wavelength axis) substantially. The ratio of the intensities of fluorescence, recorded at different wavelengths, can be used to determine the cytosolic calcium concentration (1, 2). This mode of measurement minimizes the artifacts caused by bleaching/leakage of indicator or by cellular movement. The early creative period exploiting these indicators resulted in a detailed characterization of the artifacts produced by the probes (for a detailed review of

Cell Imaging: *Methods Express* (D. Stephens, ed.)

methodology and problems of calcium measurement, see 2). One important methodological finding was that calcium indicators (including fura-2 and indo-1) and calcium buffers (including BAPTA) are antagonists of inositol trisphosphate receptors (3, 4). Another very important artifact is the formation of additional calcium buffering, which distorts and slows calcium signals. In 1992, Neher & Augustine (5) produced a detailed study of endogenous calcium buffering and contamination introduced by fura-2. In their work on chromaffin cells, the authors found that 'The low capacity, affinity and mobility of the endogenous Ca^{2+} buffer makes it possible for relatively small amounts of exogenous Ca^{2+} buffers, such as fura-2, to exert a significant influence on the characteristics of the Ca^{2+} concentration signal as measured by fluorescence ratio' (5). This lesson should not be forgotten at the present time, when calcium imaging systems have become almost as common as pH meters in physiological laboratories. Low-molecular-weight probes are not only strong calcium buffers, but are also highly mobile calcium buffers (unlike CaBPs, which diffuse more slowly in the cytosol). This will tend to distort the special properties of the calcium transients, which are particularly relevant for the imaging of fast local calcium transients. In practical terms, it is useful to minimize loading. For example, by measuring the autofluorescence level of the cell (at rest and during stimulation – the autofluorescence of intracellular NADH and FAD is of particular importance), one could estimate the minimal amount of calcium indicator that should be used during an experiment (so that the signal is not heavily contaminated by autofluorescence changes). A disadvantage of fura-2 and indo-1 is that both indicators are excitable in the UV part of the spectrum; the excitation wavelengths for fura-2 (e.g. 340 and 380 nm) and indo-1 (e.g. 350 nm) are in close proximity to the maximum of excitation of NADH. In other words, at low cytosolic concentrations of these probes, contamination by autofluorescence could be substantial.

Indicators that are excitable in the UV range are also not compatible with caged probes, as the caged probes are 'uncaged' by the same UV light as used for the excitation of fura-2 and indo-1. Examples of caged calcium probes include nitrophenyl (NP)–EGTA and DM–nitrophen – these molecules decrease their affinity to calcium upon UV illumination (6, 7). Indo-1 can be excited by a UV laser line (e.g. 351 nm) used in conjunction with laser scanning confocal microscopy (405 and 475 nm could be selected for emission channels). In contrast, the standard set of wavelengths produced by an argon UV laser is not particularly convenient for the excitation of fura-2, as there is no laser line in the 380–390 nm range, which is the part of the spectrum needed for the optimal excitation of calcium-sensitive fura-2 fluorescence). In principle, fura-2 confocal imaging using 351 and 364 nm lines is possible, but the dynamic range of fluorescence changes for such an excitation combination is very narrow (8). Therefore, fura-2 is not a convenient probe for laser scanning confocal microscopy. Currently fluo-4 and Fura Red (see *Fig. 1*) are among the frequently used indicators for confocal cytosolic calcium measurements. Fluo-4 is conveniently excitable by the 488 nm laser line. The indicator has intermediate calcium affinity (K_d = 345 nM), it increases fluorescence due to calcium binding, and it has a very substantial dynamic range (http://probes.invitrogen.com/). One useful application of fluo-4 (and other indicators excitable in the visible part of the

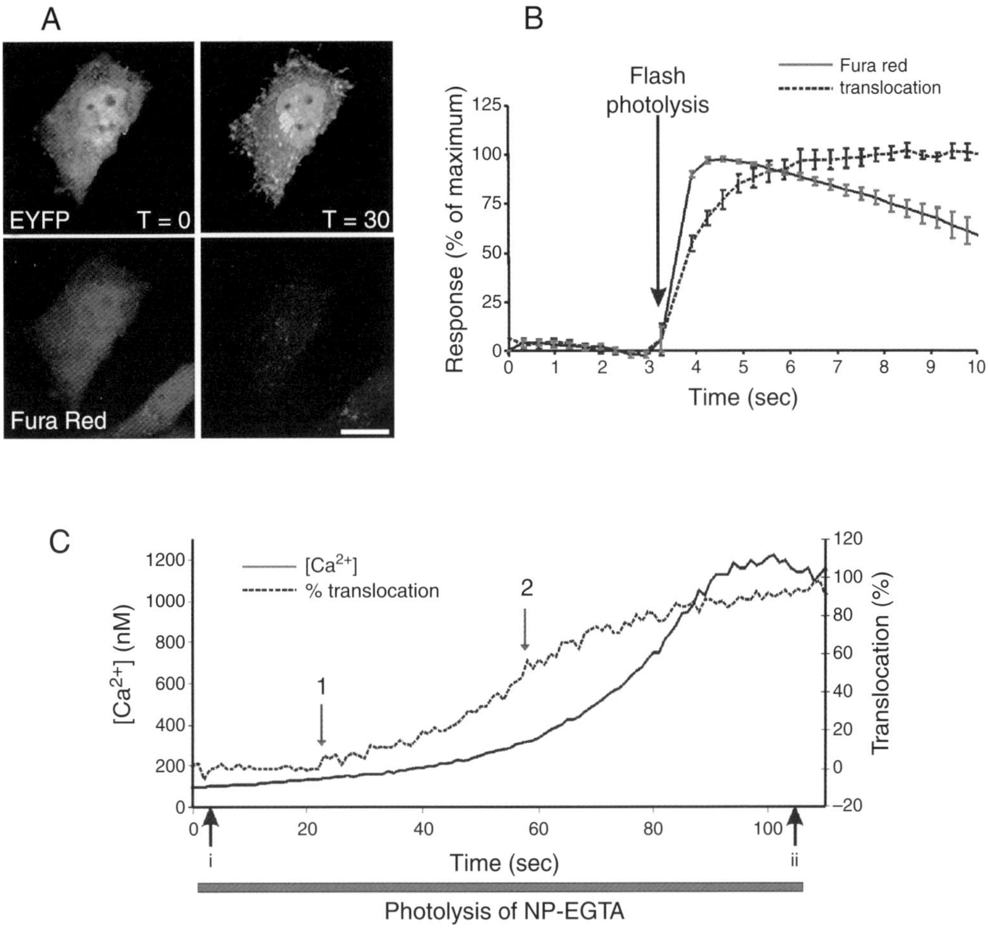

Figure 1. Simultaneous imaging of translocation of hippocalcin–EYFP and $[Ca^{2+}]$i following Ca^{2+} elevation using a variety of methods (13).
(*A*) After loading with Fura Red, live cells were imaged by confocal microscopy and $[Ca^{2+}]$i was elevated by the addition of 3 μM ionomycin. Images are shown immediately before addition ($T = 0$) or at the indicated times after addition of ionomycin. Images of Fura Red are shown beneath the corresponding images of the hippocalcin–EYFP fusion protein. Hippocalcin–EYFP was translocated from the cytosol to the Golgi near the nucleus and on to patches on the plasma membrane. Bar, 10 μm. (*B*) Rapid translocation of hippocalcin–EYFP in response to an intense flash photolysis of NP–EGTA. A HeLa cell expressing hippocalcin–EYFP was imaged for EYFP or Fura Red before and after photolysis of NP–EGTA, and fluorescence intensity was monitored in a region of interest and averaged to show the time course of changes in Fura Red and hippocalcin–EYFP fluorescence near the trans-Golgi network. Data are shown as mean ± SEM ($n = 8$ cells). (*C*) Assessment of the Ca^{2+} dependency of hippocalcin–EYFP translocation using release of Ca^{2+} by slow photolysis of NP–EGTA. Changes in hippocalcin–EYFP fluorescence were monitored in a region near the Golgi complex. Fura Red fluorescence over the cell was converted to $[Ca^{2+}]$. The initiation of translocation (labeled 1) and the time it was half maximal (labeled 2) are indicated.

spectrum) is combined fluorescence calcium measurements with simultaneous calcium uncaging (see *Protocol 1*). This technology has become particularly attractive since the introduction, by the major confocal manufacturers (e.g. Leica, Zeiss), of standard software allowing precise localized targeting of UV light to a pre-selected (on the cell image) region of the cell (often called 'region-of-interest' scanning). In the case of calcium measurements, the image is generated by exciting calcium indicators with visible light (which does not uncage the caged compound). In a recent study, Ashby and colleagues (9) used local uncaging of NP–EGTA (caged calcium) with simultaneous confocal calcium imaging (using fluo-4) to study the ability of specialized regions within polarized secretory cells to generate calcium-induced calcium release. It is also possible to use uncaging of caged neurotransmitters or hormones to stimulate cells locally. In such experiments, the UV light is targeted to the extracellular solution (near the plasma membrane) containing caged precursors of these agonists. A variation of this technique – uncaging in a cell-attached patch pipette (10) – allows local activation of small groups of receptors or even single receptors on the plasma membrane.

Fura Red is a high-affinity indicator (K_d = 140 nM), which decreases its fluorescence (see *Fig. 1A*) upon calcium binding (http://probes.invitrogen.com/). The probe can be excited by either 488 or 514 nm. Its emission can be collected at the long-wavelength part of the spectrum (e.g. above 630 nm), which makes Fura Red a very useful probe for simultaneous measurements of processes reported by more 'green' fluorescent probes. A specific example is the measurement of cytosolic calcium (with Fura Red) and mitochondrial NADH changes (11). Another application is the measurement of translocation of blue fluorescent protein (BFP)-, green fluorescent protein (GFP)- or yellow fluorescent protein (YFP)-labeled proteins with simultaneous measurement of cytosolic calcium (12, 13; see also *Fig. 1* and *Protocol 6*). A systematic study of spectral compatibility of fluorescent proteins and calcium indicators was published by Bolsover and colleagues (12). Using rather strict criteria, the authors suggest combinations of BFP with Fura Red and GFP with the calcium probe X-rhod 1. Fura Red was recently combined with the Golgi probe NBD C6-ceramide by Dolman and colleagues (14) for the simultaneous imaging of calcium signals and Golgi localization. Finally, Fura Red can be combined with caged calcium (NP–EGTA) and fluorescently labeled proteins of interest to study the calcium dependency of translocation (see *Fig. 1* and *Protocol 6*). Another approach to image calcium concentrations in the cytosol is to use calcium-sensitive fluorescent proteins, known as 'cameleons'. These proteins have been genetically engineered and contain a pair of fluorescent proteins, calmodulin and calmodulin-binding peptide. Calcium binding to the 'calmodulin' region results in a conformational change in the complex, accompanied by a change in the efficiency of fluorescence resonance energy transfer (FRET) between the two fluorescent parts. After their invention by Miyawaki and colleagues in the laboratory of R.Y. Tsien, cameleons underwent a considerable evolution, resulting in an increase in their photostability, decreased pH and Cl^- sensitivity, a reduction in their interaction with endogenous calmodulin, and an increase in the dynamic range of calcium-dependent fluorescence changes (15–18). Of particular interest with regard to cytosolic calcium imaging are cameleons with improved dynamic range, recently developed by Nagai and colleagues (17). Earlier types of cameleon were used to generate useful data in imaging experiments, but the dynamic range of calcium-dependant changes for these probes was substantially smaller than that for the majority of small fluorescent calcium indicators. It is possible that improved cameleons will become the main tool for calcium imaging experiments in cell lines and *in vivo* experiments.

2.2. Hardware for detection of calcium signals

The variability of calcium signals is substantial. Calcium signals can be very fast (elementary calcium signaling events can be only a few milliseconds), which necessitates very high rates of image acquisition. Modern confocal systems can reach rates of many tens or even hundreds of frames per second using different technologies: resonant scanning is used by Leica Microsystems, fast Nipkow disk-based scanning by PerkinElmer, and Acousto Optical Deflector technology by VisiTech International. Fast scanning confocal systems have also recently been announced by Zeiss and some other confocal manufacturers. Many of the small fluorescent indicators have a very substantial dynamic range and the fluorescence could change tens of times during a calcium response. This would necessitate the use of 10- or even 12-bit systems. Eight-bit systems, however, also have the advantage of lower memory requirements and decreased processing time (and potentially more time points available for measurements). Ideally, the digitization of an imaging system should be variable (e.g. 8- or 12-bit adjustable for an individual experiment). One important problem of fast, high-resolution imaging, as well as of prolonged imaging, is the bleaching of indicators. Systematic study of the bleaching of some common calcium indicators (including Oregon Green 488 BAPTA-1, Calcium Green-1, Fura Red, fluo-4 and Calcium Orange) revealed reasonable stability of fluo-4 and Fura Red and very impressive resistance to photobleaching of Calcium Orange (19). During the last 20 years, calcium measurements have moved from single photomultiplier-based measurements (fluorescence recorded from the whole cell) to imaging with charge-couple device (CCD) cameras, confocal microscopy, and two-photon microscopy. The popularity and utility of individual probes has changed along with these steps in the development of imaging technology. For example, fura-2 was an excellent probe for total fluorescence recording and nonconfocal imaging, but was found not to be very useful in confocal applications. Later it was recognized that fura-2 has a reasonable two-photon cross-section (20) and is therefore a useful probe for this, more novel, imaging technology. It should be noted that some other low-molecular-weight indicators and cameleons containing cyan fluorescent proteins (CFPs) and YFPs are also suitable for two-photon imaging (20–22). Recently developed 4Pi microscopy (23) offers exceptionally high axial resolution, but is still largely untested for calcium imaging experiments.

2.3. Imaging mitochondrial calcium

The bioluminescent protein aequorin has been successfully targeted to mitochondria and used for mitochondrial calcium measurements (24, 25). The low light output in such measurements makes imaging difficult (although, in principle, possible; see 26). Recent improvements in the dynamic range and targeting specificity of fluorescent calcium-sensing proteins (cameleons, pericams and camgaroos; 17, 18, 27, 28) make these very promising tools for imaging mitochondria. At the moment, however, the fluorescent probe rhod-2 (29) is probably still the most popular indicator for imaging mitochondrial calcium concentration in living cells. Rhod-2 AM has a delocalized positive charge; in this form the indicator is membrane permeable and accumulates preferentially in mitochondria because of the substantial negative voltage (with respect to the cytosol) maintained across the inner mitochondrial membrane. Following de-esterification, the indicator is retained in mitochondria. The protocol for loading rhod-2 by incubation with its AM form is described in *Protocol 2*. The indicator has intermediate calcium affinity (approximately 0.6 μM) and a very broad dynamic range – the fluorescence increases more than ten times in cellular conditions upon saturation with calcium. In many cell types, the resting mitochondrial calcium concentration is low and a low resting intensity of rhod-2 fluorescence is therefore expected. The excitation maximum for rhod-2 is approximately 550 nm and the helium neon laser line 543 nm is convenient for excitation. Emission could be collected in the waveband 560–620 nm. One of the main problems of working with rhod-2 is achieving specificity of loading. A few approaches have been developed to reduce the cytosolic component of fluorescence (i.e. the cytosolic concentration of rhod-2). Trollinger and colleagues (30) have described the procedure of selective loading (cold loading followed by warm incubation) whereby mitochondria and cytosol are loaded and then cytosol selectively unloaded (during warm incubation). Selective permeabilization of the plasma membrane (e.g. with digitonin) can be used to remove the cytosolic probe and to control the calcium concentration outside mitochondria (31). The cytosolic indicator could also be conveniently removed using a patch pipette in the whole-cell configuration (32). The mitochondrial distribution of rhod-2 fluorescence following loading could be verified by comparing the pattern of fluorescence with another mitochondrial probe (e.g. MitoFluor Green; 33). Recently another indicator from the rhod family – rhod-FF (which has a lower affinity for calcium than rhod-2) has been used to monitor mitochondrial calcium responses in exocrine cells (34). Some groups (e.g. 35) have reported mitochondrial calcium measurements using dihydro-rhod-2 AM (which apparently can be easily produced from rhod-2 AM). Some other small fluorescent calcium probes have also been shown to label mitochondria preferentially (36, 37). It is possible to combine imaging of mitochondrial calcium (using rhod-2) with imaging of NADH fluorescence (11, 35) (see *Protocol 2* and *Fig. 2* in color section). Simultaneous imaging of mitochondrial calcium (with rhod-2) and cytosolic calcium (with fluo-4) is possible (see *Protocol 2*). It is also possible to image mitochondrial and ER calcium concentrations simultaneously by combining rhod-2 and mag-fluo-4 staining (see *Fig. 3* and *Protocol 2*, and below) (32). The possibility of artifacts in such combined measurements increases substantially and it is therefore essential to carry out independent control experiments with just one indicator. Different pharmacological agents can be used to verify the specificity of rhod-2 responses.

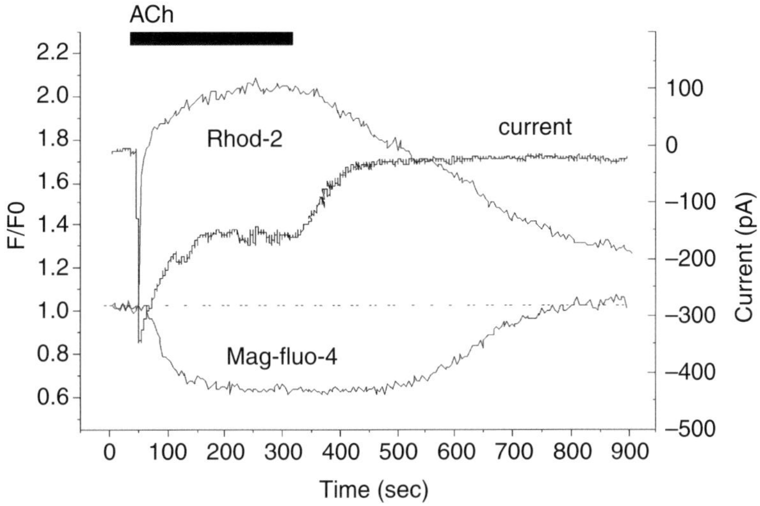

Figure 3. Simultaneous measurements of the Ca^{2+} concentration changes in the ER, mitochondria, and cytosol (32).
Time course of acetylcholine (ACh)-elicited changes in the Ca^{2+} concentrations in the ER (mag-fluo-4), in the cytosol (whole-cell Ca^{2+}-dependent current), and in the mitochondria (rhod-2).

The 'toolkit' of mitochondrial inhibitors includes protonophores (e.g. FCCP and CCCP), inhibitors of the electron transport chain (e.g. rotenone, azide, cyanide, and antimycin), an inhibitor of ATP synthase (oligomycin), an inhibitor of the Na^+/Ca^{2+} exchanger (CGP37157), and an inhibitor of mitochondria calcium influx (Ru360). A detailed review on imaging of mitochondrial functions has recently been published by Duchen and colleagues (38).

2.4. Imaging calcium in the ER lumen

Mag-fura-2 was the first probe to be successfully used for Ca^{2+} measurements in the ER lumen (39, 40). The probe has a higher affinity for Ca^{2+} (K_d = 25 μM) than for Mg^{2+} (K_d = 1.9 mM) (http://probes.invitrogen.com/). The excitation of mag-fura-2 shifts upon calcium binding and the indicator can therefore be used as a ratiometric probe. Wavelengths of 350 nm (or 340 nm if the microscope optics permit) and 385 nm can be used for the excitation, while emission can be collected at around 510 nm. There is no convenient laser line for excitation close to 385 nm, which makes confocal experiments with mag-fura-2 complicated; in some applications a 405 nm laser line (from a diode laser) is usable. The argon laser line 351 nm is close to the isosbestic point for this indicator (the wavelength at which fluorescence is independent of calcium) and this wavelength can be used locally to quench the indicator and to monitor its diffusion through the ER lumen (41). Mag-fura-2 can be loaded into the ER by incubation with the membrane-permeable AM form of the indicator. The probe is de-esterified in both the ER and the cytosol. To achieve the specificity of ER measurements, the cytosolic indicator must be removed, which could be achieved by permeabilization of the plasma membrane (e.g. with digitonin) (39) or by using a patch pipette in the whole-cell configuration (42, 43). In both cases, the cytosolic indicator is removed, while the indicator in the ER is retained. Additional tests using mitochondrial inhibitors, activators of ER calcium-releasing channels, and inhibitors of ER calcium pumps can be used to verify further the specificity of loading (40). The advantage of unloading via a patch pipette is that it allows not only removal of the contaminating probe from the cytosol (by diffusion into the large volume of the pipette solution), but also measurement of calcium and calcium-dependent currents simultaneously with ER calcium measurements (43, 44). The potential effects of Mg^{2+} on the fluorescence of the indicator can be reduced by decreasing the Mg^{2+} concentration in the perfusion/patch pipette solution. In our experience, cells withstand a reduction in Mg^{2+} concentration to 0.1 mM in patch pipette solution reasonably well (this is much lower than the K_d values for mag-fura-2), but deteriorate rapidly if the concentration is reduced further. Calibrations based on the calcium ionophore ionomycin can be used to determine the ER calcium concentration. In permeabilized cells, the calibration is straightforward; however, it is not always possible to saturate the ER probe in whole-cell patch clamp experiments, presumably due to diffusion of calcium into the pipette solution. In whole-cell patch clamp experiments, it is recommended to use high concentrations of ionomycin (e.g. 10–20 μM) and high concentrations of external calcium (e.g. 10 mM) to saturate the indicator in the ER lumen. We also found that it is possible to withdraw the patch pipette (the plasma membrane reseals in some cells) and the ER probe could then easily be saturated by treatment with ionomycin with added external calcium. The values of $[Ca^{2+}]$ in the ER are usually in the range of 100–1000 μM. Mag-fura-2 (trapped in the lumen of the ER) can be combined with high-affinity indicators fluo-3 or fluo-4 (delivered via patch pipette) for simultaneous calcium measurements in the ER lumen and the cytosol (45, 46). Mag-fluo-4 is a low-affinity Ca^{2+} probe more convenient for confocal imaging (see *Protocol 3* and *Fig. 3*). This indicator can be excited by the 488 nm argon laser line and emission can be collected in the 505–550 nm band. The indicator has a large dynamic range (the fluorescence of the probe increases upon calcium binding) and is much brighter in the ER lumen than in the cytosol (this decreases contamination by the cytosolic probe). The indicator is not ratiometric. The advantage of this probe in comparison with indicators excitable in the UV part of the spectrum is that it can be combined with caged probes (e.g. caged calcium – NP-EGTA). Mag-fluo-4 has been used in combination with whole-cell patch clamp measurements to reveal oscillations of intra-ER calcium during 'physiological' cytosolic calcium oscillations (41), and the probe was also successfully combined with NP-EGTA in the ER lumen and used to monitor the movement of calcium through the lumen of the organelle following local calcium photorelease (41). For such combined uncaging/confocal scanning experiments, both probes were loaded into the cytosol in AM form and we found that it was possible to load both probes simultaneously (see *Protocol 3*). Following incubation with AM forms of the probes and washing in indicator-free extracellular solution, the whole-cell patch clamp configuration was established and the cytosolic probes were removed by diffusion into the patch pipette. The resting calcium in the ER is sufficiently high to saturate NP-EGTA. Because of the high resting calcium in the ER, the small calcium changes due to uncaging are impossible to detect and we therefore depleted the ER of calcium using an inositol trisphosphate-producing agonist before uncaging. The local calcium transient, produced by applying UV laser light to a small region of the ER, propagates through the ER lumen and the process of such propagation can be studied using mag-fluo-4. The region of nucleoplasm (which is devoid of ER) can be used as a control for contamination by cytosolic signals. The spectral properties of mag-fluo-4 and the popular mitochondria probe rhod-2 are sufficiently different that these two indicators could be combined (see *Fig. 3* and *Protocol 2*) for simultaneous ER and mitochondrial calcium measurements. (It is important to note that excitation maxima of both probes are away from that of NADH – the main source of mitochondrial autofluorescence.) Other low-affinity calcium probes suitable for ER calcium measurements include mag-indo-1 (47) and fura-2FF (43).

Engineered fluorescent CaBPs have been successfully used for ER calcium measurements. Of particular interest for ER calcium imaging is the recently produced cameleon named Design 1 (D1; 18). The calcium dependency of fluorescence of this protein is complex and it is approximated by two dissociation constants – 0.81 and 60 μM. The presence of a low-affinity binding site (the higher K_d) makes it usable for ER calcium measurements and the value of this K_d is probably close to optimum for measuring ER calcium in the majority of cell types. The D1 cameleon was successfully targeted to the ER and used to measure the rate of calcium leakage from the ER and calcium changes in the ER lumen due to oscillatory calcium release triggered by calcium-releasing agonist (18).

2.5. Imaging calcium in the Golgi – not just yet

The acidic environment of the Golgi lumen makes it challenging to image calcium in this organelle. The majority of Golgi calcium studies have been conducted using a Golgi-targeted bioluminescent aequorin probe, which is remarkably pH resistant but difficult to image (48). A recent study by Griesbeck and colleagues (15) described a Golgi-targeted cameleon GT-YC3.3 with decreased pH and Cl^- dependency. This probe was successfully targeted to the Golgi and responded to the calcium ionophore ionomycin, which suggests that, in principle, it can be used for Golgi calcium imaging. The probe was, however, close to saturation at resting calcium levels and responded by very small changes in FRET to stimulation with a calcium-releasing agonist. Further modifications of cameleons will hopefully make the imaging of calcium in this important yet elusive organelle possible and convenient.

2.6. Imaging extracellular calcium

Calcium indicators linked to high-molecular-weight dextrans (70 or 500 kDa) can be used to image calcium in the vicinity of the cell membrane (in the extracellular solution) and to derive useful information on the distribution of calcium exit sites, the rate of calcium extrusion by the pumps in the plasma membrane, or the calcium efflux mediated by exocytosis (49). The idea behind this technique was to slow down the calcium extrusion by using an indicator that is (due to its high molecular weight) also a slowly mobile calcium buffer. Calcium Green-1 dextran was used in such experiments. It should be emphasized that the diffusion of this probe is much slower than the diffusion of free Ca^{2+} in water-based extracellular solution and also much slower than the diffusion of calcium complexes with small (nondextran-bound) calcium probes. Confocal microscopy is an essential requirement for such experiments. The changes in fluorescence in the extracellular solution are relatively small (a few percent in our confocal experiments), and in nonconfocal imaging systems, reduction of the signal-to-noise ratio (due to out-of-focus fluorescence) makes such measurements practically impossible. The imaging of extracellular calcium with Calcium Green-1 dextran can easily be combined with imaging of intracellular calcium with Fura Red, since the indicators are separated both spectrally and spatially (49). To reduce the effective diffusion coefficient of calcium sufficiently, it is necessary to use relatively high concentrations of the dextran-bound probe. We empirically found that 30–100 μM of Calcium Green-1 (bound to dextran) provides a reasonable resolution. Our experiments were conducted on nonexcitable cells; one could expect a faster extrusion rate from excitable cells where lower concentrations of dextran-bound indicator would probably be sufficient. It is important to emphasize that the concentration of the indicator is not equal to the concentration to dextran, since several molecules of indicator are bound to each molecule of dextran and the ratio of indicator to dextran varies from batch to batch of the probe. The disadvantage of this technique is that experiments are conducted at low extracellular calcium concentration; however, this removes possible contamination due to calcium influx and allows imaging of 'pure' calcium efflux. In some cases, it is necessary to reduce the calcium contamination of the extracellular solution using Calcium Sponge S (BAPTA polystyrene). To trigger calcium efflux, it is necessary to elevate intracellular calcium, for instance by uncaging caged calcium (NP–EGTA) in the cytosol of the cell or by stimulating the cells with calcium-releasing agonists (before the store is depleted). We usually apply the agonists locally via a micropipette (by pressure or iontophoretic injection). This is advantageous, since multiple cells placed in one chamber containing expensive dextran-bound fluorescent probe can be sequentially stimulated. Dextran-bound indicators can also be used to image calcium efflux not only from cells, but also from isolated cellular organelles, for example the nuclear envelope (50). An alternative strategy for imaging extracellular calcium is based on lipophilic calcium probes (e.g. Calcium Green C18 and fura-C18), which (with appropriate loading) can be connected to the outer leaflet of the plasma membrane. Recently, fura-C18 was used for imaging extracellular calcium near the plasma membrane of human embryonic kidney cells stimulated with calcium-releasing agonists (51).

2.7. From calcium signals to CaBPs

Much of our current knowledge of the Ca^{2+} affinity of CaBPs has been gained from *in vitro* studies of purified proteins (see *Protocol 4*). Calcium-sensing proteins undergo conformational changes upon binding Ca^{2+}, which can be determined by monitoring the intrinsic fluorescence in tryptophan residues in the protein (see *Protocol 5*). Most proteins contain only a few tryptophan residues and the intrinsic fluorescence of these is affected by the environment of this amino acid. Sequential addition of Ca^{2+} results in changes in the tryptophan fluorescence parameters, which mostly involve the wavelength of maximum intensity (λ_{max}) and the intensity observed at λ_{max} (52, 53). From the data, a plot of change in fluorescence vs. calculated Ca^{2+} concentration can be generated (see *Fig. 4*). The example shown in *Fig. 4* is of a decrease in emission intensity following Ca^{2+} elevation with the neuronal Ca^{2+} sensor protein K^+ channel-interacting protein 1 (KChIP1).

Small fluorescence molecules (e.g. fluorescein derivatives) have been successfully used to label CaBPs and image their translocation in living cells (54, 55). The advantage of such probes is that they do not affect the diffusion properties of the proteins, which is particularly relevant for studies of nuclear

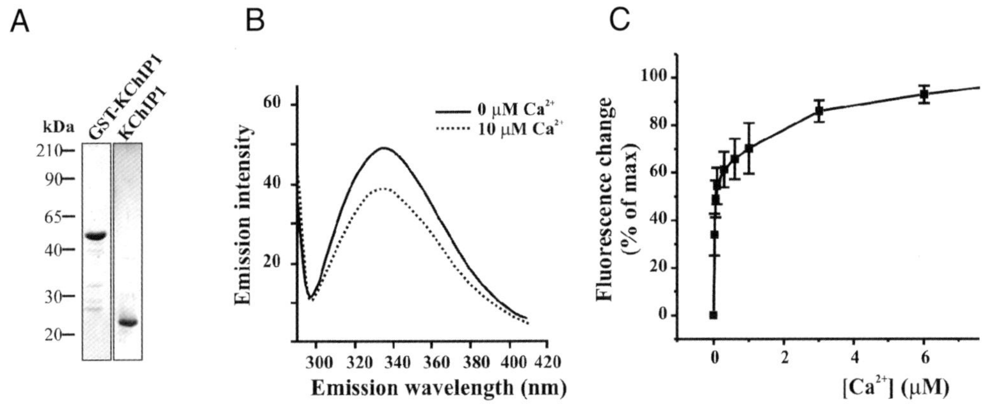

Figure 4. Analysis of expressed KChIP1 and the Ca^{2+}-induced conformational changes in tryptophan fluorescence.
The affinity of KChIP1 for free Ca^{2+} is unknown and so we examined this by determining the conformational change on Ca^{2+} binding through monitoring the intrinsic fluorescence of the protein's single tryptophan at position 129. (*A*) Coomassie brilliant blue-stained SDS gel of the bacterially expressed and purified GST–KChIP1 and the cleaved and purified free KChIP1. (*B*) Tryptophan fluorescence emission spectra from cleaved 1 μM KChIP1 excited at 280 nm with emission fluorescence collected between 290 and 410 nm. The two traces shown were obtained before and after the addition of 10 μM Ca^{2+}. Sequential addition of Ca^{2+} resulted in a decrease in the peak tryptophan fluorescence. (*C*) Emission spectra were measured over a range of free Ca^{2+} concentrations in three experiments and the mean changes in peak fluorescence were plotted as a function of calculated free Ca^{2+} concentration. The conformational change was half maximal at around 60–100 nm free Ca^{2+}.

translocation. The disadvantage is that such labeled proteins need to be delivered into the cells by invasive techniques (e.g. pressure injection, patch clamp, reversible permeabilization).

GFP from the jellyfish *Aequorea victoria* (56) has proved to be an extremely useful tool in biological research, mainly as a protein marker/tracker (57). Engineering of the original wild-type GFP sequence has resulted in the development of enhanced GFP (EGFP), optimized for imaging, and the spectrally different EGFP variants including the most commonly used ECFP and EYFP (57, 58). In addition, there are also a number of reef coral-derived fluorescent proteins (59) that are commercially available. These include red variants such as DsRed and HcRed. cDNAs encoding fluorescent proteins are available from a wide range of suppliers including Clontech, MBL International, Evrogen, and Invitrogen. The availability of a range of colors is useful, as it allows imaging of several different proteins at once, for instance of a CaBP and its target protein, or the simultaneous use of Ca^{2+} indicator fluorescent probes (see, for example, the examination of hippocalcin–EYFP translocation and the use of Fura Red in *Fig. 1* and *Protocol 6*). Other applications include FRET, which can be used to detect protein interactions (60). In general, to produce fluorescently tagged fusion proteins, the gene of interest (e.g. a CaBP) is inserted in frame into the expression vector encoding the fluorescent protein. The vector is transfected into cells and specifically labeled protein in live cells is expressed and can be imaged. The fluorescent protein may be tagged to the N or C terminus of the protein of interest and the version resulting in the least disruption to the protein's structure and function by the tag is usually chosen. A worthy supplement to the existing GFP variants is the more recently developed photoactivatable GFP (PA–GFP; 61), which has had a great impact on studies where the tracking of movements of proteins inside cells as they occur is required. This approach relies on the fact that PA–GFP is only dimly fluorescent until excited by a pulse of blue light and thus it is possible to photoactivate and reveal fusion proteins within a region of interest in the cell and then follow them as they move around in the cell away from the place of activation. This provides the opportunity to study the pathways and movement velocities of the protein of interest precisely. GFP has also proved to be a powerful fusion tag for measuring the dynamics of translocation events, for instance of CaBPs (an example is shown in *Fig. 1*). A particularly vivid illustration of the usefulness of this technology is the study of calcium-dependent translocation of protein kinase C (PKC)-γ (62). Conserved domains of PKC-γ (C1 and C2) are the interaction sites for diacylglycerol and Ca^{2+}, respectively, and monitoring the membrane translocation of the GFP-tagged C1 or C2 domains allows the study of signal transduction mechanisms and spatio-temporal differences for different receptor stimuli (62). Recently, GFP tagging was used for studies of the translocation of another C2 domain-containing CaBP, calcium-promoted Ras inactivator (63). Clearly, GFP and its variants are immensely valuable for research in the Ca^{2+} signaling field, as they not only provide the opportunity to visualize and track CaBPs in a spatial and temporal manner, but also allow measurement of their activities and interactions, which can often be Ca^{2+} dependent. Most importantly, these methods can be applied to living cells and organisms.

3. RECOMMENDED PROTOCOLS

Protocol 1

Imaging cytosolic calcium with simultaneous uncaging of caged calcium (NP–EGTA)

(a) Intact cells

Equipment and Reagents

- Fluo-4 AM (Invitrogen Molecular Probes)
- NP–EGTA AM (Invitrogen Molecular Probes)
- Confocal microscope equipped with UV and visible lasers (in our laboratory, we use Leica TCS SP2, Leica SP2 AOBS, or Zeiss 510 for such combined experiments)
- Standard Na-HEPES-buffered extracellular solution (140 mM NaCl; 4.7 mM KCl; 1.13 mM $MgCl_2$; 10 mM glucose; 1 mM $CaCl_2$; 10 mM HEPES; adjusted to pH 7.2 with NaOH; all from Sigma)

Method

1. Incubate cells in a standard Na-HEPES-buffered solution containing 2.5–5 μM fluo-4 AM and 5 μM NP–EGTA AM in the dark for 25 min at room temperature. Wash by centrifugation in indicator-free solution. Incubate in indicator-free solution for 20–30 min before beginning the experiment.
2. Use a 488 nm laser line to excite fluo-4 fluorescence. Collect emission at 505–550 nm. Fluo-4 is a nonratiometric 'single wavelength' indicator. Fluorescence values for this probe can be converted into Ca^{2+} concentrations using the standard equation:

   $$[Ca^{2+}] = K_d \times (F - F_{min})/(F_{max} - F)$$

   where F_{min} corresponds to the fluorescence of the calcium-free form of the indicator, F_{max} represents indicator saturated by calcium, and F is the current value of fluorescence (for which the calcium concentration is calculated). The values of F_{min} and F_{max} can be determined by calibration using the calcium ionophore ionomycin[a].
3. Select a specific region for uncaging (i.e. a region for targeting UV light) on the fluo-4 image of the cell (this can be any identifiable part of the cell, such as the nucleus or secretory granule region). Use a combination of laser lines (351 and 364 nm) for fast local uncaging. Continue recording fluo-4 fluorescence with 488 nm excitation[b].

Notes

[a]We used 10 or even 20 μM ionomycin (Invitrogen Molecular Probes) in our experiments on primary isolated cells. We found that lower concentrations of ionophore could be used on cell lines (e.g. 3 μM; see *Protocol 6*). To determine F_{min}, incubate in a solution containing 10 μM ionomycin and 1–5 mM EGTA for 20–30 min (note that a relatively long time is required to attain F_{min}). To determine F_{max}, incubate in a solution containing 10 μM ionomycin and 1–5 mM calcium. The saturation with calcium is usually achieved rapidly and is frequently followed by disintegration of the cells and loss of fluo-4 fluorescence. The equation should be applied to the specific regions on the image. In some cases, such local calibrations are not possible because ionomycin in combination with high calcium induces substantial changes in cell shape and movement of the regions of interest. A detailed study on the behavior of calcium indicators in different cellular compartments, which also contains

useful suggestions for *in situ* and *in vivo* calibrations (and discrepancy between the two calibrations), has recently been published by Thomas and colleagues (19).
[b]Fluo-4 does not absorb UV light effectively, but during the application of the powerful pulse of UV light, the fluorescence of the indicator will increase locally due to the additional excitation (not only due to a calcium rise).

(b) Patched cells

Equipment and Reagents

- Fura Red, tetrapotassium salt (cell impermeant; Invitrogen Molecular Probes)
- NP–EGTA, tetrapotassium salt (cell impermeant; Invitrogen Molecular Probes)
- Confocal microscope equipped with UV and visible lasers
- Patch clamp set-up (EPC-8 from HEKA is used in our laboratory)
- Standard K-HEPES-buffered patch pipette solution (130–140 mM KCl; 20 mM NaCl; 1.13 mM $MgCl_2$; 2 mM Na_2ATP; 10 mM HEPES; adjusted to pH 7.2 with KOH; all from Sigma)

Method

1. Use 50–150 μM Fura Red in the patch pipette solution. Add 1 mM NP–EGTA and 0.8 mM $CaCl_2$ to the patch pipette solution.
2. Establish a whole-cell patch configuration and wait for 2–3 min before the confocal imaging/uncaging experiment.
3. Use a 488 nm laser line to excite Fura Red fluorescence and a long-pass 560 nm filter or band-pass 580–680 nm filter to collect the fluorescence of Fura Red.
4. Use a combination of 351 and 364 nm laser lines to induce fast uncaging of NP–EGTA.

Protocol 2

Imaging mitochondrial calcium with rhod-2

This can be combined with NADH measurements, NP–EGTA uncaging, cytosolic calcium measurements using fluo-4, or ER calcium measurements using mag-fluo-4.

Equipment and Reagents

- Rhod-2 AM (Invitrogen Molecular Probes)
- Pluronic F-127 (Invitrogen Molecular Probes)
- NP–EGTA (Invitrogen Molecular Probes)
- Fluo-4 AM (Invitrogen Molecular Probes)
- Ionomycin (Invitrogen Molecular Probes)
- Confocal microscope (equipped with UV and visible lasers) and patch clamp set-up (the equipment used in our laboratory is described in *Protocol 1*)
- Standard extracellular solution (see *Protocol 1*)
- Patch pipette solution (see *Protocol 1*)

Method

1. Load the cells with rhod-2 AM by incubation in a solution containing 3–5 μM indicator and 0.01% Pluronic F-127. Incubation can be conducted at 37°C for 20–30 min (other authors have reported loading at room temperature, or even at 4°C). It is important to adjust the loading protocol in such a way that mitochondria are not overloaded. Mitochondria should become bright and clearly visible at the peak of calcium accumulation, but not necessarily at resting calcium levels.

2. Remove the indicator and Pluronic F-127 by centrifugation in indicator-free solution. Incubate cells in indicator-free solution for 10–20 min before beginning the experiment.
3. Use a 543 nm laser line to excite the fluorescence of rhod-2. Emission can be collected with a long-pass 560 nm filter.
4. For combined measurements of mitochondrial calcium (with rhod-2) and NADH changes use a 351 nm laser line to excite the NADH and collect NADH fluorescence in the range of 390–470 nm.
5. For combined measurements of mitochondrial calcium (with rhod-2) and cytosolic calcium uncaging, use a patch pipette to add NP-EGTA to the cytosol (an added bonus is that it reduces the concentration of contaminating cytosolic rhod-2). The patch pipette solution should contain NP-EGTA (1–2 mM in our experiments) and added calcium at a concentration slightly lower than the concentration of NP-EGTA (e.g. 0.9–1.8 mM). Use 351 or 364 nm laser lines (or combined) to trigger calcium uncaging. UV light can be targeted to the part of the cell devoid of mitochondria[a].
6. For combined measurements of mitochondrial and cytosolic calcium, load the cells with rhod-2 as described above. Wash once by centrifugation and then incubate cells with fluo-4 AM (2.5 μM for 20 min at room temperature). Use a 488 nm laser line for the excitation of fluo-4 and collect emission light between 505 and 550 nm from the regions of the cytosol free from mitochondria. Use a 543 nm laser line to excite the fluorescence of rhod-2. Collect emission with a long-pass 560 nm filter (or longer) from mitochondria-enriched regions. Both fluo-4 and rhod-2 are nonratiometric indicators, strongly increasing fluorescence upon calcium binding. It is impossible to calibrate both indicators simultaneously. Calibrations should be conducted separately and could be used for estimation of calcium concentrations in combined experiments.
7. For combined measurements of mitochondrial and ER calcium (see *Fig. 3*), load the cells with rhod-2 as described above. Wash once by centrifugation and then incubate cells with mag-fluo-4 (4 μM mag-fluo-4 AM and 0.01% Pluronic F-127 for 20–30 min at 37°C). Use a whole-cell patch clamp configuration to remove cytosolic indicators. The patch pipette should contain a low concentration of calcium buffer to allow calcium exchange between the ER, cytosol, and mitochondria. This is also beneficial because the low calcium buffering allows detection of cytosolic calcium changes by recording calcium-dependent currents (see *Fig. 3*). Choose a region enriched in mag-fluo-4 fluorescence to record ER calcium changes. Choose a region with strong rhod-2 fluorescence to record mitochondrial calcium changes. In such combined measurements, only qualitative analyses of calcium responses in three cellular compartments is possible.

Notes

[a]Pancreatic acinar cells, for which we developed these protocols, have a considerable calcium-binding capacity (64) and it is probable that for other cell types, lower concentrations of the NP-EGTA/calcium mix will be sufficient.

Protocol 3

Imaging calcium and uncaging caged calcium in the ER lumen

Equipment and Reagents

- Mag-fluo-4 AM (Invitrogen Molecular Probes)
- NP-EGTA AM (Invitrogen Molecular Probes)
- Pluronic F-127 (Invitrogen Molecular Probes)
- Confocal microscope (equipped with UV and visible lasers) and patch clamp set-up (the equipment used in our laboratory is described in *Protocol 1*)
- Standard extracellular solution (see *Protocol 1*)
- Patch pipette solution (see *Protocol 1*)

Method

1. Incubate cells in solution containing 4–6 μM mag-fluo-4 AM and 0.01% Pluronic F-127 for 20–30 min at 37°C. Wash once by centrifugation. Incubate in solution containing 20–25 μM NP-EGTA AM and 0.01% Pluronic F-127 for 30 min at 37°C in the dark. Wash once by centrifugation.
2. Place cells in a perfusion chamber on the table of a confocal microscope. Establish a whole-cell patch clamp configuration. Cytosolic indicators are removed by diffusion into the patch pipette. The patch pipette solution in our experiments contained 10 mM BAPTA and 2 mM Ca^{2+}.
3. Reversibly deplete the ER store by applying an appropriate calcium-releasing agonist. The agonist could be applied by perfusion in calcium-free extracellular solution.
4. Remove the agonist from the extracellular solution.
5. Image the distribution of mag-fluo-4 fluorescence using a 488 nm laser line for excitation. Emission can be collected using a band-pass 505–550 nm filter.
6. Select the region for uncaging of NP-EGTA using the image of mag-fluo-4 distribution.
7. Apply UV light (351 or 364 nm or a combination of the two lines) to uncage calcium locally in the selected region of the ER. Image the ER calcium changes using excitation and emission wavelengths appropriate for mag-fluo-4. To evaluate the level of contamination from cytosolic signals, the region of interest can be placed over the nucleoplasm (which is diffusionally connected to the cytosol via nuclear pore complexes and deprived of ER).

Protocol 4

CaBP production and purification

Production and purification of CaBPs involves a series of procedures, which vary depending on the unique properties of each protein. The following protocol can be used as a guideline.

Equipment and Reagents

- pGex-6P-1 vector (Amersham Biosciences)
- *Escherichia coli* BL21 (Gene Therapy Systems)
- Superbroth (0.5% (w/v) NaCl; 1.5% (w/v) tryptone; 2.5% (w/v) yeast extract)
- Ampicillin (use at 100 μg/ml)
- Incubator (37°C) containing a shaker (250 r.p.m.)
- Isopropyl-1-thio-β-D-galactopyranoside (IPTG)
- Breaking buffer (500 mM KCl; 100 mM HEPES; 5 mM ATP; 5 mM $MgCl_2$; 2 mM 2-mercaptoethanol; pH 7.0)
- Protease inhibitor cocktail, lysozyme, and DNase I (Sigma)
- Sonicator
- Centrifuges
- Phosphate-buffered saline (PBS) (137 mM NaCl; 2.7 mM KCl; 10 mM Na_2HPO_4; 2 mM NaH_2PO_4; pH 7.4)
- Glutathione–Sepharose 4B (Amersham Biosciences)
- PreScission protease (Amersham Biosciences)
- Cleavage buffer (50 mM Tris-HCl; 150 mM NaCl; 1 mM EDTA; 1 mM dithiothreitol; pH 7.0)

Method

1. Clone the CaBP sequence into the multiple cloning site of the pGex-6P-1 vector to obtain an N-terminal glutathione *S*-transferase (GST) fusion-tagged CaBP plasmid.
2. Perform a transformation of *E. coli* BL21 cells.
3. Inoculate 100 ml of superbroth containing ampicillin with a single colony and grow overnight at 37°C with shaking at 250 r.p.m.
4. The following morning, add the 100 ml of culture to 900 ml of superbroth with ampicillin for a 1 l culture, shake at 37°C for 1.5 h, and induce by adding IPTG to a final concentration of 1 mM (OD at this point is between 1.5 and 2.5).
5. Shake at 37°C for another 3 h and recover the cells by centrifugation at 4000 r.p.m. for 20 min at 4°C.
6. Discard the supernatant and resuspend the pellets in 20 ml total volume of pre-chilled breaking buffer. Add 200 μl of protease inhibitor cocktail and swirl the samples immediately. Freeze the suspension at −80°C until further use.
7. Thaw the cell suspension, add lysozyme to a concentration of 1 mg/ml and incubate for 30 min on ice, followed by sonication for further disruption.
8. Add DNase I to a concentration of 2 μg/ml to the sonicated sample and incubate on ice for 15 min. Pass sample through a hypodermic needle several times to reduce viscosity.
9. Centrifuge the sample for 1 h at 4°C at 100 000 *g*.
10. Incubate the supernatant with 1 ml of glutathione–Sepharose slurry pre-washed in PBS for 1 h at 4°C with end-over-end mixing.
11. Wash the Sepharose beads in PBS (three 10 ml washes)[a] and in cleavage buffer (three 10 ml washes).

12. To 1 ml of the washed Sepharose, add 1 ml of cleavage buffer containing 80 units of PreScission protease and incubate for 4 h (or overnight) at 4°C with end-over-end mixing[b]. The supernatant will contain GST-free CaBP[c].

Notes

[a]At this stage, if desired, the GST-tagged version of the protein can be obtained by elution with reduced glutathione.
[b]Fusion proteins will be cleaved between the GST moiety and the fusion partner.
[c]The yield can be determined by a Bradford assay. It is advisable to check correct protein production by SDS–PAGE coupled to Western blot analysis using an antibody to detect the CaBP of interest.

The Ca^{2+}-induced conformational changes in calcium-sensing proteins can be monitored by tryptophan fluorescence measurements as described in *Protocol 5.*

Protocol 5

Trytophan fluorescence measurements of conformational change on binding of Ca^{2+} to CaBPs

Equipment and Reagents

- Luminescence spectrometer
- Buffer A (20 mM HEPES; 139 mM NaCl; 5 mM EGTA; 5 mM nitrilotriacetic acid; pH 7.4)
- GST-free CaBP dialyzed against Buffer A
- 1 M $CaCl_2$

Method

1. Place a cuvette containing 3 ml of Buffer A in the luminescence spectrometer and run an emission spectrum to obtain a blank reading (excitation at 280 nm, emission collected between 290 and 410 nm)[a].
2. Add 1 μM CaBP to 3 ml of Buffer A and obtain a tryptophan fluorescence emission spectrum in the absence of Ca^{2+}.
3. Add $CaCl_2$ stepwise to the sample to give calculated free Ca^{2+} concentrations ranging from 0.03 to 10 μM and obtain emission spectra for each addition. (To 3 ml of Buffer A add sequentially 3.49, 2.07, 1.94, 3.84, 1.71, 0.93, 1.49, 1.05, and 1.05 μl of 1 M $CaCl_2$ to obtain free [Ca^{2+}] (μM) of 0.03, 0.06, 0.1, 0.3, 0.6, 1, 3, 6, and 10, respectively.)
4. The percentage change in peak fluorescence intensity can be calculated and plotted as a function of free [Ca^{2+}].

Notes

[a]It is advisable to perform control readings in the absence of protein to check that the mere addition of Ca^{2+} to Buffer A does not affect the emission spectrum.

Protocol 6

Simultaneous imaging of CaBP translocation and $[Ca^{2+}]i$

Equipment and Reagents

- Laser scanning confocal microscope (e.g. Zeiss LSM 150 confocal microscope or Leica TCS-SP-MP microscope) with a 63× water-immersion objective with a 1.2 numerical aperture
- Cells
- Fura Red (Molecular Probes)
- NP-EGTA (Molecular Probes)
- Krebs–Ringers solution (145 mM NaCl; 5 mM KCl; 1.3 mM $MgCl_2$; 1.2 mM NaH_2PO_4; 10 mM glucose; 20 mM HEPES; pH 7.4) with 3 mM $CaCl_2$
- Plasmid encoding EYFP-tagged construct of CaBP

Method

1. Transfect cells with the appropriate transfection reagent for the cell type using the manufacturer's recommended protocol.
2. At 24–72 h post-transfection, load cells with Fura Red by incubation in 5 μM of the AM, and with NP-EGTA by incubation in 10 μM of the AM in growth medium for 30 min.
3. Place cells on the microscope stage, bathed in Krebs–Ringers solution, and excited at 488 nm. Light is collected at 625–725 nm for Fura Red emission and at 525–590 nm for EYFP fusion protein emission.
4. Add ionomycin, when used, to the bath solution to a final concentration of 3 μM.
5. Photolysis of NP-EGTA is carried out by illumination with a 360 nm laser light at full power for rapid photolysis and at 6% power for generation of a $[Ca^{2+}]$ ramp.
6. To calibrate Fura Red fluorescence and calculate $[Ca^{2+}]$, treat cells with 3 μM ionomycin at the end of the experiment in the presence of 3 mM EGTA to determine F_{min}. F_{min} represents the calcium-free form of the indicator (referring to the minimal concentration of calcium). Since Fura Red decreases its fluorescence (for excitation at 488 nm) upon calcium binding, F_{min} is larger than F_{max}. The ratio of the fluorescence intensities, $\alpha = (F_{min}/F_{max})$, is determined in separate experiments dedicated to calibration. In such a calibration experiment, F_{min} is achieved by incubating cells in solution containing ionomycin and EGTA (no added calcium), while F_{max} is determined by the subsequent application of solution containing ionomycin with 1–5 mM Ca^{2+} added. $[Ca^{2+}]$ is calculated using a K_d of 140 nM from the equation:

   $$[Ca^{2+}] = K_d \times (F_{min} - F)/(F - F_{max})$$

   or from the equation:

   $$[Ca^{2+}] = K_d \times (F_{min} - F)/(F - F_{min}/\alpha)$$

   where $\alpha = F_{min}/F_{max}$ is determined in a separate experiment.

4. TROUBLESHOOTING

Problems frequently occur during the calibration of fluorescence responses of cytosolic indicators (see *Protocol 1*). In our experience, it usually takes a substantial amount of time to achieve F_{min} values by treatment with the solution containing ionomycin and EGTA. Ionomycin not only transports calcium from the cytosol (assuming very low extracellular calcium concentration), but also releases calcium from internal stores. The recovery from this release (and consequently elevated cytosolic [Ca^{2+}]) could be slow, particularly at concentrations close to the resting cytosolic [Ca^{2+}]. Negative values of [Ca^{2+}] in recalculations of experimental data (from fluorescence to [Ca^{2+}]) could indicate that insufficient time was spent on incubation in ionomycin/EGTA solution. It is also useful to remember that calcium contamination of nominally calcium-free solutions could be up to 50–100 μM; therefore, a robust EGTA concentration of a few millimolar should be used to attain sufficiently low calcium in extracellular solution for F_{min} calibration.

When loading mitochondria with rhod-2 (see *Protocol 2*), it is very easy to overload the mitochondria with this probe, in which case cells display a bright 'mitochondrial' pattern of fluorescence, but the indicator fails to respond to calcium changes. It is useful to adjust the loading protocol in such a way that the resting fluorescence of mitochondrial rhod-2 is weak (e.g. only two to three times higher than the autofluorescence) and strong mitochondrial fluorescence develops only during the calcium response.

When the cell is stimulated by calcium-releasing agonists, the direction of fluorescence changes of cytosolic and ER-trapped indicators are opposite. If you attempt to measure ER calcium concentration (see *Protocol 3*) and see a biphasic response (in the case of mag-fluo-4, an initial rise followed by a delayed decline), this suggests that the cytosolic contamination is not sufficiently suppressed and either more time is necessary for removal of the cytosolic probe (e.g. by diffusion into the patch pipette), or a higher concentration of the cytosolic buffer needs to be used (e.g. increasing BAPTA concentration in the patch pipette solution).

One of the key determinants of successful protein production (see *Protocol 4*) is the induction with IPTG. The concentration of IPTG and time point of induction given in the protocol can be used as a guideline, but this step is also one of the first points to optimize when no protein is being produced. Depending on the individual protein's needs, a slower induction may be desirable, in which case the concentration of IPTG should be lowered and the incubation temperature reduced. For measuring tryptophan fluorescence changes upon Ca^{2+} binding of a CaBP (see *Protocol 5*), the main advice to be given in advance is to check that the protein of interest actually contains tryptophan residues. A good control for this experiment that can be done and is often requested is also to carry out tryptophan fluorescence measurements on a mutant form of the protein (for instance, to disable the EF-hands) that is unable to bind Ca^{2+}.

Acknowledgements

We would like to thank our former and current collaborators who have participated in the development of techniques/protocols described in this chapter: Michael C. Ashby, Pavel V. Belan, Nicholas Dolman, Oleg and Julia Gerasimenko,

Hideo Mogami, Dermott W. O'Callaghan, Myoung K. Park, Ole H. Petersen, and Svetlana G. Voronina. The technical help of Mark Houghton is gratefully acknowledged.

5. REFERENCES

1. **Grynkiewicz G, Poenie M & Tsien RY** (1985) *J. Biol. Chem.* **260**, 3440–3450.
★★ 2. **Takahashi A, Camacho P, Lechleiter JD & Herman B** (1999) *Physiol. Rev.* **79**, 1089–1125. – *Useful review of the technology of calcium measurements, containing instructive information about the calibration of ratiometric and single wavelength indicators and a detailed description of the artifacts of calcium measurements.*
3. **Morris SA, Correa V, Cardy TJ, O'Beirne G & Taylor CW** (1999) *Cell Calcium,* **25**, 137–142.
4. **Richardson A & Taylor CW** (1993) *J. Biol. Chem.* **268**, 11528–11533.
★★★ 5. **Neher E & Augustine GJ** (1992) *J. Physiol.* **450**, 273–301. – *Probably the most detailed numerical account of endogenous calcium-binding capacity and the additional calcium-binding capacity introduced by a calcium indicator.*
★ 6. **Ellis-Davies GC & Kaplan JH** (1994) *Proc. Natl. Acad. Sci. U. S. A.* **91**, 187–191. – *Description of NP-EGTA, which is probably the most popular caged calcium probe.*
7. **Ellis-Davies GC** (2003) *Methods Enzymol.* **360**, 226–238.
8. **Nitschke R, Wilhelm S, Borlinghaus R, Leipziger J, Bindels R & Greger R** (1997) *Pflugers Arch.* **433**, 653–663.
9. **Ashby MC, Craske M, Park MK, *et al.*** (2002) *J. Cell Biol.* **158**, 283–292.
10. **Ashby MC, Camello-Almaraz C, Gerasimenko OV, Petersen OH & Tepikin AV** (2003) *J. Biol. Chem.* **278**, 20860–20864.
11. **Voronina S, Sukhomlin T, Johnson PR, Erdemli G, Petersen OH & Tepikin A** (2002) *J. Physiol.* **539**, 41–52.
★ 12. **Bolsover S, Ibrahim O, O'luanaigh N, Williams H & Cockcroft S** (2001) *Biochem. J.* **356**, 345–352. – *Systematic study of the compatibility of calcium indicators and fluorescent proteins.*
13. **O'Callaghan DW, Tepikin AV & Burgoyne RD** (2003) *J. Cell Biol.* **163**, 715–721.
14. **Dolman NJ, Gerasimenko JV, Gerasimenko OV, Voronina SG, Petersen OH & Tepikin AV** (2005) *J. Biol. Chem.* **280**, 15794–15799.
15. **Griesbeck O, Baird GS, Campbell RE, Zacharias DA & Tsien RY** (2001) *J. Biol. Chem.* **276**, 29188–29194.
★★★ 16. **Miyawaki A, Llopis J, Heim R, *et al.*** (1997) *Nature* **388**, 882–887. – *Study describing the first generation of cameleons.*
★ 17. **Nagai T, Yamada S, Tominaga T, Ichikawa M & Miyawaki A** (2004) *Proc. Natl. Acad. Sci. U. S. A.* **101**, 10554–10559. – *Description of the introduction of cameleons with a substantially improved dynamic range.*
18. **Palmer AE, Jin C, Reed JC & Tsien RY** (2004) *Proc. Natl. Acad. Sci. U. S. A.* **101**, 17404–17409.
★ 19. **Thomas D, Tovey SC, Collins TJ, Bootman MD, Berridge MJ & Lipp P** (2000) *Cell Calcium* **28**, 213–223. – *Detailed study of the behavior of calcium indicators in live cells, including information on the affinity of indicators in different compartments of the cell, the region-specific dynamic range of fluorescence changes, and photostability of the indicators.*
20. **Xu C, Zipfel W, Shear JB, Williams RM & Webb WW** (1996) *Proc. Natl. Acad. Sci. U. S. A.* **93**, 10763–10768.
21. **Fan GY, Fujisaki H, Miyawaki A, Tsay RK, Tsien RY & Ellisman, MH** (1999) *Biophys. J.* **76**, 2412–2420.
22. **Mainen ZF, Malinow R & Svoboda K** (1999) *Nature,* **399**, 151–155.
23. **Egner A & Hell SW** (2005) *Trends Cell Biol.* **15**, 207–215.
24. **Chiesa A, Rapizzi E, Tosello V, *et al.*** (2001) *Biochem. J.* **355**, 1–12.
25. **Rizzuto R, Simpson AW, Brini M & Pozzan T** (1992) *Nature,* **358**, 325–327.

26. **Rutter GA, Burnett P, Rizzuto R, *et al.*** (1996) *Proc. Natl. Acad. Sci. U. S. A.* **93**, 5489–5494.
27. **Filippin L, Abad MC, Gastaldello S, Magalhaes PJ, Sandona D & Pozzan T** (2005) *Cell Calcium,* **37**, 129–136.
28. **Rudolf R, Mongillo M, Magalhaes PJ & Pozzan T** (2004) *J. Cell Biol.* **166**, 527–536.
29. **Minta A, Kao JP & Tsien RY** (1989) *J. Biol. Chem.* **264**, 8171–8178.
30. **Trollinger DR, Cascio WE & Lemasters, JJ** (1997) *Biochem. Biophys. Res. Commun.* **236**, 738–742.
31. **Szabadkai G, Pitter JG & Spat A** (2001) *Pflugers Arch.* **441**, 678–685.
32. **Park MK, Ashby MC, Erdemli G, Petersen OH & Tepikin AV** (2001) *EMBO J.* **20**, 1863–1874.
33. **Boitier E, Rea R & Duchen MR** (1999) *J. Cell Biol.* **145**, 795–808.
34. **Bruce JI, Giovannucci DR, Blinder G, Shuttleworth TJ & Yule DI** (2004) *J. Biol. Chem.* **279**, 12909–12917.
35. **Hajnoczky G, Robb-Gaspers LD, Seitz MB & Thomas AP** (1995) *Cell,* **82**, 415–424.
36. **Csordas G & Hajnoczky G** (2003) *J. Biol. Chem.* **278**, 42273–42282.
37. **Ricken S, Leipziger J, Greger R & Nitschke R** (1998) *J. Biol. Chem.* **273**, 34961–34969.
★★ 38. **Duchen MR, Surin A & Jacobson J** (2003) *Methods Enzymol.* **361**, 353–389. – *Very useful review, describing measurements of different mitochondrial functions, including measurements of mitochondrial calcium concentration.*
★★★ 39. **Hofer AM & Machen TE** (1993) *Proc. Natl. Acad. Sci. U. S. A.* **90**, 2598–2602. – *Original and detailed study describing the technique of ER calcium measurement with a trapped small fluorescent indicator.*
40. **Hofer AM & Machen TE** (1994) *Am. J. Physiol.* **267**, G442–G451.
41. **Park MK, Petersen OH & Tepikin AV** (2000) *EMBO J.* **19**, 5729–5739.
42. **Lomax RB, Camello C, van Coppenolle F, Petersen OH & Tepikin AV** (2002) *J. Biol. Chem.* **277**, 26479–26485.
43. **Mogami H, Tepikin AV & Petersen OH** (1998) *EMBO J.* **17**, 435–442.
44. **Hofer AM, Fasolato C & Pozzan T** (1998) *J. Cell Biol.* **140**, 325–334.
45. **Chatton JY, Liu H & Stucki JW** (1995) *FEBS Lett.* **368**, 165–168.
46. **Solovyova N, Veselovsky N, Toescu EC & Verkhratsky A** (2002) *EMBO J.* **21**, 622–630.
47. **Tse FW, Tse A & Hille B** (1994) *Proc. Natl. Acad. Sci. U. S. A.* **91**, 9750–9754.
48. **Pinton P, Pozzan T & Rizzuto, R** (1998) *EMBO J.* **17**, 5298–5308.
49. **Belan PV, Gerasimenko OV, Berry D, Saftenku E, Petersen OH & Tepikin AV** (1996) *Pflugers Arch.* **433**, 200–208.
50. **Gerasimenko OV, Gerasimenko JV, Tepikin AV & Petersen OH** (1995) *Cell,* **80**, 439–444.
51. **de Luisi A & Hofer AM** (2003) *J. Cell Sci.* **116**, 1527–1538.
52. **Cox JA, Durussel I, Comte M, *et al.*** (1994) *J. Biol. Chem.* **269**, 32807–32813.
53. **McFerran BW, Weiss JL & Burgoyne RD** (1999) *J. Biol. Chem.* **274**, 30258–30265.
54. **Craske M, Takeo T, Gerasimenko O, *et al.*** (1999) *Proc. Natl. Acad. Sci. U. S. A.* **96**, 4426–4431.
55. **Torok K, Wilding M, Groigno L, Patel R & Whitaker M** (1998) *Curr. Biol.* **8**, 692–699.
56. **Shimomura O, Johnson FH & Saiga Y** (1962) *J. Cell Comp. Physiol.* **59**, 223–239.
57. **Lippincott-Schwartz J & Patterson G H** (2003) *Science* **300**, 87–91.
58. **Zhang J, Campbell RE, Ting AY & Tsien RY** (2002) *Nat. Rev. Mol. Cell Biol.* **3**, 906–918.
59. **Matz MV, Fradkov AF, Labas YA, *et al.*** (1999) *Nat. Biotechnol.* **17**, 969–973.
60. **Sekar RB & Periasamy A** (2003) *J. Cell Biol.* **160**, 629–633.
★ 61. **Patterson GH & Lippincott-Schwartz J** (2002) *Science* **297**, 1873–1877. – *Description of the new technology of PA-GFP, which can be used for dynamic studies of protein translocation.*
62. **Oancea E & Meyer T** (1998) *Cell* **95**, 307–318.
63. **Lockyer PJ, Kupzig S & Cullen PJ** (2001) *Curr. Biol.* **11**, 981–986.
64. **Mogami H, Gardner J, Gerasimenko OV, Camello P, Petersen OH & Tepikin AV** (1999) *J. Physiol.* **518**, 463–467.

CHAPTER 8
Total internal reflection fluorescence microscopy

David Zenisek

1. INTRODUCTION

Total internal reflection fluorescence microscopy (TIRFM) has gained popularity in recent years in the study of cellular processes and single biological molecules. As a technique, it is relatively easy and inexpensive to implement compared with scanning techniques, such as laser scanning fluorescence confocal and two-photon fluorescent microscopy. Despite its simplicity, it provides remarkably good optical sectioning equal to or better than that of these other techniques. The utility of this technique is limited, however, to the study of fluorescent molecules within 100 nm of a coverslip. This chapter provides an overview of the theory of TIRFM and a practical overview of the different ways in which TIRFM can be implemented. In addition to this chapter, there are a number of good review articles on the subject of TIRFM (1–6).

1.1. General properties of reflection and refraction at an interface

At most angles, light traveling from a substance of one refractive index (n_1) to one of a different refractive index (n_2) is both refracted and reflected. If the refractive index of the second substance is less than that of the first, the refracted light is bent towards perpendicular to the interface between the two media, whereas light traveling from a higher to a lower refractive index is bent away from perpendicular (see *Fig. 1*). The degree to which the light is refracted is described by Snell's law:

$$n_1\sin(\alpha_1) = n_2\sin(\alpha_2) \qquad \text{(Equation 1)}$$

where n_1 and n_2 are the refractive indices of the incident and refracted light, respectively, and α_1 is the angle between the incident light and a line drawn perpendicular to the interface of the two substances. The angle α_2 is the same as

Cell Imaging: ***Methods Express*** (D. Stephens, ed.)

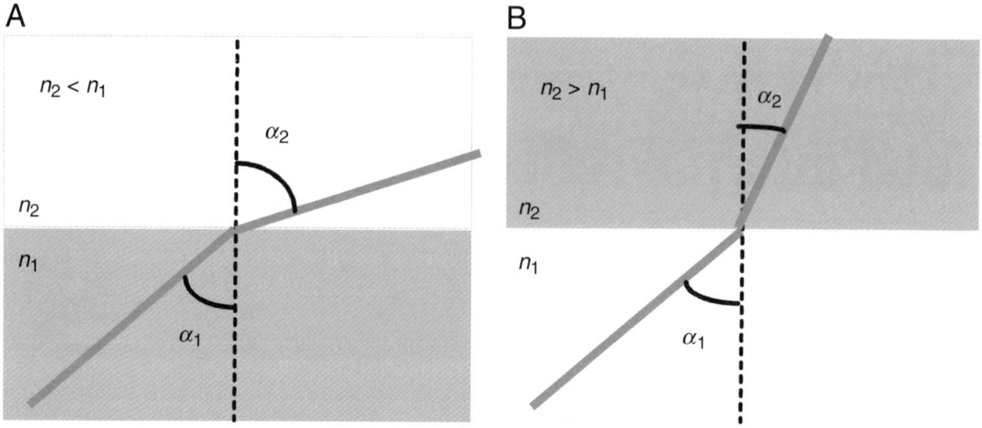

Figure 1. Refraction at an interface between two media.
Light traveling from a medium of one refractive index to another of a different refractive index is refracted depending on their refractive indices. (*A*) Incident light traveling in a substance with a refractive index (n_1) which is greater than the substance it is meeting at the interface (n_2) will be refracted away from perpendicular to the interface. (*B*) As in (*A*), but for $n_2 > n_1$. Note that both conditions obey Snell's law (Equation 1).

α_1 for the refracted light (see *Fig. 1*). One can rearrange Snell's law to solve for the angle of refraction (α_2):

$$\sin(\alpha_2) = (n_1/n_2)\sin(\alpha_1) \qquad \text{(Equation 2)}$$

From Equation 2, one can see that when $n_1 > n_2$, $\sin(\alpha_2) > 1$ for incident angles where

$$\sin(\alpha_c) > n_2/n_1 \qquad \text{(Equation 3)}$$

This angle (α_c) is the so-called critical angle, beyond which no light is refracted. Instead, the light is totally internally reflected and the energy from the incident light remains within the more refractive substance. Nonetheless, the electric field does penetrate the lower refractive index medium to generate an exponentially decaying 'evanescent field' with an intensity distribution (I) described by Equations 4 and 5:

$$I = I_o e^{(-z/d)} \qquad \text{(Equation 4)}$$

$$d = (\lambda/4\pi)(n_1^2 \sin^2\alpha - n_2^2)^{-1/2} \qquad \text{(Equation 5)}$$

where z is the distance from the interface, d is the penetration depth, and λ is the wavelength of light (see *Fig. 2A*). In TIRFM, this evanescent field is used to excite fluorophores in the lower refractive index medium.

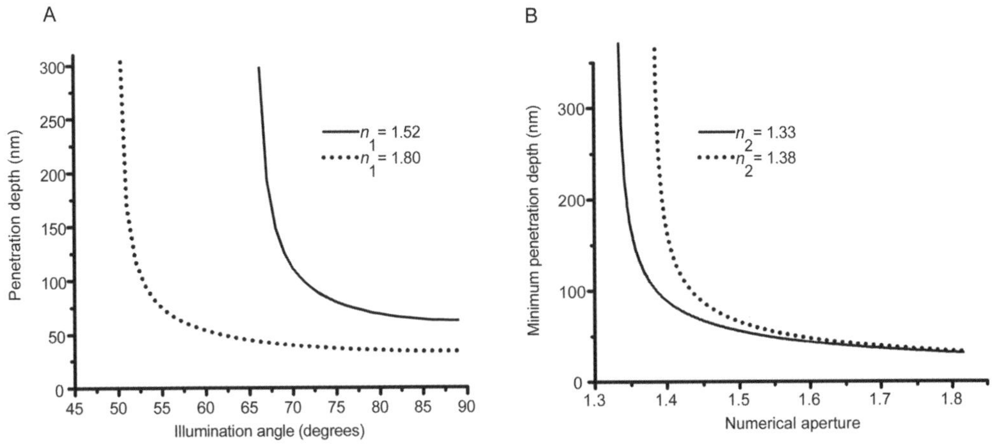

Figure 2. Relationship between excitation angle and evanescent field depth.
(*A*) Plot of penetration depth (*d* in Equations 5 and 8) as a function of excitation angle for light traveling into a medium of refractive index 1.38. This plot is a graphical representation of Equation 5. The solid line indicates penetration depth of the evanescent field calculated for excitation light traveling through regular glass ($n = 1.52$) and the dotted line shows the same calculation for light traveling in highly refractive medium of $n = 1.80$. (*B*) Graphical representation of Equation 8. The plot shows the relationship between numerical aperture and penetration depth for light traveling into water (solid line) or a cell with a refractive index of 1.38 (dotted line).

1.2. Principles of TIRFM

The evanescent field generated in the lower refractive index medium during total internal reflection is exploited in TIRFM to excite fluorescent molecules. The light from the fluorescent molecules can then be collected by an objective lens and recorded using a detector, usually a camera, as one would do for epifluorescence. The chief advantage of this method is the thinness and exponential decay of the illumination. From Equation 5 and *Fig. 2*, it can be seen that the exponential penetration depth of the evanescent field can be considerably smaller than the wavelength of incident light. As an example, regular glass has an index of refraction of 1.52 and water 1.33 for 488 nm light. Light shone at a 70° angle is totally internally reflected and generates an evanescent field with a length constant of 74.6 nm. Given this thin layer of illumination, under optimal conditions, there will be very little out-of-focus fluorescent light. In practice, light scattering at interfaces between substrates with different refractive indices (for example, at a lens surface) somewhat degrades the signal away from those expected under optimal conditions. Even with this caveat, TIRFM provides far superior optical sectioning to conventional epifluorescence microscopy. It achieves this optical sectioning without using a pinhole to reject out-of-focus light as in confocal laser scanning fluorescence microscopy. Thus, much more of the fluorescent light can be collected without a loss in resolution.

TIRFM also gives information about movement towards and away from the interface. Fluorescence intensity is exquisitely sensitive to displacement from the interface in TIRFM. As a fluorescent object moves towards an interface, it experiences a more intense electromagnetic field and thus fluoresces more intensely, as long as the fluorophores are not approaching ground state saturation. Taking the same example as above, an object in an evanescent field with a length constant of 74.6 nm increases its fluorescence emission twofold as it moves 50 nm towards the glass–water interface. This change in fluorescence is often used as an index of movement towards and away from the coverslip, and multiple-angle TIRFM has been used to estimate objects' absolute distance from the coverslip (7, 8). It should be noted, however, that due to near-field effects, emitted light also depends on distance from the interface. At distances close to the interface, more light is emitted towards the higher refractive medium than at further distances. Thus, the fluorescence–distance relationship deviates somewhat from purely exponential (9).

1.3. Applications of TIRFM

In general, TIRFM is a technique applicable to studying anything that can be brought close to a coverslip. Specifically, in biology, TIRFM has been used to study single molecules near coverslips and activity near the surface of acutely dissociated or tissue-cultured cells. A number of applications are listed in *Table 1*.

Table 1. Applications of TIRFM

Application	Reference
Cell-surface adhesion	16
3D imaging of single-molecule dynamics	17
Single-molecule FRET of epidermal growth factor (EGF) binding to EGF receptor	18
Single-molecule detection of motor protein-driven transport	1, 19–22
Visualization of synaptic ribbons	13
Dense core granule trafficking	23
Monitoring the cytoskeleton in living cells	24, 25
Membrane curvature using polarized TIRFM	26
Monitoring synaptic vesicle trafficking	27
Constitutive vesicle trafficking and exocytosis	28, 29
Clathrin-mediated endocytosis	30
Near-membrane calcium transients	31–33
Calcium entry through single calcium channels	34
Simultaneous recording of optical and electrical properties of ion channels	35, 36
Monitoring DNA hybridization	37

2. METHODS AND APPROACHES

2.1. Techniques for TIRFM

Although there are numerous methods for achieving total internal reflection for TIRFM, these methods can be broken down into two categories: prism-type and objective-type TIRFM. Each category has its own advantages and disadvantages.

2.1.1. Prism-type TIRFM

Prism-type TIRFM makes use of a prism (usually trapezoidal or hemi-cylindrical) made of higher refractive material than the biological sample. Light shone at an acute angle travels through the prism, a refractive index-matched immersion fluid, and then the coverslip to the interface between the sample and the coverslip, where it undergoes total internal reflection. Since the prism is on the coverslip, prism-type TIRFM usually (with one exception discussed below) requires the illumination (and therefore the sample) and the light collection to come from opposite sides of the microscope. One way to implement prism-type TIRFM is using an upright fluorescence microscope with excitation light introduced from below the sample (see *Fig. 3A*). Since most biological samples are studied in aqueous buffers, a water (or air)-immersion objective lens must be used to collect the fluorescent light, if one wants access to the sample. Because water has a lower refractive index than glass, water-immersion lenses have a lower numerical aperture (NA) than oil-immersion lenses (NA is proportional to the refractive index), thus providing lower resolution and poorer light collection. Furthermore, objective working distance is inversely related to NA, exacerbating the cost in resolution and light collection for some applications (e.g. where patch clamp is desired). A variation of prism-type TIRFM with an upright microscope is to place the sample between two coverslips, which limits access to the sample, but allows the use of oil-immersion objectives. Similarly, an inverted microscope can be used for prism-type TIRFM by placing the sample between two coverslips. If the sample is a cell culture, then this requires adherence of the cells to the top piece of glass, upside down.

An alternative to using a prism to introduce light from the opposite side of the sample to the objective is to use the coverslip itself as a light guide to bring totally internally reflected light to the sample (see *Fig. 3B*) (1). To do this, a prism is brought into optical contact (via immersion oil) to the bottom surface of a glass coverslip on an inverted microscope outside the field of view of the objective. Light can then be shown at supercritical angles through the prism into the coverslip and the light bounces off both the bottom and top surfaces of the coverslip, thus generating an evanescent field at both surfaces. This allows one full access to the sample from the top.

2.2.2. Objective-type TIRFM

A second method, objective-type TIRFM (10), has become more popular in recent years since the introduction of very high-NA objective lenses by several

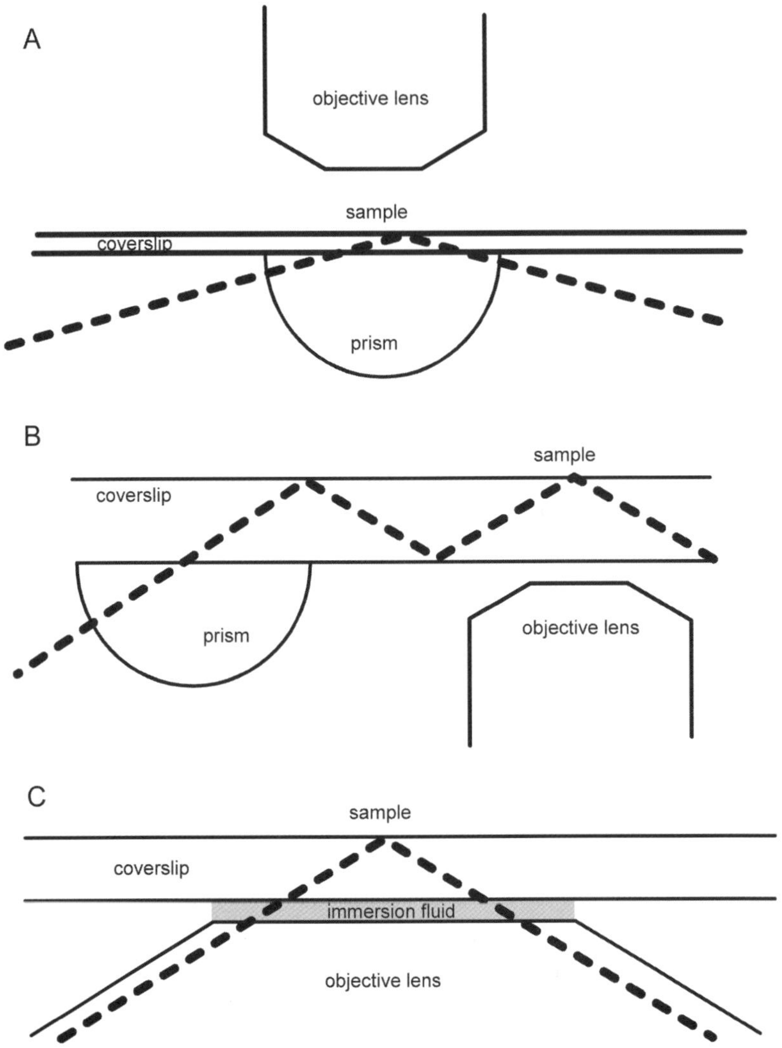

Figure 3. Different methods of achieving TIRFM illumination.
(*A*) Diagram of the light path for TIRFM illumination using a prism on an upright microscope. A prism placed beneath the sample allows one to achieve supercritical angles for TIRFM illumination. Fluorescent light is collected by an objective above the sample. The dashed line indicates the laser light path. (*B*) Diagram showing the use of the coverslip as a waveguide to deliver totally internally reflected light. As in (*A*), excitation light is indicated by a dashed line. Note that the coverslip is shown larger than scale for illustration purposes. (*C*) Diagram showing the light path for objective-type TIRFM illumination.

microscope companies. Objective-type TIRFM uses a high-NA objective lens instead of a prism to introduce light at a supercritical angle (see *Fig. 3C*). Equation 6 describes the definition of NA:

$$NA = n_1 \sin(\alpha_{max}) \qquad \text{(Equation 6)}$$

where n_1 is the refractive index ($n = 1.52$ for oil-immersion lenses) and α_{max} is the maximum illumination angle (also collection angle) achievable with the objective. By substituting Equation 6 into Equation 3 and rearranging, one sees that, in order to get total internal reflection, the NA must be greater than the refractive index of the sample:

$$NA \equiv n_1 \sin(\alpha) > n_2 \quad \text{(Equation 7)}$$

For water, the refractive index is 1.33 and for a cell it is approximately 1.36 (11). The minimum penetration depth achievable for a given objective can be derived by substituting Equation 6 into Equation 5:

$$d = (\lambda/4\pi)\,(NA^2 - n_2^2)^{-1/2} \quad \text{(Equation 8)}$$

Fig. 2(B) plots the relationship between the NA and the minimum penetration depth of evanescent field achievable. It is apparent from Equation 8 and *Fig. 2(B)* that the penetration depth achievable has an inverse relationship to NA. In practice, higher-NA objectives, at the shallowest angles, also produce less propagated light than lower-NA objectives. In recent years, Olympus, Zeiss, and Nikon have all developed high-NA objective lenses that are useful for objective-type TIRFM. Olympus and Nikon have 60× objectives with an NA of 1.45 and Zeiss has a 100× 1.45 NA objective, which use standard glass and immersion oil (n = 1.515 at 488 nm). In addition, Olympus has a 1.65 NA objective, which uses highly refractive glass (n = 1.78 for 488 nm light) and index-matched immersion oil (diiodomethane with sulphur). The Olympus 1.65 NA objective provides superior optical sectioning than the others, with a penetration depth that can be less than 50 nm. This improvement comes at a cost, however, both financial and in terms of hassle. The high-NA coverslips are very expensive and the immersion oil is volatile and leaves a yellow (sulphur) residue. Moreover, the immersion oil and objective are opaque to UV light. I recommend using the 1.65 NA objective only if the thinner penetration depth is absolutely necessary. For most applications, the 1.45 NA objectives would suffice.

To achieve total internal reflection using an objective-type system, one must avoid illuminating the sample at angles that are propagated through the sample rather than being totally internally reflected. One way to achieve this is to use an opaque disk or annulus in the conjugate back focal plane of the microscope, which is used to block all subcritical angles of light. Such a method can be used with either laser or arc lamp illumination (1). Alternatively, total internal reflection can be achieved by focusing a laser beam off-axis towards the back focal plane of the objective lens. Light focused at the back focal plane is directed towards the sample with angles dependent upon the position of the focused beam, with angles of excitation increasing with distance from the center of the back focal plane. Hence, the angle of excitation can be decreased by moving the focused beam towards the edge of the back focal plane and increased by moving it towards the center. Since the penetration depth of the evanescent field is dependent upon the angle of excitation, this allows one to control the depth of the evanescent field. By contrast, the use of an opaque disk does not allow this flexibility.

2.2.3. Prism-type vs. objective-type TIRFM

The prism-type and objective-type TIRFM systems each offer their own advantages and disadvantages. Prisms are much cheaper than high-NA objectives, making prism-type TIRFM the less-expensive option. Additionally, since microscope objectives are made up of many smaller lenses, each with the potential to scatter light, the evanescent field is cleaner with prism-type systems than it is with objective-type systems. Furthermore, using high refractive index prisms, one has the potential to reach shallower penetration depths with a prism-type system than with objective-type systems. These advantages come at a cost in light collection and access to the cell. As mentioned above, prism-type TIRFM requires either the use of a longer working distance water-immersion objective, which limits light collection, or sandwiching the sample between two pieces of glass, which sacrifices access to the sample. The sacrifice in light collection is inversely related to the working distance of the objective: longer working distance objectives have lower NAs and the brightness of fluorescent light is proportional to the fourth power of the NA (NA^4). It is important to note that even using the coverslip as a waveguide requires the use of a water- or air-immersion objective, since the refractive index of oil will cause the light to be propagated to the objective rather than totally internally reflected.

2.2.4. Commercially available TIRFM systems

Currently, a number of TIRFM systems are available for purchase. Olympus, TILL Photonics, and Nikon have commercially available TIRFM systems that are designed to attach to the epi-illumination port on inverted microscopes. These systems are all objective-type TIRFM systems using high-NA objectives, which are coupled via fiber optics to their laser illumination source. Light emerging from the fiber is collimated, reflected, and then focused off-axis to the back focal plane of the objective.

3. RECOMMENDED PROTOCOLS

Protocol 1

Measuring excitation angle for TIRFM

Note that this protocol describes the procedure for measuring excitation angle for an objective-type inverted TIRF microcsope. A similar procedure can also be used to measure excitation angle for other TIRFM configurations.

Equipment and Reagents

- Inverted microscope configured with a light path for TIRFM illumination with a 1.45 or 1.6 NA objective lens
- Prism (hemi-cylindrical, trapezoidal, or rectangular; available from ThorLabs, Newport, or most other optics companies)
- Fluorescent beads (e.g. 100 nm fluorescent microspheres; Invitrogen)
- Ruler or tape measure

The microscope must be set up carefully to optimize its use for TIRFM. Since the penetration depth of the evanescent field is dependent upon the angle of excitation, it is often important to measure this angle. Once measured, one can calculate the penetration depth of the evanescent field using Equation 5 above. This is useful for estimating the distance traveled by objects during experiments.

Method

1. Align the microscope using fluorescent beads. First place a drop of undiluted fluorescent beads on an appropriate refractive index coverslip and place the coverslip on the microscope. Bring the focus of the microscope to the plane of the coverslip. Under conditions where light is propagated, fluorescence will appear mostly diffuse and individual beads will be hard to distinguish. By adjusting the excitation angle (how this is done will depend on your specific microscope) so that the excitation light is totally internally reflected, one will be able clearly to see beads that are immobilized on the coverslip. As the penetration depth is increased, one can see more beads above the coverslip undergoing Brownian motion. Adjust the angle to the desired depth.
2. Turn off the excitation light and, without changing the angle, remove the coverslip and beads and place a drop of immersion oil on to the objective lens.
3. Carefully place the flat surface of a prism on top of the oil droplet. The glass coverslip has a higher refractive index than the aqueous solution and light will no longer be totally internally reflected, but instead propagated through the prism.
4. Project the light towards a sheet of paper and measure the vertical and horizontal displacement to determine the excitation angle. If one is using a hemi-cylindrical or hemi-spherical prism of the same refractive index as the immersion oil (and the light is entering the center of the prism), then the exit angle is equal to the angle of the propagated light. For other prism geometries or refractive indices, then the angle must be back calculated using Snell's law (Equation 1), taking into account each interface between two substances of different refractive indices (the oil-to-prism and prism-to-air interfaces).

As described earlier in this chapter, TIRFM can be used for a wide range of experiments. *Protocol 2* gives one example of its use: imaging intracellular vesicles.

Protocol 2

Imaging FM1-43-labeled retinal bipolar cell vesicles

Equipment and Reagents

- Retinal bipolar cells prepared as described previously (see 12–14)
- Enzymes and materials for tissue preparation (see 13–15)
- Standard buffer (120 mM NaCl; 2.5 mM KCl; 2.5 mM $CaCl_2$; 1 mM $MgCl_2$; 10 mM HEPES; adjusted to pH 7.2 with NaOH)
- 5 μM FM1-43 (Molecular Probes) in depolarizing buffer (95 mM NaCl; 25 mM KCl; 2.5 mM $CaCl_2$; 1 mM $MgCl_2$; 10 mM HEPES; adjusted to pH 7.2 with NaOH)
- Low-calcium buffer (120 mM NaCl; 2.5 mM KCl; 0.5 mM $CaCl_2$; 1.25 mM EGTA; 1 mM $MgCl_2$; 10 mM HEPES; adjusted to pH 7.2 with NaOH)
- System for local perfusion (e.g. SF-77B Perfusion Fast-Step; Warner Instruments, or other)

Method

1. Plate the retinal bipolar cells on washed and rinsed coverslips. Bipolar cells are identifiable by their distinct morphology, containing a cell body with a single axon terminating in a single, large (~10 μm diameter), nearly spherical synaptic terminal.
2. After allowing the cells to settle, search the coverslip for bipolar cells with tightly adherent synaptic terminals. To determine whether a bipolar cell is well attached, it is often helpful to tap the microscope stage lightly; an adherent cell moves very little when the microscope stage is tapped.
3. Superfuse attached bipolar cells with 5 μM FM1-43 in depolarizing buffer using a local perfusion pipette aimed directly at the cell for 10 s at room temperature. Resist the temptation to use higher concentrations of FM1-43, since higher concentrations (as low as 10 μM) are toxic to the cells and may permeate the membrane (unpublished observation).
4. Using a local perfusion pipette, wash the cells with dye-free, low-calcium buffer for more than 30 min to remove fluorescent dye from the bipolar cell plasma membrane and from the coverslip (FM1-43 and other styryl dyes adhere tightly to glass; 15). This long wash period is also necessary to ensure that synaptic vesicles are labeled rather than large endosomes or cisternae.
5. Replace the low-calcium buffer with standard buffer.
6. Image the cells using TIRFM. To visualize vesicles, it is helpful to use an intensified camera; either a standard image intensifier or a electron multiplying camera will work for this purpose. An argon ion laser-based system excites FM1-43 well, but many other lasers will also excite FM1-43 due to its extremely broad excitation spectrum.

4. REFERENCES

★★ 1. **Axelrod D** (2001) *Traffic*, **2**, 764–774. – *Excellent review describing, among other things, many of the different configurations that can be used for TIRFM.*

★★ 2. **Steyer JA & Almers W** (2001) *Nat. Rev. Mol. Cell Biol.* **2**, 268–275. – *Another good review dealing with many of the practical aspects of TIRFM.*

3. **Toomre D & Manstein DJ** (2001) *Trends Cell Biol.* **11**, 298–303.
4. **Thompson NL, Pearce KH & Hsieh HV** (1993) *Eur. Biophys. J.* **22**, 367–378.
5. **Axelrod D, Thompson NL & Burghardt TP** (1983) *J. Microsc.* **129**, 19–28.
6. **Oheim M, Loerke D, Chow RH & Stuhmer W** (1999) *Philos. Trans. R. Soc. Lond. B Biol. Sci.* **354**, 307–318.
7. **Burmeister JS, Truskey GA & Reichert WM** (1994) *J. Microsc.* **173**, 39–51.
8. **Olveczky BP, Periasamy N & Verkman AS** (1997) *Biophys. J.* **73**, 2836–2847.
9. **Axelrod D & Hellen EH** (1989) *Methods Cell Biol.* **30**, 399–416.
10. **Stout AL & Axelrod D** (1989) *Applied Optics*, **28**, 5237–5242.
11. **Curl CL, Bellair CJ, Harris T, *et al.*** (2005) *Cytometry A*, **65**, 88–92.
12. **Zenisek D, Steyer JA, Feldman ME & Almers W** (2002) *Neuron*, **35**, 1085–1097.
13. **Zenisek D, Horst NK, Merrifield C, Sterling P & Matthews G** (2004) *J. Neurosci.* **24**, 9752–9759.
14. **Heidelberger R & Matthews G** (1992) *J. Physiol.* **447**, 235–356.
15. **Rouze NC & Schwartz EA** (1998) *J. Neurosci.* **18**, 8614–8624.
16. **Todd I, Mellor JS & Gingell D** (1988) *J. Cell Sci.* **89**, 107–114.
17. **Dickson RM, Norris DJ, Tzeng YL & Moerner WE** (1996) *Science*, **274**, 966–969.
18. **Sako Y, Minoghchi S & Yanagida T** (2000) *Nat. Cell Biol.* **2**, 168–172.
19. **Vale RD, Funatsu T, Pierce DW, Romberg L, Harada Y & Yanagida T** (1996) *Nature*, **380**, 451–453.
20. **Quinlan ME, Forkey JN & Goldman YE** (2005) *Biophys. J.* **89**, 1132–1142.
21. **Kural C, Kim H, Syed S, Goshima G, Gelfand VI & Selvin PR** (2005) *Science*, **308**, 1469–1472.
22. **Inoue Y, Iwane AH, Miyai T, Muto E & Yanagida T** (2001) *Biophys. J.* **81**, 2838–2850.
23. **Steyer JA, Horstmann H & Almers W** (1997) *Nature*, **388**, 474–478.
24. **Krylyshkina O, Anderson KI, Kaverina I, *et al.*** (2003) *J. Cell Biol.* **161**, 853–859.
25. **Lanni F, Waggoner AS & Taylor DL** (1985) *J. Cell Biol.* **100**, 1091–1102.
26. **Sund SE, Swanson JA & Axelrod D** (1999) *Biophys. J.* **77**, 2266–2283.
27. **Zenisek D, Steyer JA & Almers W** (2000) *Nature*, **406**, 849–854.
28. **Schmoranzer J, Goulian M, Axelrod D & Simon SM** (2000) *J. Cell Biol.* **149**, 23–32.
29. **Toomre D, Steyer JA, Keller P, Almers W & Simons K** (2000) *J. Cell Biol.* **149**, 33–40.
30. **Merrifield CJ, Feldman ME, Wan L & Almers W** (2002) *Nat. Cell Biol.* **4**, 691–698.
31. **Omann GM & Axelrod D** (1996) *Biophys. J.* **71**, 2885–2891.
32. **Zenisek D, Davila V, Wan L & Almers W** (2003) *J. Neurosci.* **23**, 2538–2548.
33. **Becherer U, Moser T, Stuhmer W & Oheim M** (2003) *Nat. Neurosci.* **6**, 846–853.
34. **Demuro A & Parker I** (2003) *Cell Calcium*, **34**, 499–509.
35. **Yokota H, Kaseda K, Matsuura H, *et al.*** (2004) *J. Nanosci. Nanotechnol.* **4**, 616–621.
36. **Ide T, Takeuchi Y, Aoki T & Yanagida T** (2002) *Jpn. J. Physiol.* **52**, 429–434.
37. **Yao G, Fang X, Yokota H, Yanagida T & Tan W** (2003) *Chemistry*, **9**, 5686–5692.

CHAPTER 9

An overview of three-dimensional and four-dimensional microscopy by computational deconvolution

José-Angel Conchello

1. INTRODUCTION

Three-dimensional microscopy by computational deconvolution has steadily gained acceptance as an alternative and complement to purely optical methods such as confocal scanning fluorescence microscopy and multi-photon fluorescence excitation microscopy. Several methods have been developed for computational optical sectioning microscopy. The goal of these methods is to undo the degradation introduced by the process of image formation and recording. To achieve this goal, these methods require a mathematical model for this process. In general, more exact models lead to better results but at the expense of increased computer time and memory. This article gives an overview of the existing computational deconvolution methods.

1.1. General principles

Three-dimensional fluorescence microscopy is often done by the method of optical sectioning. In this method, the microscope is focused at some plane in the specimen and a 2D image is collected. The microscope is then focused at a different plane and another 2D image is collected. The process of focusing and collecting 2D images continues until all the structures of interest have been imaged. The 2D images are called optical slices (as opposed to physical slices obtained with the microtome). Each optical slice contains the in-focus information that appears in clear focus plus a large contribution of out-of-focus material that obscures the in-focus information and reduces contrast. In fact, for a wide-field microscope, usually more than 90% of the light that reaches the detector is out-of-focus light. For example, the depth of focus of a 1.4 numerical

Cell Imaging: *Methods Express* (D. Stephens, ed.)

aperture (NA) objective lens is only about 0.3 μm, whereas cells are about 10 μm in diameter. Thus, for this objective, 97% of such a cell is out-of-focus; however, the fluorescent light from the whole cell reaches the detector. The relative contribution of in-focus light to the total amount of light decreases with an increasing NA and with increasing specimen thickness.

There are several methods to avoid or ameliorate the detrimental effect of out-of-focus light on the recorded image. The methods can be classified as: (i) optical methods, such as confocal scanning microscopy and multi-photon fluorescence excitation microscopy; (ii) computational methods (usually grouped under the name *deconvolution*); and (iii) hybrid methods, i.e. those that require an optical and a computational component, such as structured illumination microscopy and aperture correlation microscopy. Optical methods are covered elsewhere in this volume. This chapter concentrates on computational methods for 3D microscopy. The basis of all computational methods is first to develop a mathematical description of the process of image formation and recording and then to use this description or model to undo the degradations caused by this process, i.e. to calculate the distribution of fluorescent dye that can give rise to the recorded image and is consistent with the model of image formation and recording. An exact mathematical description of the process of image formation and recording is not only virtually impossible to obtain but also computationally intractable. Thus, all models used for computational 3D microscopy use simplifying assumptions, and models that use different simplifications lead to different algorithms or methods. The general rule of thumb is that more precise models give better results but have a higher computational burden, in terms of computer time, memory requirements, or both. This article presents algorithms for 3D computational microscopy arranged by the increasing complexity of the mathematical model they are based on. There are few mathematical details provided in this chapter. The more mathematically oriented reader is encouraged to read the references given for the different methods.

The methods are presented in increasing degree of complexity. Sections 2.1 and 2.2 present algorithms based on the simplest models for image formation. These models lead to algorithms that have very low computational demands and yet can be applied to certain types of specimens. Section 2.3 covers methods based on modeling the microscope from a frequency components standpoint. Before covering the algorithms, the section introduces the important concepts of frequency components and frequency response. Section 2.4 presents methods that exploit knowledge about the specimen to greatly improve the frequency response of the algorithms. Section 2.5 covers algorithms based on models of image formation and recording that explicitly take into account the unpredictable variations in the image caused by the random behavior of photon absorption, emission, collection, and detection. Section 2.6 presents methods that can be used when there is not enough information about the imaging conditions and thus they have to be estimated from the recorded image and prior knowledge about the specimen and the microscope. Finally, section 3 presents a brief list of sources of degradation that are not taken into account on any of the existing algorithms.

2. METHODS AND APPROACHES

2.1. Simple models for image formation

The most simplified model for image formation is to consider the image as a blurred version of the distribution of fluorescent dye in the specimen. The inversion of this model is simply to sharpen the recorded image with an arbitrary sharpening filter. These methods do not require sophisticated software. Sharpening utilities are included in many software packages for generic image manipulation such as PHOTOSHOP (Adobe Systems), METAMORPH (Universal Imaging), and IMAGEJ (National Institutes of Health, Bethesda, MD). These greatly simplified methods, however, have severe limitations for dealing with out-of-focus light and thus we will not consider them further.

2.2. Pseudo-3D models – nearest-neighbors subtraction

The next level of sophistication is what I call pseudo-3D methods. These are methods based on models that only partially consider the out-of-focus light present in images of 3D specimens. These methods take into account the fact that the blur observed in an image is caused by diffraction. Due to diffraction, the image of a point source of light is not a point. When the point source is in focus, the image is a small, bright patch of light surrounded by alternating dark and bright concentric circles. The brightness of the concentric rings decreases from one ring to the next as the square of the distance from the center of the image. As the point source moves out of focus, the size of the central patch of light increases and its brightness decreases. At the same time, the brightness of the first secondary ring increases. At some out-of-focus distance, the central spot vanishes completely. The structure of the out-of-focus image then becomes more complicated, but with less contrast. The image of a point source of light is referred to as the point spread function (PSF) of the microscope and is used to characterize its imaging properties.

The earliest model that takes into account the PSF is as follows (1). The light recorded in a given optical slice consists of the in-focus light plus out-of-focus contributions from the optical slices immediately above and below the optical slice of interest, i.e. the mathematical model assumes that only the two nearest neighbors contribute out-of-focus light. The contributions of the two nearest neighbors to the optical slice of interest are blurred by a 2D PSF at an out-of-focus distance equal to the depth separation between the optical slices.

To invert this model of image formation, each optical slice of the restored image is calculated by taking the corresponding optical slice of the recorded image and its two nearest neighbors. The two neighbors are computationally blurred by a 2D PSF calculated at an out-of-focus distance equal to the depth difference between the optical slices. The blurred neighbors are then subtracted from the optical slice of interest. The result is optionally sharpened by a 2D Wiener filter (described below) that uses the in-focus PSF (2). Although the method is quite simplistic, it can provide better resolution and contrast than the original images in cases when the specimen is made of punctate and filamentous structures. This is because, for such specimens, the out-of-focus light decreases rapidly away from focus and thus the simplified model for image formation applies (see *Fig. 1*). When the specimen includes fluorescently tagged membranes or volumes significantly larger than the diffraction-limited spot, the model does not describe the actual conditions for image formation and thus the algorithm fails. When the method can be applied, it presents significant advantages over other more sophisticated methods because it requires very little computer memory and central processing unit time. At any given time, three planes of the input image, one of the output image, and two 2D filters need to be in memory. Because of its simplicity and low computational load, the method is generally included in many commercial deconvolution packages.

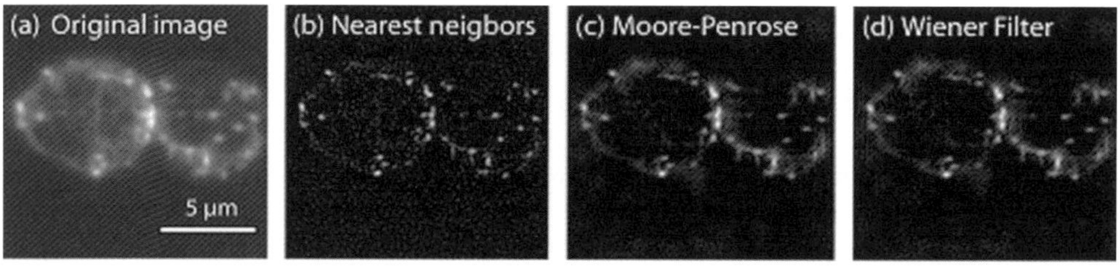

Figure 1. Image and restoration by nearest-neighbors subtraction and using Moore–Penrose pseudo-inverse and Wiener–Helstrom filters.
An optical slice through a 3D through-focus series of fluorescently labeled actin filaments collected with a 100×/1.3 NA objective is shown. (*a*) Recorded image; (*b*) nearest-neighbors subtraction; (*c*) Moore–Penrose pseudo-inverse filter; and (*d*) Wiener–Helstrom filter. Bar, 5 μm. (Original stack courtesy of Dr. Tatiana Karpova, National Cancer Institute, Bethesda, MD, USA).

2.3. Methods based on the frequency response

2.3.1. Frequency components

Before we go into a series of more sophisticated and more powerful methods, it is necessary to introduce an alternative description of the process of image formation, namely, the frequency characterization of the microscope. For this it is necessary to introduce the concept of frequency components of a specimen and image. Frequency components in 1D are wave-like shapes of different frequencies. In 2D, they are patterns that alternate between dark and bright, as shown in *Fig. 2(A)*. If we define any two axes in the plane of the specimen, the frequency components along those axes will be patterns that change only along one of the axes. For example, in *Fig. 2(A)* for the horizontal and vertical axes, we take the frequency components along those axes given in the top row and left column, respectively. In addition to the components along the two chosen axes, there are components that combine the frequencies of those two axes. In *Fig. 2(A)*, these are the boxes at the intersection of the column of the vertical component and the row

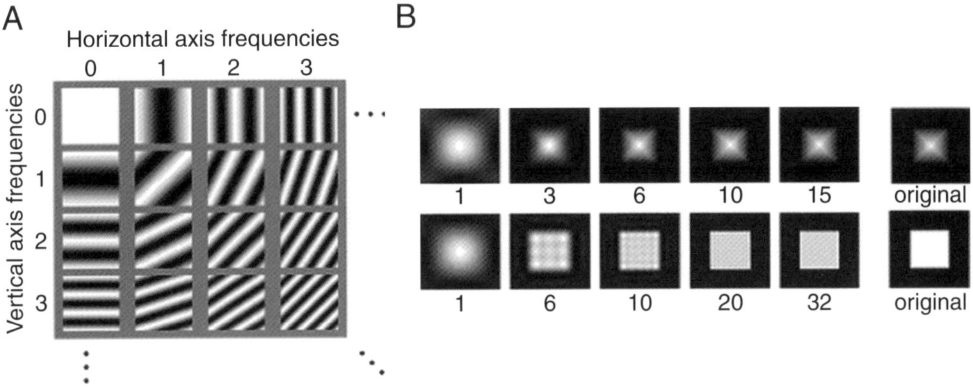

Figure 2. Two-dimensional frequency components.
(*A*) Frequency components in two dimensions are patterns that alternate between dark and bright. The more cycles that fit in a given space, the higher the frequency of that component. (*B*) Synthesis of an image from its frequency components. The left-most column shows the shape being synthesized. The numbers under the frames are the number of cycles over the field of view of the highest frequency used. Top: a shape whose brightness decreases gradually from the center to the edges. Bottom: a figure with the same shape, but the brightness is uniform across the shape and goes to zero sharply at the edges.

of the horizontal component. For example, the component with vertical frequency of two cycles per field of view and horizontal frequency of three cycles per field of view is the box in the row labeled 2 and column labeled 3. An important characteristic to note is that anywhere we trace a vertical line in this box, the line goes over two cycles of the pattern of dark and bright. Likewise, if we trace a horizontal line, the line will go over exactly three cycles, regardless of where in the box we trace the line. By combining frequency components of the appropriate amplitude and phase, any shape can be synthesized. Conversely, any image can be analyzed into its frequency components by an operation called the Fourier transform that computers perform very rapidly. *Fig. 2*(*B*) shows two simple images and their synthesis with a subset of their frequency components. The top row shows an image of a square whose brightness decreases linearly from the center to the edges (right-most frame). The number under the frames indicates the highest-frequency component used to synthesize the image in that frame. For example, the frame labeled '3' uses the constant term and the components that have one, two, or three cycles within the field of view. Because of the size of the image, the highest frequency that any component may have is 127 cycles vertically and horizontally across the field of view. From the top row of *Fig. 2*(*B*) it is clear that the figure can be synthesized to a very good approximation using only a small set of all its frequency components. The bottom row shows a similar example, except that the brightness of the square is constant and sharply goes to zero at the edges of the square. Although the square shape can be approximated with a few frequency components, the uniform brightness across the square cannot. Even with components up to 32 cycles per frame, there is a wavy look to the square. This figure shows the relationship between the description of an image from its pixel values and from its frequency components. Namely, coarse detail (e.g. the square shape) and smooth detail (e.g. the smooth transition from black to white) can be synthesized from lower-frequency components, whereas fine detail, such as edges (i.e. the sharp transition from black to white) requires higher-frequency components.

2.3.2. Frequency response of the microscope

So, how do frequency components help us restore an image? To answer this question, it is necessary to introduce the concept of the frequency response of the microscope, i.e. how the microscope passes the frequency components from the specimen to the image. Due to diffraction and the finite size of the objective lens, the microscope does not pass all the frequency components of the specimen, and from those that it passes, the lower frequencies are passed better than the higher ones. *Fig. 3* shows a plot of the frequency response of the microscope in any of the lateral directions. In 2D, the function has circular symmetry (obtained by rotating *Fig. 3* around the vertical axis). This function is called the optical transfer function (OTF) because it describes how the microscope transfers the frequency components of the specimen to the image and it is customary to normalize it so that the zero frequency passes unaltered, i.e. the OTF is 1.0 at zero frequency. In 2D, there are two frequencies, one along each of two arbitrary axes, usually chosen to be the horizontal and vertical axes. More precisely, let ξ and η be frequencies in the horizontal and vertical directions, respectively. If $S(\xi,\eta)$ describes the frequency component of the specimen at those frequencies and $H(\xi,\eta)$ is the OTF, i.e. it describes how the microscope transfers the frequency component of the specimen to the image, then the frequency components of the

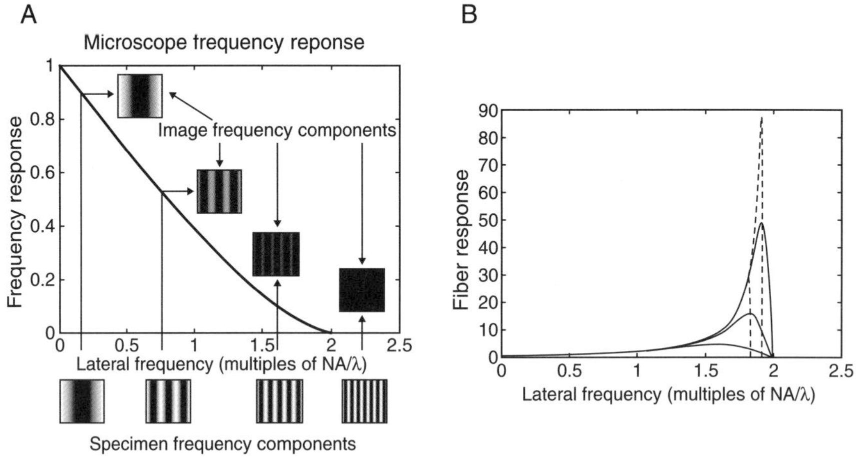

Figure 3. Frequency responses.
(*A*) Frequency response of the microscope. The higher the frequency, the more it is attenuated. Frequencies higher than $2NA/\lambda$ are not passed by the objective. (*B*) Frequency response of the Moore–Penrose pseudo-inverse filter (dashed line) for $\varepsilon =$ 0.01 and 0.03, and the Wiener–Helstrom filter (solid line) for $\varepsilon = 0.01$, 0.001, and 0.001. Smaller values of ε give taller frequency response plots for both filters.

image $G(\xi,\eta)$ are the frequency components of the specimen multiplied by the OTF:

$$G(\xi,\eta) = H(\xi,\eta)S(\xi,\eta) \quad \text{(Equation 1)}$$

Knowing the OTF and frequency components of the image, one would be tempted to obtain the frequency components of the specimen by simply solving this equation for the specimen frequency components, i.e.:

$$S(\xi,\eta) = G(\xi,\eta)/H(\xi,\eta) \quad \text{(Equation 2)}$$

This equation is called inverse filtering and would be very easy to implement in a computer. One needs to give the image and the PSF of the microscope to the computer; the computer then calculates their Fourier transforms and divides them to give the frequency components of the specimen. These are then inverse Fourier transformed by the computer to give the restored specimen. This approach, however, would not work. This is because the OTF, the frequency response of the microscope, is zero except for frequencies smaller than $2NA/\lambda$, where λ is the wavelength of fluorescent light. Thus, the inverse filter divides by zero, an operation that is not allowed. Ideally, the image would not have components at frequencies higher than the cut-off frequency of $2NA/\lambda$. In practice, however, the image is contaminated by noise resulting from the random nature of photon emission and from noise in the detector. This noise invariably introduces components at frequencies not passed by the microscope. In addition, as seen in *Fig. 3*, the OTF becomes small for frequencies close to but smaller than the cut-off frequency. At frequencies where the OTF is small, the noise is usually larger than the signal (the Fourier transform of the image) and thus dividing by the OTF greatly increases the amount of noise in the restored image. Thus, it is necessary to modify Equation 2 to avoid the division by zero or by a small number. These modifications are presented in the next section.

2.3.3. Truncated inverse (Moore–Penrose pseudo-inverse) filter

One widely used method to avoid division by zero or by a small number is to ignore the components of the image and the specimen at those frequencies where the OTF is zero or a small number. Thus, instead of Equation 2, the frequency components are obtained as:

$$\begin{aligned} S(\xi,\eta) &= G(\xi,\eta)/H(\xi,\eta) \text{ if } |H(\xi,\eta)| > \varepsilon \\ S(\xi,\eta) &= 0 \text{ otherwise} \end{aligned} \quad \text{(Equation 3)}$$

where ε is a small positive number. Although the idea looks quite simple, it has a solid mathematical foundation (3, 4). Because the image can have frequency components where the OTF is zero, Equation 1 is inconsistent in the sense that it does not have a solution, i.e. there is no specimen that would satisfy this equation. When faced with an inconsistent equation, one might try to find a solution that is optimal in some sense. One widely used criterion for optimality is to use least squares, i.e. to look for a solution to Equation 1 for which the squared error

$$|G(\xi,\eta) - H(\xi,\eta)S(\xi,\eta)|^2$$

added for all frequency components is minimized. It turns out that Equation 3 is such a solution. Because Equation 3 truncates the frequency components of the specimen, it is sometimes called a truncated inverse filter. Other names used for this filter are linear least squares and Moore–Penrose pseudo-inverse (MPPI). The latter name, which we will use in this chapter, comes from the mathematical derivation to minimize the squared error. The selection of ε permits controlling the trade-off between the achievable resolution and the amount of noise in the restored image. If ε is large (i.e. close to 1.0), the restored image will contain only those few frequencies at which the OTF is above this value, which are usually low frequencies. In this case, the restored specimen function will have only coarse and smooth detail. If ε is decreased, the restored specimen function will contain more frequency components, usually at higher frequencies, and thus will show finer detail. Using very small values of ε, however, brings into the specimen estimate components at those frequencies where the OTF is small and the recorded image is likely to be dominated by noise. Thus, a small value of ε usually results in the amplification noise, usually high-frequency noise. *Fig. 4* shows the results of deconvolving a 3D image of *Histoplasma capsulatum* (courtesy of Dr. William Goldman, Washington University School of Medicine, USA). In this image, only the membrane is fluorescently labeled. Therefore, there should be no fluorescent light inside the membrane. *Fig. 4* shows a section of the optical slice that had the most light in the stack. In the original image (left frame), there is a substantial amount of out-of-focus light. The nearest-neighbors subtraction somewhat reduces this, but at the expense of a salt-and-pepper background and a disjoint membrane. The MPPI filter with $\varepsilon = 0.01$, on the other hand, removes a significant amount of out-of-focus light while preserving a continuous membrane. Reducing ε, however, increases the amount of background noise and results in a disjoint membrane. The MPPI method is available in AXIOVISION (Carl Zeiss) and in the freely distributed XCOSM deconvolution package (www.omrfcosm.omrf.org).

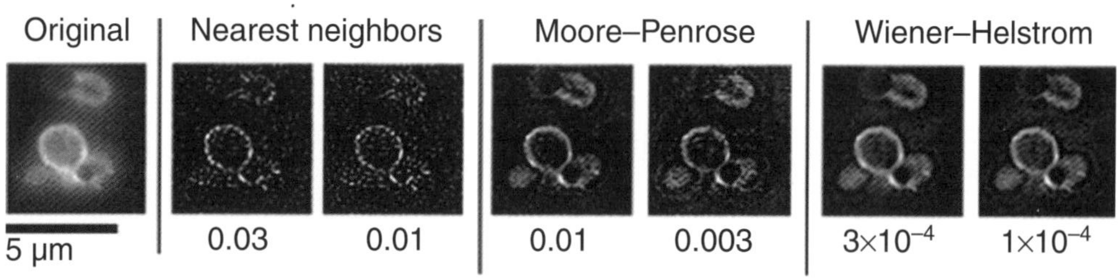

Figure 4. A plane through a 3D image of *Histoplasma capsulatum* whose membrane was fluorescently stained.
The different values of ε are shown beneath each image. (3D stack courtesy of Dr. William Goldman, Washington University School of Medicine.)

Wiener–Helstrom filter

Another method commonly used to avoid the division by zero in Equation 1 is to add a small positive number to the denominator of Equation 2. However, the OTF cannot be guaranteed to be non-negative. In fact, the OTF usually has complex values. Thus, the idea of simply adding a small positive number to the denominator must be modified to ensure that the denominator is never zero. The most widely used method to do this is to use a filter of the form:

$$S(\xi,\eta) = G(\xi,\eta)H^*(\xi,\eta)/[|H(\xi,\eta)|^2 + \varepsilon] \qquad \text{(Equation 4)}$$

where the * superscript and the bars |•| denote the conjugate and the modulus, respectively, of a complex-valued quantity. This expression is known as the Wiener filter (or, more correctly, Wiener–Helstrom filter (WHF); see reference 5, pp. 206–210). In the original derivation, the constant ε is the noise-to-signal ratio at the corresponding frequencies (ξ,η) and, to obtain Equation 4, it is assumed to be the same for all frequencies. The latter restriction, however, seldom applies to microscopic images and thus, for microscopy, Equation 4 is not strictly a WHF but a Wiener–Helstrom-like filter. In this article, however, we will use the shorter WHF name. As with the MPPI or truncated inverse filter, the value of ε controls the trade-off between the achievable resolution and the amount of noise in the restored specimen. From *Fig. 3*, it is clear that when the OTF is much larger than ε (low frequencies in *Fig. 3*), the responses of the WHF and MPPI filter are similar. However, at those frequencies where the OTF is comparable to or smaller than ε, the two filters have different behaviors. The MPPI filter keeps increasing as the OTF approaches ε and then cuts off sharply to zero. The WHF, on the other hand, stops growing when the OTF decreases so its square is equal to ε and monotonically decreases to zero at the point where the OTF becomes zero. Because of this behavior, the WHF usually requires a smaller value of ε than the MPPI filter to achieve a similar amount of detail. In addition, the latter usually shows high-frequency oscillations about the edges. Although the WHF can show these oscillations too, they have smaller amplitude than those produced by the MPPI filter. The two left-most frames of *Fig. 4* show two different WHF deconvolutions of the *H. capsulatum* image for different values of ε. For this image, the MPPI filter provides better results, although this is not true for all images. For some images, the WHF gives better results and in many cases the MPPI filter and WHF give very similar results. The WHF is part of many deconvolution packages including DELTAVISION (Applied Precision), AUTODEBLUR (AutoQuant Imaging), and HUYGENS (Scientific Volume Imaging).

Fig. 5 illustrates a deficiency of the methods presented so far. The specimen is a 10 μm diameter latex microsphere that has a layer of fluorescence between 1 and 2 μm thick. The microsphere was embedded in optical cement to prevent it from moving and imaged with a 40×/1.0 NA oil-immersion objective (original image courtesy of Dr. James G. McNally, National Cancer Institute, USA). Two sections through the center of the microsphere are shown, a horizontal (top row) and a vertical or axial (bottom row) section. The left-most column shows the

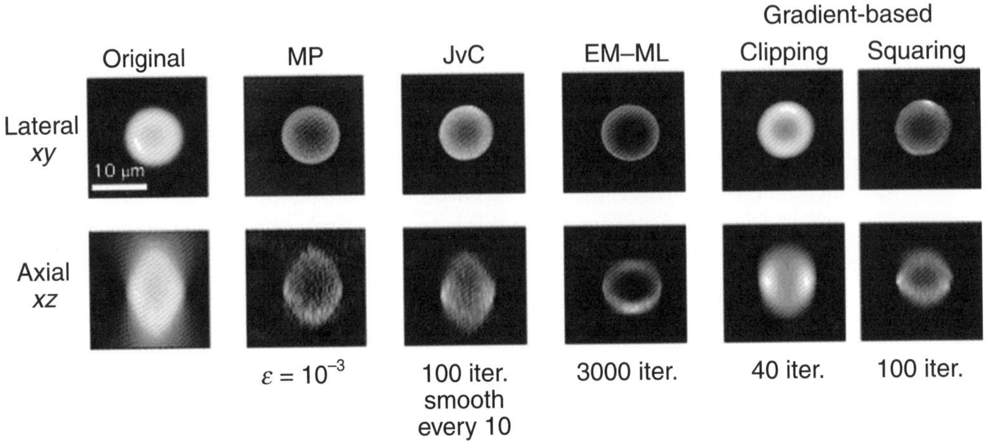

Figure 5. Horizontal and vertical sections through the center of the image and restoration of a spherical object.
The diamond shape in the vertical (or axial) section is due to the missing cone. MP, Moore–Penrose pseudo-inverse filter; JvC = Jansson–van Cittert method of repeated convolution; EM–ML, expectation-maximization maximum likelihood. For the gradient-based maximum likelihood, clipping indicates setting negative pixels to zero at every iteration, and squaring indicates defining the specimen to be the square of another quantity and estimating that quantity. (Original images courtesy of Dr. James G. McNally, National Cancer Institute, Bethesda, MD.)

recorded image. The image greatly departs from what is known about the specimen. First, the image does not show the dark core devoid of fluorescence, and secondly, the vertical section does not look like a spherical object. This section is diamond shaped. The second column shows the deconvolution obtained with the MPPI filter. The lateral section shows the dark core of the microsphere; however, the vertical section still looks highly elongated. The image was measured to be 17 μm (full width at half maximum) along the vertical axis. The PSF of the 40×/1.0 NA can explain an elongation of less than 2 μm above and below the microsphere.

The reason a spherical object has a diamond shape with a longer axis along the optical axis of the microscope (the depth direction) has to do with the frequency response of the microscope in 3D. The plot shown in *Fig. 3* depicts only the lateral, in-focus frequency response of the microscope. Although this is not enough to describe the 3D imaging characteristics of the microscope, it is useful to illustrate the behavior of the WHF and MPPI filter. As mentioned above, the frequency response of the microscope has cylindrical symmetry. This holds in focus, as well as out of focus. However, in 3D the OTF is nonzero over a more restricted volume of the frequency space or domain as shown in *Fig. 6* (6). An important feature to notice is that the bow-tie shape of this area excludes all the frequency components along the depth axis ($\zeta = 1/z$) and a wedge-shaped region about this axis. In 3D, this wedge-shaped region becomes a cone-shaped volume of frequencies that the microscope blocks. This region, commonly called the *missing cone*, is the reason for the diamond shape of the image of a spherical object. Unfortunately, none of the methods described thus far can recover the frequency components within the missing cone.

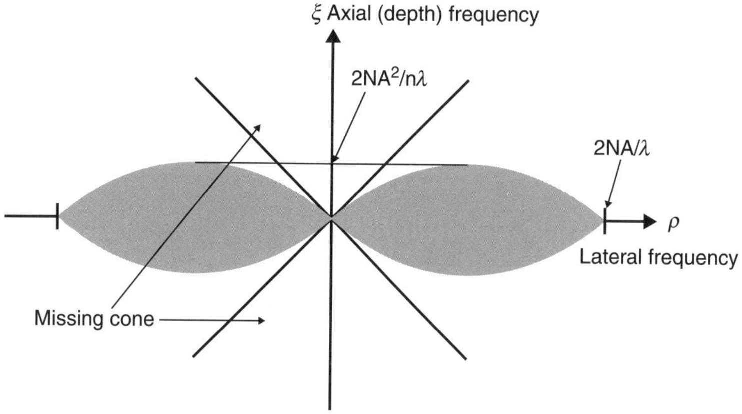

Figure 6. Region of the frequency domain where the OTF of the wide-field microscope is not zero.
The actual frequency response is the 3D shape that results from rotating the figure about the vertical axis. There is a conic region about this axis where there are no frequencies passed by the microscope.

2.4. Constrained deconvolution

The methods described thus far cannot recover frequency components that are not passed by the microscope. These frequency components, however, are sometimes necessary for recovering even coarse detail, such as the shape of large objects such as blood cells. Unfortunately, the image and the PSF (or, equivalently, the OTF) do not provide enough information to recover any of the frequency components not passed by the microscope. To be able to recover some of the lost frequency components, any algorithm needs to incorporate information about the specimen. There are several examples of information that can be used: (i) the concentration of fluorescent dye is never negative; (ii) the concentration of fluorescent dye is always finite; (iii) the specimen is never infinitely large. These are properties that apply to all fluorescent specimens. Constraining the restored specimen function to satisfy one or more of these characteristics leads to a group of algorithms called constrained deconvolution. With very few exceptions (7), constrained deconvolution algorithms are iterative, i.e. they perform a series of steps over and over until a certain criterion is satisfied. The general form of the iterative deconvolution algorithm is as follows:

1. Estimate (guess) the specimen function.
2. Use the model of image formation and recording to calculate the image of the specimen function estimate.
3. Compare the estimated image with the one recorded from the microscope.
4. Use the result of this comparison to update the estimated specimen function.
5. Enforce constraints.

6. Reduce unwanted behavior.
7. Repeat steps 2–6 until the estimated specimen function is satisfactory (in some pre-determined sense).

Note that steps 4–6 need not be carried out in the order indicated above, or even as separate steps. Depending on the algorithm, the order of these steps may be different and/or two or more steps might be combined into a single one. Several methods for iterative deconvolution have been developed. The most significant ones are listed in the following sections.

2.4.1. The Jansson–van Cittert method of repeated convolution

The oldest method for iterative deconvolution to be applied to fluorescence microscopy is the Jansson–van Cittert (JvC) method of repeated convolution. In this method, the initial guess for the specimen function is usually the recorded image. This image is convolved in the computer with the PSF of the microscope. The calculated image is subtracted from the recorded one and the difference is used to correct the estimated specimen function. In the original van Cittert method of repeated convolution, the difference was added to the current specimen estimate to obtain the new specimen estimate. Doing so, however, often results in negative pixels. Jansson modified the method to avoid negative pixels (8), namely, by multiplying the difference by a factor that depends on the value of the pixel being updated. If the pixel value is close to zero or to a maximum allowed pixel value, the factor is small. The factor increases towards the center of the range of allowed pixel values. In particular, Jansson proposed the weight factor shown as a dashed line in *Fig. 7*. Later, Agard (1) suggested the weight factor shown by the solid line in this figure. Notice that in the JvC algorithm, instead of enforcing the pixels to be between two known values, it prevents them from taking values outside the allowed range. One disadvantage of this approach is the need to know an upper limit, s_{max} on the allowed pixel values. This value usually has to be estimated and adjusted during the deconvolution process. This value, however, has a strong effect on the convergence rate of the algorithm. If s_{max} is set to a value much larger than it actually needs to be, the vast majority of the pixel values will fall in the lower end of the weight factor (close to $s = 0$). As a result, the weight factor is small and so is the correction term added to the pixel value. Thus, at each iteration, the specimen estimate receives very small corrections and therefore the algorithm requires a large number of iterations to converge. On the other hand, if s_{max} is set smaller than it needs to be, there will be specimen values that are larger than s_{max}. Because the weight factor is zero for pixels with values greater than s_{max}, these pixels are never corrected.

Another modification to the JvC method is to prevent unwanted behavior of the restored image (step 6 of the generic iterative deconvolution). To counteract the amplification of high-frequency noise with iteration, a smoothing filter is applied to the restored image every few iterations. Typically, the smoothing filter has a Gaussian shape. The JvC algorithm is the basis for the constrained iterative deconvolution in DELTAVISION (Applied Precision) and MICROTOME CI (VayTek and Scanalytics) although the algorithm has been extensively modified and improved

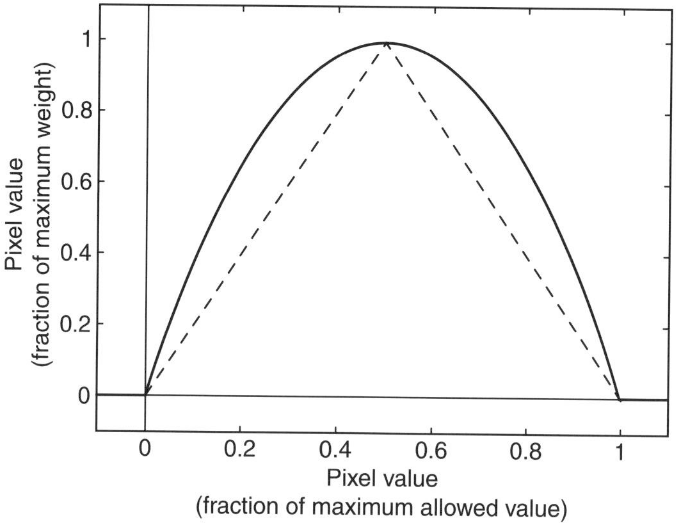

Figure 7. Weight factors for the Jansson–van Cittert algorithm as proposed by Jansson (dashed line) and by Agard (solid line).

by Drs. David Agard and John Sedat and their group (University of California, San Francisco, CA).

2.5. Statistical image estimation

One problem of the JvC algorithm is that it deals with noise in a somewhat *ad hoc* manner by simply smoothing the estimate after an arbitrary number of iterations. There are, however, several methods that are derived from models for image formation and recording that explicitly take into account the random nature of photon absorption, emission, collection, and detection, as well as the presence of noise in the detector electronics. These algorithms approach deconvolution as a statistical estimation problem – knowing the probability distribution of the image when the specimen is known and knowing the PSF, these methods try to estimate the specimen function that best matches the statistical description of the process of image formation and recording and the recorded image. In principle, any of the approaches to statistical estimation could be used for statistical image estimation. The ones that have found most applicability are those based on maximum-likelihood and, to a lesser extent, maximum-entropy estimation. It is worth mentioning that, although there have been methods published with the name of maximum *a posteriori* probability (MAP) (9–11), these are, strictly speaking, maximum penalized likelihood (MPL) methods. MAP estimation requires knowing the probability distribution of the specimen *a priori*, i.e. prior to the collection of an image, whereas MPL methods require only specifying undesired characteristics of the estimated specimen function.

2.5.1. Maximum likelihood (ML) and MPL

The most widely used statistical model for image formation and recording assumes that the dominant source of random variability in microscopic images is photon noise that results from the random absorption, emission, collection, and detection of photons. Within certain limits that often apply, photon noise follows a Poisson probability distribution. The oldest method for maximum likelihood estimation based on Poisson statistics was originally derived for astronomical images by Richardson (12) and independently by Lucy (13). The algorithm was then independently re-derived for ML estimation of microscopic images using the expectation-maximization (EM) formalism of Dempster *et al.* (14) by Holmes *et al.* (15, 16), Conchello *et al.* (6, 17), and Snyder *et al.* (18). In its most basic form the algorithm is as follows:

1. Estimate (guess) the specimen function.
2. Using the model for image formation and recording, calculate the image of the estimated specimen function.
3. Compare the recorded and calculated images by dividing the former by the latter.
4. 'Project' the ratio from image space (the image side of the objective) to object space (the specimen side of the objective) by convolving with the 'mirror image' of the PSF.
5. Multiply the projected ratio by the current specimen function estimate to obtain an improved specimen function estimate.
6. Repeat steps 2–5 as necessary.

One strong advantage of the algorithm above is that the non-negativity constraint is implicit to the algorithm. To see this, consider an initial guess that is non-negative. The image of a non-negative distribution of fluorescent dye is also non-negative. Because both the recorded and calculated images are non-negative, their ratio is non-negative. This ratio is projected by a non-negative PSF into the specimen side of the objective and thus the projection is non-negative. Finally, the non-negative specimen estimate is updated by multiplying it by the projected ratio and thus the result is non-negative. *Fig. 5* shows the results of deconvolving the 10 μm microsphere with the EM–ML algorithm. Clearly, the algorithm restores the spherical shape of the specimen, implying that it recovers frequencies within the missing cone (for a more mathematical demonstration, see 19).

The algorithm described above has been modified (10, 15) to incorporate the possibility of missing pixels (e.g. due to charge-coupled device (CCD) wells that are defective or when not enough out-of-focus planes are collected). In the modification by Holmes & Liu (15), the ratio of the recorded to the calculated image in step 3 of the algorithm is set to 1.0 for those pixels that are defective or were not recorded. Conchello & McNally modified the algorithm to compensate for large CCD wells (20). In this modification, the model for image formation in step 2 and the projection in step 4 are modified to account for the fact that the CCD well integrates the light falling over an area comparable to or larger than the diffraction-limited spot. For details, see the references cited.

ML estimation has two serious drawbacks: first, as the iteration progresses and resolution is gained, noise is also amplified, especially high-frequency noise (6). Thus, even if the imaging conditions are exactly known, the iteration has to be stopped before noise dominates the restored image – a problem common to most iterative deconvolution algorithms. The second drawback is that ML estimation has an inherent tendency to increase the brightness of isolated bright spots in the image at the expense of reducing contrast elsewhere in the restored specimen function. One widely used method to deal with both of these drawbacks is to use MPL. In MPL, the likelihood or log-likelihood function is modified to include a term that causes the merit function to decrease with the increase of some unwanted characteristic of the specimen, such as high-frequency noise or overly bright isolated spots. The most common form of MPL is to subtract from the log-likelihood a term that increases with the unwanted characteristics. This method has been used to avoid the isolated bright spots (21) and to prevent the amplification of high-frequency noise (10, 21). In the former case, the penalty term increases with the brightness of the specimen and is usually called intensity penalty. To prevent high-frequency noise, the penalty term increases with large changes between neighboring pixels, i.e. with the roughness of the specimen estimate. This term is therefore called roughness penalty and was originally introduced by Good & Gaskins (22).

2.5.2. Gradient-based ML and MPL algorithms

The algorithm described above and its variants have proven to be robust for the estimation of frequency components not passed by the microscope (19). However, the algorithm has a very slow convergence rate that sometimes makes it impractical for processing images in more than 3D (e.g. time series of 3D images at multiple wavelengths). For this reason, other ML estimation algorithms have been derived based on different maximization methods. In addition, when the pixel values are larger than a few tens, the Poisson distribution can be safely approximated by a Gaussian or normal distribution. In this case, ML estimation is equivalent to least squares estimation. Various investigators (9, 11, 23–29) have derived faster algorithms for ML using gradient-based maximization either for Poisson or for Gaussian statistics. The most successful of these methods use different maximization approaches at different stages of the deconvolution. In one example (26), a simple steepest-ascent method was used for the first few iterations (less than five or ten) and the conjugate-gradient approach was used thereafter. One problem of gradient-based methods is that non-negativity has to be explicitly enforced. There are two widely used methods to enforce non-negativity in gradient-based deconvolution algorithms. In the first method, at every iteration, any pixel that becomes negative is set to zero. Although this might sound simple, it is actually the procedure obtained from applying the Kuhn–Tucker conditions of constrained maximization to the deconvolution process and was the method used by Carrington & Fogarty (23). The second approach is to define the specimen function s to be the square of another (real) quantity c, i.e. $s = c^2$, and to estimate c instead of s. This ensures that the specimen function s is always positive. It has been shown that the time required by the EM–ML algorithm is proportional to the square of the time required by the gradient methods to achieve similar results (26).

Maximum likelihood deconvolution is available in HUYGENS (Scientific Volume Imaging), AUTODEBLUR (AutoQuant Imaging), SLIDEBOOK (Intelligent Imaging Innovations), AXIOVISION (Karl Zeiss), and XCOSM (www.omrfcosm.omrf.org).

2.6. Blind deconvolution

2.6.1. Obtaining the PSF

All the algorithms described above depend on having a description of the process of image formation and recording. The process of image formation is give by the PSF of the objective lens, i.e. the image of a point source of light. Alternatively, image formation can be described by the OTF, the Fourier transform of the PSF. There are two ways to obtain the PSF or OTF prior to deconvolution. One is to calculate the PSF from a mathematical physics model (e.g. 30–32, among others). These models require precise knowledge of the refractive indices and thicknesses of all the media between the specimen and the objective lens. The second method is to measure the PSF from the microscope. This is done by imaging one or more specimens whose geometry is known. The most widely used specimens are fluorescently labeled microspheres smaller than the diffraction-limited spot. To collect the images of the microspheres, it is necessary to know and replicate the conditions under which the specimen is imaged. If these microspheres can be embedded within and imaged together with the specimen, replicating the imaging conditions is obviated. Both theoretically computed and experimentally measured PSFs have advantages and disadvantages. Theoretical computation is noise free and can be carried out over a large volume that includes several microns above and below the center of the PSF. However, the imaging conditions might not be known with enough precision for this calculation. For experimental measurement, microspheres smaller than the diffraction-limited spot weakly emit light and thus the PSF can be measured only for a small volume around the location of the microsphere. Even in that small volume, the PSF is contaminated by photon and detector noise. In addition, if the microspheres cannot be imaged at the same time as the specimen, replicating the imaging conditions can be difficult, in particular, the depth at which the specimen was under the coverslip. A third method of obtaining the PSF is to estimate it at the same time as the specimen – a process called blind deconvolution.

2.6.2. Blind deconvolution

In blind deconvolution, the PSF is estimated at the same time as the specimen during the deconvolution process. As mentioned above, deconvolution is an underdetermined problem in that there is not enough information in the image and PSF to recover the frequency components blocked by the microscope, components that can be necessary to recover the coarse shape of the specimen. Blind deconvolution is even more underdetermined. Thus, blind deconvolution algorithms require even more additional information, namely information about the PSF, the OTF, or both. Currently, there are two methods for blind deconvolution, one developed by Holmes and co-workers (33–35) and the other by Markham & Conchello (36, 37). Each of these blind deconvolution approaches uses a different set of constraints in the PSF to avoid the underdeterminacy of the problem. Holmes and co-workers (33–35) constrained the PSF to be non-negative and cylindrically symmetrical, and the OTF to be nonzero only over a well-known region of the frequency domain and to have the cylindrical symmetry that results from that of the PSF. Markham & Conchello assumed that the PSF follows a well-defined mathematical expression (e.g. 30–32). The mathematical expression depends on a small number of parameters (usually less than 20), and estimating these parameters leads to the estimation of the complete PSF. Both methods use ML estimation and alternate between estimating the specimen function assuming the PSF is known, and estimating the PSF assuming the specimen function is known – an approach called grouped-coordinates optimization. Both methods are successful in estimating both the specimen function and the PSF. However, in its current version, the method of Markham & Conchello is slower. *Fig. 8* compares the results of ML deconvolution knowing the PSF with those obtained with blind deconvolution. Although the latter has little information about the imaging conditions, the results are very similar. ML blind deconvolution is available in AUTODEBLUR (AutoQuant Imaging).

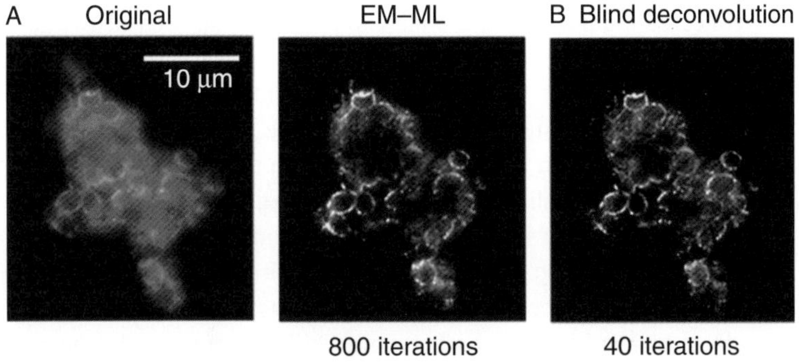

Figure 8. Blind deconvolution image restoration.
An optical slice through a 3D image of *Histoplasma capsulatum.* (A) Deconvolution using a known PSF and the expectation-maximization algorithm for maximum likelihood. (*B*) Blind deconvolution result. (Original stack courtesy of Dr. William Goldman, Washington University School of Medicine, USA. Blind deconvolution programmed and tested by Ms. Joanne Markham, Washington University School of Medicine.)

3. RECOMMENDED PROTOCOLS

3.1. Acquiring images to correct for background and a nonuniform flat-field response

The methods presented in section 2 compensate mostly for the compensations introduced by the process of image formation and to a much lesser extent for those introduced by the process of image detection and recording. In particular, they do not compensate for the variations across the field of view caused by nonuniform illumination, by the heterogeneous sensitivity of the CCD detector, and by the background introduced by the CCD. These degradations are sometimes called *structured noise* because they change very little from one 2D image to another. All three forms of structured noise are detrimental to quantitative measurements. In addition, the heterogeneous sensitivity of the CCD camera has the form of high-frequency noise, the kind that is typically amplified by the deconvolution process. To correct for these three sources of noise, the image has to be *pre-corrected* before the deconvolution process (one exception is the algorithm derived by Joshi & Miller (10), which can compensate for the heterogeneous sensitivity of the CCD camera). This pre-correction requires collecting images of a uniform field of fluorescence. To do this, a drop of fluorescent dye is placed in the slide and covered with a coverslip. To estimate the background level, a series of images is collected with the excitation and detection (camera) shutters closed. Each image is collected with the same exposure time as each of the optical slices in the 3D through-focus series. Any readings obtained from these images are due to the dark current and electronic noise in the detector. Averaging these images gives an estimate of the background present in all other images. To correct for the nonuniform illumination and the heterogeneous response of the CCD camera, another series of images called *flat-field* images is collected (again, with the same exposure time as each of the 2D optical slices in the through-focus stack). Each of these images is corrected for background noise by subtracting the background estimate from each of the flat-field images. The resulting images are then averaged and the average is normalized so that it has a maximum equal to 1.0. The 3D stack is then pre-corrected by dividing each optical slice by the normalized average of the background-corrected flat-field image. It is worth mentioning that the averaging, normalization, and division all have to be done with floating-point numbers, not with integers, as the image is collected. Because the pre-correction greatly reduces the structured noise, deconvolution algorithms can achieve better results with pre-corrected images.

3.2. Acquiring a suitable image stack for PSF calculation

As mentioned above, one way to obtain the PSF necessary for deconvolution is to measure it from an object of known geometry. Unfortunately, there are not many microscopic fluorescent objects whose geometry is precisely known. In fact, the only ones are fluorescently stained microspheres made of latex or polystyrene. The most widely used method of measuring the PSF is to collect the images of fluorescently labeled microspheres smaller than the diameter of the diffraction-limited spot (which is $1.22\lambda/NA$). These microspheres can be obtained, for example, from Molecular Probes. How small, however, requires a trade-off between having the correct shape for the PSF compared with having one that is reasonably free of noise. More precisely, for the PSF to have the correct shape, the microsphere should approximate a point source of light. However, the fluorescent irradiance from a microsphere is proportional to its volume and therefore proportional to the cube of its diameter. This means that reducing the size of the microsphere by a factor of two reduces the irradiance it produces by a factor of eight. The weak fluorescence has several detrimental effects. First, the brightest measurements possible (those at and close to the location of the microsphere) are at best weak and thus have low signal-to-noise ratio. Measurements away from the location of the microsphere fall below the noise floor and thus are of no use. On the other hand, microspheres larger or comparable to the size of the diffraction-limited spot broaden the PSF and hide the ring structure predicted by diffraction. It has been found experimentally that, for MPPI, deconvolution, microspheres with diameters of approximately 30–40% of the diameter of the diffraction-limited spot perform adequately. It is possible, however, to obtain better-quality PSFs by using several steps in the measurement and computation that reduces the amount of noise in the PSF.

Prepare a suspension of microspheres that ensures a mean distance, D, between the microspheres between $5z_o$ and $10z_o$ (suggested $D \approx 8z_o$), where $z_o = 2n\lambda/\mathrm{NA}^2$ is the distance from the peak of the theoretical PSF to the first null on the depth direction and n is the refractive index of the immersion medium. To achieve this degree of separation, the concentration of microspheres should be $1/D^3$. For example, for an 1.4 NA oil-immersion objective and a fluorescent wavelength of 480 nm, z_o = 1.42 μm, so for a separation of $10z_o$, the concentration of microspheres should be about $1/(14.2\ \mu\mathrm{m})^3 \approx 350$ microspheres/picoliter. This suspension should be in a medium whose refractive index is very close to that of the specimen mounting medium. (This is a serious drawback when the mounting medium has low viscosity, e.g. water or saline. The microspheres suspended in water undergo Brownian motion and make it impossible to record a 3D stack.) If using optical cement, put a drop of the suspension of microspheres in optical cement directly on to a coverslip and cure the optical cement as required by the manufacturer. If using a liquid, place a drop of the suspension on to a microscope slide and cover it with the coverslip. Under the microscope, search for a region that has a few microspheres and focus on an arbitrary one. Adjust the exposure time so that the image of that microsphere is not too dim but does not saturate the detector. Sometimes aiming for a reading of about 200 at the brightest pixel is bright enough to collect a PSF but not to bleach the microspheres.

Move the microscope stage to find another region that has a number of microspheres and collect a 3D stack that captures all the microspheres within the field of view. Correct the image for background and for the heterogeneous response of the CCD camera (for small objects, nonuniform illumination has little effect). From axial or x-z sections locate the image of a microsphere that is at about the same depth as the specimen you are studying. To do this, make a maximum-intensity axial projection of the stack. In this projection, the bottom of the coverslip (top for inverted microscopes) appears as a horizontal line. From this line, measure the approximate depth of the specimen (which can be obtained in a similar way, i.e. by looking for the coverslip and measuring the depth from there to the center of the specimen). Once you have located the specimen depth on the 3D stack of microspheres, look for the image of a microsphere that is closest to this depth and does not have another microsphere nearby (i.e not closer than about $8z_0$). Using a program that permits extracting subsets from 3D images (e.g. MATLAB, IPLAB SPECTRUM, METAMORPH, IMAGEJ), extract the volume around the location of the microsphere in all three dimensions. Using an image manipulation program, re-arrange the pixels in the image of the microsphere to place the brightest pixel at the 'origin' of the array (the origin is the pixel with the lowest indices; this is either (0, 0, 0) or (1, 1, 1), depending on the program you are using to manipulate the image). Re-arranging the pixels is done by cyclically shifting them, i.e. when shifting to the right; the pixels that 'fall off' at the right boundary of the volume are placed into the left-most pixels of the image. This shifted image can be used as a PSF for deconvolution.

As mentioned above, when the medium in which the microspheres are suspended has low viscosity (close to the viscosity of water or saline), Brownian motion makes it impossible to collect a 3D stack that does not suffer from motion blur. In this case, it is possible to measure the PSF only at zero depth under the coverslip. When this measure is satisfactory it can be obtained with less noise than in the case of a PSF at an arbitrary depth.

Prepare a suspension of microspheres with a mean separation between microspheres of between $50z_0$ and $100z_0$ (suggested $D \approx 80z_0$). As before, the density of microspheres per volume is $1/D^3$. Place a thin layer of the suspension on a coverslip and leave to dry in a dark place. The drying will deposit the microspheres on the surface of the coverslip. Once the liquid has evaporated, place a drop of optical cement on top of the microspheres and cure it as required by the manufacturer. The optical cement has two functions: (i) it holds the microspheres in place; and (ii) it locks oxygen out, preventing the fluorescent dye from oxidizing and thus bleaching. Place the coverslip under the microscope, locate a field that has a number of microspheres, and focus the microscope at the plane of the microspheres (which are now all in the same plane). Adjust the exposure time so that the maximum brightness is above 250 for a camera with 10-bit or more gray-level resolution, to about 200 for an 8-bit detector. Move the stage to locate another field that has a number of microspheres and collect a 3D stack that covers from about $10z_0$ above to $10z_0$ below the plane of the microspheres. Correct the stack for dark current and for heterogeneous CCD sensitivity.

From this 3D stack, there are several ways to get the PSF. The simplest one is to simply search for a microsphere whose nearest-neighbor microspheres are as far away as possible and extract the microsphere and the volume around it. Then shift the pixels so that the brightest spot is at the origin of the array and use the result as the PSF.

An experimentally measured PSF has noise that can be reduced. One way to reduce the noise is to use the images of several microspheres. More precisely, from the 3D stack corrected for dark current and for heterogeneous CCD sensitivity, extract the volume around a number of microspheres. Using an image manipulation program, move the centroid of each image to the center of the corresponding volume and average all the images of the different microspheres. The resulting image has much less noise than any of individual images. Shift the pixels of this average so that the brightest pixel is at the origin of the volume and use it as the PSF for the deconvolution.

An alternative to using subresolved microspheres is to use a larger object of known geometry – usually a microsphere larger than the diffraction-limited spot. Images of these objects are less affected by noise and can be collected for larger distances above and below the location of the object than the subresolved microspheres. Once the 3D image of the large object is collected, it is deconvolved using the known geometry of the object as a PSF in a constrained deconvolution method. For example, using a 5 μm diameter microsphere, the 3D image of the microsphere is collected and corrected for nonuniform illumination, dark current, and heterogeneous CCD sensitivity. Using an image manipulation program, a synthetic image of a spherical object of uniform intensity and 5 μm diameter is generated. The pixels of the synthetic image are shifted so that the center of the sphere is at the origin of the synthetic 3D image. This is used to deconvolve the image collected from the microscope. The result of the convolution is a PSF that can be used in deconvolution of other images collected under similar conditions. One potential disadvantage of this method is that our knowledge of the geometry of the object of known geometry might not match that of the actual object. For example, in the case of a 5 μm diameter microsphere, if the concentration of fluorescent dye throughout its volume is not uniform, the microsphere has defects (e.g. cracks or bubbles), or there is a tolerance in the diameter comparable to or larger than the size of the diffraction-limited spot, the PSF would be artifactual.

As mentioned above, the PSF can also be calculated from a mathematical expression or estimated at the same time as the specimen.

4. CONCLUSIONS AND CURRENT RESEARCH

From the several algorithms that exist for 3D microscopy, the ones that use a model for image formation that better fits the actual imaging characteristics lead to better results. Unfortunately, more precise models for image formation and recording lead to methods that are computationally more demanding. In some cases, the additional computation is justified due to the potential for artifacts resulting from inadequate models. However, for certain specimens (e.g. those dominated by puncta or filaments), the less demanding methods often lead to satisfactory results.

The problem of image estimation for fluorescent microscopy is still an active area of research. There are several sources of degradation that are not included in any of the methods available (commercially or otherwise). These include the fact that, for thick specimens, the imaging conditions change significantly throughout the depth of the specimen, leading to a model of image formation in which the PSF changes with the depth at which the microscope is focused. There have been few attempts to compensate for this degradation due to the computational complexity of the problem (38, 39). Another source of degradation not accounted for in any existing model is the blur caused by scattering as light propagates through thick, dense specimens. This problem requires a new model for image formation.

It is worth reminding the reader that although this chapter has focused on deconvolution of wide-field fluorescent images, all the methods presented here can and have been applied to images from confocal scanning fluorescence microscopes and multi-photon fluorescence excitation microscopes. The only change is in the PSF that defines the image formation.

Acknowledgements

I want to thank the many people who have contributed to the continued development of the research presented here. Because this research spans several years and institutions the list is long: Dr. Eric Hansen (Dartmouth College); Debbie Petterson, Leslie Nelson, Shirley Gonzales-Rubio, Stan Phyllips, Gerald Johns (WUIBC); Joanne Markham, Troy Jones, John Jenkins, Lucile Miller, Ann Olendorf, Dr. William Goldman (WUMS); Dr. Chrysanthe Preza (WU School of Engineering); Dr. James G. McNally, Dr. Tatiana Karpova (NCI); Dr. Jeff W. Lichtman (Harvard University); Danny Smith, Mary Flynn, Barbara Irish, Brad Pazoureck, Les Cummings (OMRF).

The research that led to the development, implementation, testing, and dissemination of the algorithms presented here has been supported by a large number of grants from the National Institutes of Health (Bethesda, MD), the National Science Foundation (Arlington, VA), the Thayer School of Engineering at Dartmouth College (Hanover, NH), the Oklahoma Medical Research Foundation (Oklahoma City, OK), and the University of Oklahoma (Norman, OK).

5. REFERENCES

★★★ 1. **Agard DA** (1984) *Annu. Rev. Biophys. Bioeng.* **13**, 191–219. – *A good introduction to computational optical sectioning microscopy (or deconvolution microscopy) that includes the earlier methods.*
2. **Monok JR, Oberhauser AF, Keating TJ & Fernandes JM** (1992) *J. Cell Biol.* **116**, 745–759.
3. **Joyce LS & Root WL** (1984) *J. Opt. Soc. Am. A Opt. Image Sci. Vis.* **1**, 149–168.
★ 4. **Preza C, Miller MI, Thomas LJ, Jr & McNally JG** (1992) *J. Opt. Soc. Am. A Opt. Image Sci. Vis.* **9**, 219–228. – *For the mathematically oriented reader, this paper provides a more thorough description of MPPI deconvolution.*
★ 5. **Frieden BR** (1983) *Probability, Statistical Optics, and Data Testing.* Springer-Verlag, Berlin, Germany. – *Mathematical-based description of Wiener–Helstrom deconvolution.*
6. **Conchello J-A, Kim JJ & Hansen EW** (1994) *Appl. Opt.* **33**, 3740–3750.
7. **Walsh DO & Nielsen-Delaney PA** (1994) *J. Opt. Soc. Am. A Opt. Image Sci. Vis.* **11**, 572–579.
8. **Frieden BR** (1975) In: *Picture Processing and Image Filtering*, pp. 177–249. Edited by TS Huang. Springer-Verlag, New York, NY.
9. **Verveer PJ & Jovin TM** (1997) *J. Opt. Soc. Am. A Opt. Image Sci. Vis.* **14**, 1696–1706.
10. **Joshi S & Miller MI** (1993) *J. Opt. Soc. Am. A Opt. Image Sci. Vis.* **10**, 1078–1085.
11. **Verveer PJ, van Kempen GMP & Jovin TM** (1997) In: *Three-dimensional Microscopy: Image Acquisition and Processing* IV, pp. 125–135. Edited by J-A Conchello, CJ Cogswell & T Wilson. SPIE Press, Bellingham, WA.
12. **Richardson WH** (1972) *J. Opt. Soc. Am.*, **62**, 55–59.
13. **Lucy LB** (1974) *Astronom. J.* **76**, 745–754.
14. **Dempster AP, Laird NM & Rubin DB** (1977) *J. R. Stat. Soc. B*, **39**, 1–38.

★ 15. **Holmes TJ & Liu Y-H** (1989) *Appl. Opt.* **28**, 4930–4938. – *Mathematical description of MPPI deconvolution, Wiener–Helstrom deconvolution, and maximum-likelihood deconvolution methods.*

★ 16. **Holmes TJ** (1988) *J. Opt. Soc. Am. A Opt. Image Sci. Vis.* **5**, 666–673. – *Mathematical-based description of MPPI deconvolution, Wiener–Helstrom deconvolution, and maximum-likelihood deconvolution methods.*

★★ 17. **Conchello J-A & Hansen EW** (1990) *Appl. Opt.* **29**, 3795–3804. – *A thorough mathematical-based description of the ML deconvolution method.*

18. **Snyder D, Hammoud AM & White RL** (1993) *J. Opt. Soc. Am. A Opt. Image Sci. Vis.* **10**, 1014–1023.

19. **Conchello J-A** (1998) *J. Opt. Soc. Am. A Opt. Image Sci. Vis.* **15**, 2609–2619.

20. **Conchello J-A & McNally JG** (1997) In: *Three-dimensional Microscopy: Image Acquisition and Processing* IV, pp. 164–174. Edited by J-A Conchello, CJ Cogswell & T Wilson. SPIE Press, Bellingham, WA.

21. **Markham J & Conchello J-A** (1997) In: *Three-dimensional microscopy: Image Acquisition and Processing* IV, pp. 136–145. Edited by J-A Conchello, CJ Cogswell & T Wilson. SPIE Press, Bellingham, WA.

22. **Good IJ & Gaskins RA** (1971) *Biometrika*, **58**, 255–277.

23. **Carrington WA & Fogarty KE** (1991) Iterative Image Restoration Device. US patent number 5047968, granted 10 September, 1991.

24. **Carrington WA, Lynch RM, Moore EDW, Isengerg G, Fogarty KE & Fay FS** (1995) *Science*, **268**, 1483–1487.

25. **Rizzuto R, Carrington WA & Tuft R** (1998) *Trends Cell Biol.* **8**, 288–292.

26. **Markham J & Conchello J-A** (2001) *J. Opt. Soc. Am. A Opt. Image Sci. Vis.* **18**, 1062–1071.

27. **Verveer PJ & Jovin TM** (1998) *J. Opt. Soc. Am. A Opt. Image Sci. Vis.* **15**, 1077–1083.

28. **van Kempen GMP, van Vliet LJ & Verveer PJ** (1997) In: *Three-dimensional microscopy: Image acquisition and processing* IV, pp. 114–124. Edited by J-A Conchello, CJ Cogswell & T Wilson. SPIE Press, Bellingham, WA.

29. **Verveer PJ, Gemkov MJ & Jovin TM** (1999) *J. Microsc.* **193**, 50–61.

30. **Gibson SF & Lanni F** (1991) *J. Opt. Soc. Am. A Opt. Image Sci. Vis.* **8**, 1601–1613.

31. **Haeberlé O** (2003) *Opt. Commun.* **216**, 55–63.

32. **Török P & Varga P** (1997) *Appl. Opt.* **36**, 2305–2312.

33. **Holmes TJ, Bhattacharyya S, Cooper JA, Hanzel D, Krishnamurthi V, Lin W-C, Roysam B, Szarowski DH & Turner JN** (1995) In: *Handbook of Biological Confocal Microscopy*, pp. 389–402. Edited by JB Pawley. Plenum Press, New York.

★ 34. **Holmes TJ** (1992) *J. Opt. Soc. Am. A Opt. Image Sci. Vis.* **9**, 1052–1061. – *For the mathematically oriented reader, this paper provides a more thorough description of the blind deconvolution method.*

35. **Battacharyya S, Szarowski DH, Turner JN, O'Connor N & Holmes TJ** (1996) In: *Three-dimensional microscopy: Image acquisition and processing* III, pp. 175–186. Edited by CJ Cogswell, GS Kino & T Wilson. SPIE Press, Bellingham, WA.

★ 36. **Markham J & Conchello J-A** (1999) *J. Opt. Soc. Am. A Opt. Image Sci. Vis.* **16**, 2377–2391. – *Mathematical-based description of the blind deconvolution method.*

37. **Markham J & Conchello J-A** (1998) In: *Three-dimensional and Multidimensional Microscopy: Image Acquisition and Processing* V, pp. 38–49. Edited by J-A Conchello, CJ Cogswell & T Wilson. SPIE Press, Bellingham, WA.

38. **Preza C & Conchello J-A** (2004) *J. Opt. Soc. Am. A Opt. Image Sci. Vis.* **21**, 1593–1601.

39. **Preza C & Conchello J-A** (2003) In: *Three-dimensional and Multidimensional Microscopy: Image Acquisition and Processing* X, pp. 135–142. Edited by J-A Conchello, CJ Cogswell & T Wilson. SPIE Press, Bellingham, WA.

CHAPTER 10

Correlative microscopy using Tokuyasu cryosections: applications for immunolabeling and *in situ* hybridization

Miguel R. Branco, Sheila Q. Xie, Sonya Martin and Ana Pombo

1. INTRODUCTION

The study of subcellular compartments emerged over a century ago and has evolved significantly since the early days of immunocytochemistry. Nowadays, researchers are interested not only in the localization of proteins or nucleic acids to a particular organelle, but also in their distribution pattern within that compartment and in the spatial relationships between different molecules. Such studies require methods of visualization that are sensitive enough to detect most molecules of interest and imaging devices that provide enough spatial resolution for subcellular localization. For example, nascent transcripts are detected inside mammalian nuclei in thousands of very small sites (~50 nm), such that, under a conventional light microscope, they can produce a diffuse pattern with only a small number of brighter foci (1, 2; see *Fig. 1A*). Optical sectioning of specimens with confocal laser scanning microscopes improves the axial and (to a lesser extent) lateral imaging resolution. Although several closely spaced sites can be resolved (see *Fig. 1B*), patterns can still be complicated by 'out-of-focus' flare from sites above and below the focal plane. The process of deconvolution has been used to subtract such flare, but the algorithms involved require subtraction of 'background' levels that may lead to loss of low-level signals and underestimation of the number of sites (3). Physical sectioning of frozen, fixed cells (Tokuyasu cryosections; 4, 5) can improve the axial resolution of light microscopy (LM) from >500 nm to ~100 nm (i.e. the section thickness), simplify patterns, and minimize chromatic aberration under the fluorescence microscope (6; see *Fig. 1C* and *D*). Importantly, cryosections also improve probe accessibility to their targets (see

Cell Imaging: *Methods Express* (D. Stephens, ed.)

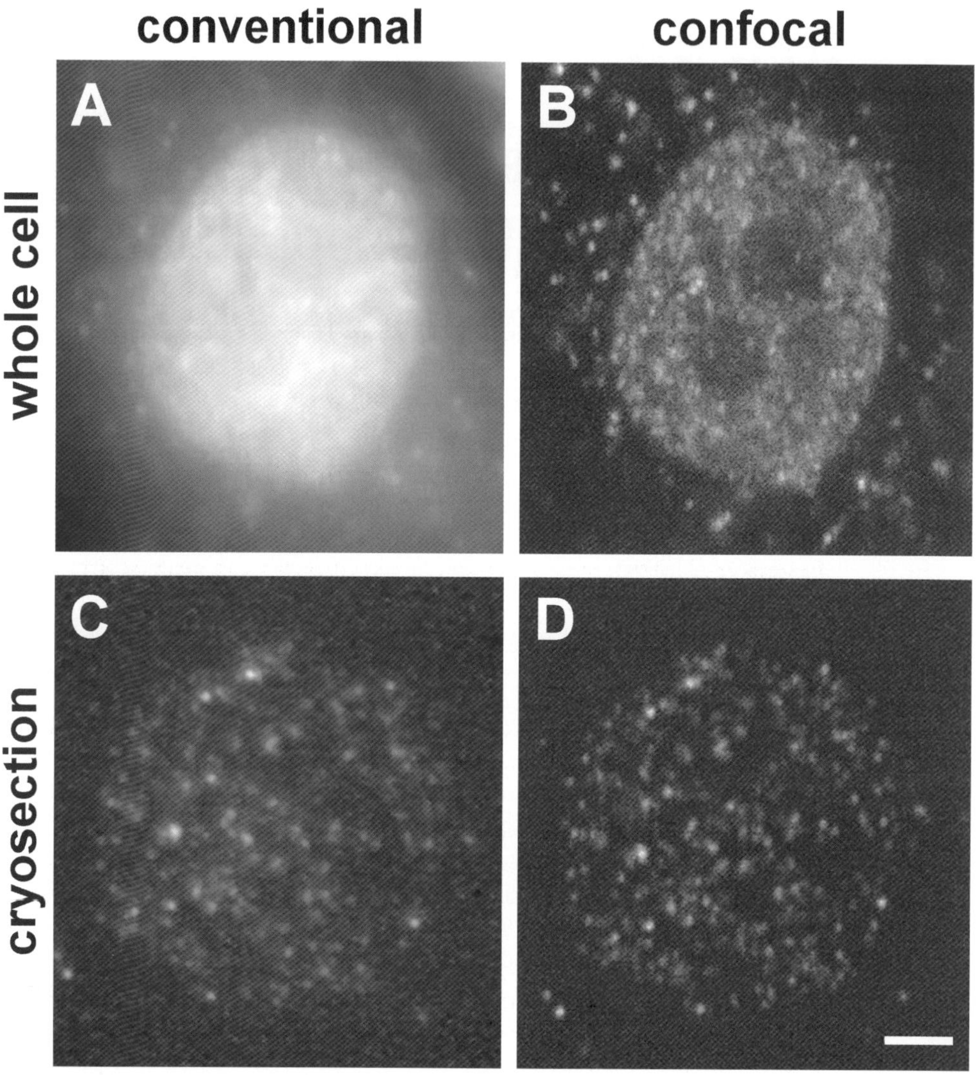

Figure 1. Transcription sites in HeLa cells imaged in whole cells or in cryosections using conventional or confocal microscopy.
Cells were permeabilized in a physiological buffer, nascent transcripts were allowed to extend in the presence of bromouridine triphosphate (BrUTP) for 15 min, and cells were then fixed. BrUTP-labeled RNA (BrRNA) was indirectly immunolabeled with fluorescein in whole cells or in cryosections. (*A, B*) Two views of BrRNA in a whole cell. (*C, D*) Two views of BrRNA in a cryosection (~150 nm) of a different cell. Bar, 2.5 μm. Reproduced, with permission from The Histochemical Society, from (6).

below), while using strong fixatives that do not compromise the preservation of cellular ultrastructure (7; see *Fig. 3D, F*, also available in color section).

This chapter concentrates on the application of thin Tokuyasu cryosections for subcellular localization of proteins and RNA. The main advantages of cryosectioning relative to other sectioning strategies, such as those involving resin embedding, are speed of preparation, the improved sensitivity of detection, and avoidance of artifacts caused by harsh organic solvents and resin polymerization. Cryosections are suitable for imaging on light and electron microscopes as appropriate to the problem in hand, allowing the same structures to be visualized in a correlative manner. Stereological methods that allow quantitative analysis of numbers or dimensions of subcellular features (e.g. 8) are beyond the scope of this chapter.

1.1. Antibody accessibility through cryosections

The ultrastructural analysis of well-fixed samples is often associated with low sensitivity of detection by probes that do not penetrate sections, such as antibodies and particularly the gold conjugates used for electron microscopy (EM). As a consequence, only structures at the section surface are labeled, giving lower sensitivity of detection, often associated with poor signal-to-background ratios.

Elegant work on Tokuyasu cryosections has shown that they are weakly penetrated by antibodies, probably due to cross-linking of cell components during fixation (9). However, we have found that accessibility can easily be achieved by pre-treating sections with a mild detergent and increasing antibody incubation and wash times (6). Antibody penetration through the sections can be verified by varying section thickness and counting discretely labeled structures that are uniformly distributed across a particular compartment. For example, the active form of RNA polymerase II (labeled by antibody H5) is known to be uniformly distributed in discrete sites throughout the nucleoplasm of HeLa cells (10, 11) and so can be used as a test marker for antibody penetration. H5 antibody is an IgM (~900 kDa molecule with radius 20 nm; 12), which is known to have limited penetration through whole cells, unless weak fixatives are used (7). If the H5 antibody only labeled the surface or a little way into the section, one would expect that the number of sites per unit area would remain constant with increasing section thickness (see *Fig. 2A, B*) and that the number per unit volume would decrease, resulting in underestimation of site number for thicker sections. Alternatively, if H5 can penetrate and label throughout the section thickness, one would expect the number of sites per unit area to increase proportionally with increasing thickness (see *Fig. 2A, C*) and the number per unit volume to remain constant. *Table 1* contains the results obtained using the EM labeling procedure described in *Protocol 3*, clearly showing that probes as large as IgMs are able to penetrate Tokuyasu cryosections of at least 220 nm. The number of H5 sites per unit area increases with increasing section thickness and the number of sites per unit volume does not change. Surprisingly, the 100 nm sections showed fewer sites per unit volume. However, this is consistent with reports of cryosection stretching (a commonly cited disadvantage of cryosections), which is also observed in the general appearance of ultrathin sections stained with uranyl acetate. Section stretching can be minimized using different section-retrieval solutions (13) and technical expertise. Nevertheless, our results show that cryosections can be used with minimal distortions and maximum labeling sensitivity if their thickness lies between 140 and 220 nm.

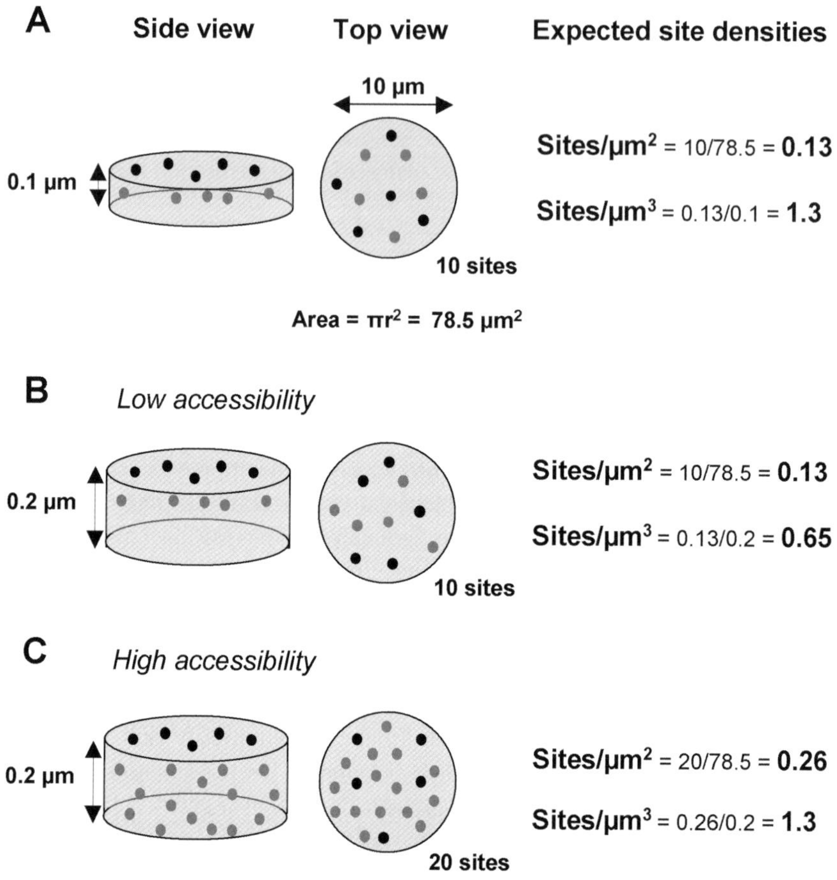

Figure 2. Testing antibody accessibility in cryosections.
Schematic diagram showing sections of different thickness, with antibodies (spots) labeling a discrete, uniformly distributed structure. (*A, B*) If antibodies cannot penetrate throughout the thickness of the section, then the number of sites per unit area will remain the same with increasing section thickness, whilst the number per unit volume will decrease. (*A, C*) If antibodies can penetrate throughout the thickness of the section, then the number of sites per unit area will increase with increasing section thickness, whilst the number per unit volume remains the same.

2. METHODS AND APPROACHES

2.1. Cell fixation, embedding, and cryosectioning

Tokuyasu, sucrose-embedded cryosections allow probe accessibility to subcellular structures under conditions that preserve cellular ultrastructure and overcome the need to optimize different fixation/permeabilization protocols for every antibody of interest (e.g. 2, 7). They are ideally suited for analysis of membranous compartments that are easily destroyed by organic solvents and resin embedding

Table 1. Antibody penetration of Tokuyasu cryosections

Sections of different thickness were cut and labeled for EM with H5 antibody and 5 nm gold particles using *Protocol 3*. Thickness was estimated from interference color, except for the ~300 nm sections, which have no discernable color; their thickness was considered to be that of the microtome setting used. The increase in density per unit area of H5-positive sites corresponded to the increase in thickness of the section, while the density per unit volume remained the same, implying that the antibodies were able to penetrate sections of at least 220 nm. The value for pale yellow sections was lower, probably due to section stretching. For each thickness, n=10 images.

Section color	Thickness (nm)	Sites/area (μm^2)	Sites/volume (μm^3)
Pale yellow	~100	5.6 ± 2.8	39 ± 19
Yellow/pink	~140	9.9 ± 3.8	54 ± 21
Blue	~220	14.4 ± 4.7	54 ± 18
Not applicable	~300	18.4 ± 5.8	53 ± 17

and can be applied to whole tissues or cells in culture (see *Protocol 1*). For a comparison of different protocols of fixation including the one presented here, see (7). A detailed review of embedding and section retrieval, with comments about artifacts such as section stretching and compression, can be found in (13), and for a more recent review, see (14).

2.2. Immunolabeling for LM and EM

LM and EM provide complementary information about cellular structure and compartmentalization. Fluorescence immunolabeling allows a fast assessment of the structures of interest by LM and an understanding of the cell population variability, and constitutes a fast pre-EM survey. EM gives the ultimate resolution required for localization of biomolecules within the cell, but is often perceived as a time-consuming, specialized, and highly skilled technique.

We have developed simple immunolabeling protocols that are reliable with most antibodies for both LM and EM (see *Protocols 2* and *3*), whilst preserving cellular ultrastructure (7). For any new antibody, we first optimize its ideal dilution by immunofluorescence on cryosections, including, where possible, negative control samples that do not contain the epitope of interest (see *Table 2*). We next optimize secondary (see *Table 3*) and gold conjugate (see *Table 4*) dilutions that will be used for EM by indirect immunofluorescence labeling of the gold conjugates and analysis by LM. Immunogold labeling for EM is only done afterwards, requiring only minor fine-tuning of dilutions. *Tables 2–4* exemplify a typical optimization for antibody H5, an IgM, which detects a hyper-phosphorylated form of RNA polymerase II; negative control samples that do not contain the hyperphosphorylated epitope can be obtained by pre-treatment of sections with alkaline phosphatase (15).

Table 2. Primary antibody optimization

Sample	Alkaline phosphatase	H5 (IgM)	FITC-labeled donkey anti-mouse Ig	Outcome
FITC-labeled anti-mouse control[a,b]	–	–	+	No signal
H5 dilution test[b]	–	+	+	Maximize signal; minimize background
H5 specificity test[b]	+	+	+	No signal

[a]Include a positive control for FITC-labeled anti-mouse, using a pre-tested primary antibody.
[b]Try a range of dilutions of primary antibody and choose the dilution that gives maximum signal and minimum background.

Table 3. Secondary antibody optimization

Sample	H5	Rabbit anti-mouse Ig	FITC-labeled donkey anti-rabbit Ig	Outcome
FITC anti-rabbit control[a]	–	–	+	No signal
Rabbit anti-mouse control[b]	–	+	+	No signal
Rabbit anti-mouse dilution test[b,c]	+	+	+	Maximize signal; minimize background

[a]Include a positive control for FITC-labeled anti-rabbit, using a pre-tested primary antibody.
[b]Try a range of dilutions of rabbit anti-mouse antibody and choose the dilution that gives maximum signal and minimum background.
[c]Run a positive control for H5 in parallel (e.g. with FITC-labeled anti-mouse), to help interpret lack of staining or excess background when using rabbit anti-mouse.

Table 4. Gold-conjugated antibody optimization

Sample	H5	Rabbit anti-mouse Ig	Gold (5 nm)-labeled goat anti-rabbit Ig	FITC-labeled donkey anti-goat Ig	Outcome
FITC-labeled anti-goat control[a]	–	–	–	+	No signal
Gold (5 nm)-labeled anti-rabbit control	–	–	+	+	No signal
Gold (5 nm)-labeled anti-rabbit dilution test[b]	+	+	+	+	Maximize signal; minimize background

[a]Include a positive control for FITC-labeled anti-goat, using a pre-tested primary antibody.
[b]Run in parallel a positive control for H5 (e.g. with FITC-labeled anti-mouse), to help interpret lack of staining or excess background with rabbit anti-mouse or 5 nm gold-labeled anti-rabbit.

2.3. RNA *in situ* hybridization

Fluorescence *in situ* hybridization (FISH) has been accomplished as a fundamental approach in genome analysis since its development over 20 years ago. Detection of RNA in the cell can be used as a single-cell assay to measure expression levels of a particular gene, providing information on the biological variance across the cell population (16–18). Furthermore, visualization of newly synthesized RNA, preferably by use of probes that detect introns only, is indicative of active transcription at a certain point in time and can be used to study spatial relationships between a particular transcription site and subnuclear structures (19, 20).

These studies can be carried out with higher resolution and sensitivity in thin Tokuyasu cryosections. For example, mature mRNAs can be detected by hybridizing their poly(A)$^+$ tails with FITC-labeled poly(dT) oligonucleotide on to cryosections (see *Fig. 3A*, also available in color section, and *Protocol 4*). The use of cryosections provides the most efficient detection of the poly(A)$^+$ RNA signal, when compared with whole cells after either a mild or strong fixation (see *Fig. 3*). Although a strong fixation preserves nuclear structure (see *Fig. 3D*), it decreases nuclear accessibility (see *Fig. 3C*). On the other hand, mild fixatives provide accessibility (see *Fig. 3E*), but disrupt ultrastructure (*see Fig. 3F*). Cryosections are well fixed and still allow probe accessibility (see *Fig. 3A*), as discussed previously. Therefore, FISH on ultrathin Tokuyasu cryosections provides an important tool to localize RNA accurately within cellular compartments for functional genomic research. Current developments in our laboratory show that cryosections are also suitable for DNA FISH.

2.4. Correlative microscopy

Imaging of the same sample by both LM and EM can provide more information on the structure to be analyzed (21–23). Correlative microscopy is particularly useful when studying structures that require the resolution of EM but are rarely found in sections, either because they are infrequent within each cell or because they are not present in all cells of the population. Examples are nuclear bodies such as Cajal bodies (one to three per cell) or single gene loci (two in each diploid cell). Whereas the lower throughput of EM usually makes such studies time-consuming, the association of LM and EM provides a way of quickly finding a section with the desired structure by LM, followed by relocation of the same cell section by EM.

To image the same sample by LM and EM one requires: (i) a labeling strategy that enables imaging on both microscopes (e.g. with fluorochromes and gold particles); and (ii) a sample processing method for mounting for LM (on a glass coverslip) and EM (on a metal grid).

As labeling of structures for LM is normally done by fluorochrome-conjugated antibodies, these can be further visualized by EM after indirect immunolabeling of the fluorochromes with gold particles (antibodies are now available against FITC, Alexa Fluor 488, rhodamine, Texas Red and cyanine dyes). Alternatively, secondary

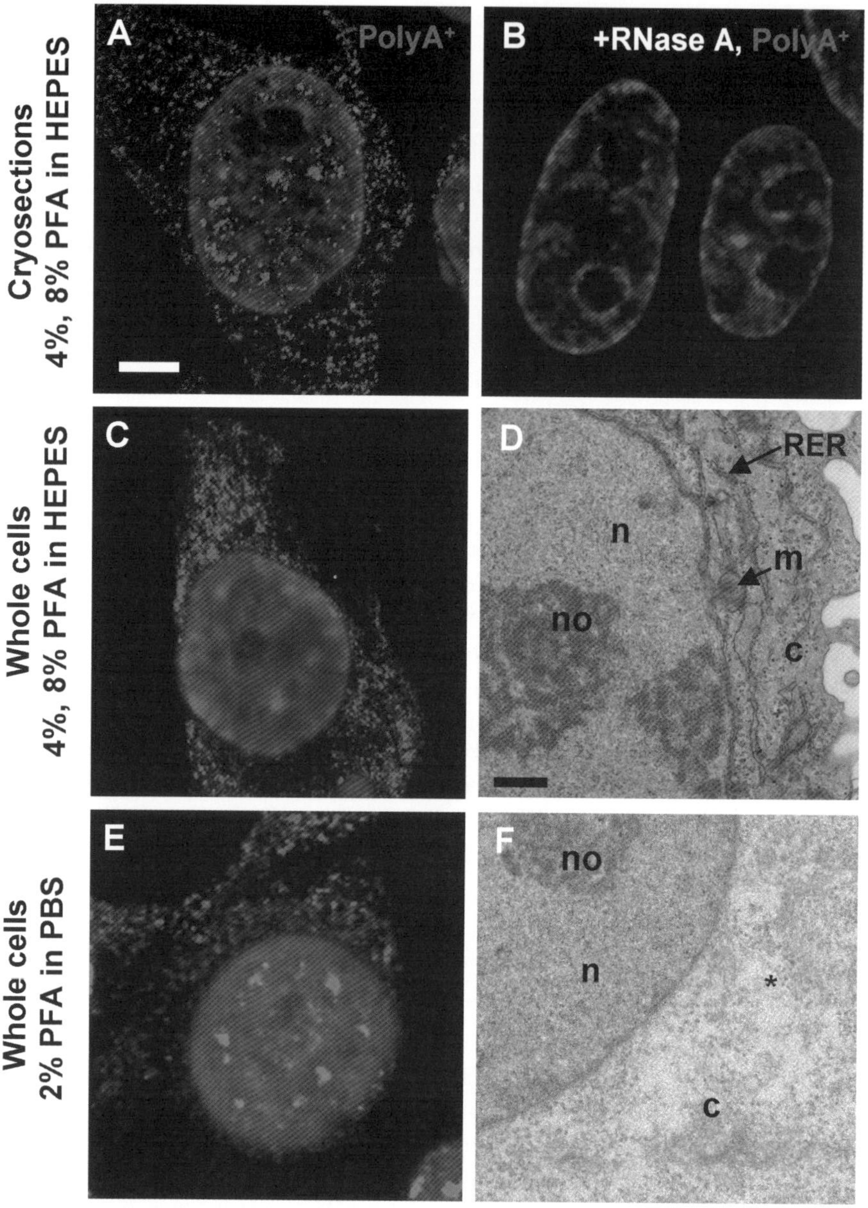

Figure 3. RNA FISH in HeLa cells (see page xxv for color version).
(*A*, *B*) HeLa cells were fixed in 4 and 8% paraformaldehyde (PFA) in 250 mM HEPES, embedded in sucrose, cryosectioned (100–120 nm thick; see *Protocol 1*), and permeabilized with 0.1% Triton X-100 for 10 min before RNA FISH using an FITC–poly(dT) oligonucleotide probe, as described in *Protocol 4*. Both nuclear and cytoplasmic pools of poly(A)$^+$ RNA are efficiently detected (*A*), and treatment of sections with RNase A shows the specificity of detection (*B*). (*C*)–(*F*) HeLa cells were grown on coverslips, fixed as for cryosections, and permeabilized with 0.1% Triton X-100 (*C*, *D*) or fixed in 2% PFA in PBS for 15 min (28) and permeabilized with 0.5% Triton X-100 for 5 min (*E*, *F*). After fixation, cells were processed for RNA FISH according to *Protocol 4* (*C*, *E*) or for ultrastructural analysis by EM (*D*, *F*) (7). Cells fixed in a stronger fixative showed good preservation of cellular morphology (*D*), but resulted in reduced probe accessibility, such that nuclear RNAs were not detected (*C*). Cells fixed in a weaker fixative provided improved probe accessibility (*E*), but exhibited poor preservation of ultrastructure (*F*), such as disrupted rough endoplasmic reticulum (RER) and mitochondria (*). n, Nucleus; no, nucleolus; c, cytoplasm; m, mitochondria. Bars, 4 μm (LM) and 1 μm (EM).

antibodies that carry both fluorochrome and gold particles can be used, avoiding the need for additional immunolabeling steps after LM imaging. For example, FITC–Protein A conjugated with 6 nm colloidal gold particles has been used to label transcription sites indirectly in HeLa cells (6). FluoroNanogold secondary antibodies provide a similar detection system and are commercially available, but the gold particles are 1.4 nm in size, requiring silver enhancement (24). More recently, luminescent semi-conductor quantum dots have been developed for use in cell imaging. These 10–15 nm ligands have a high quantum yield, high resistance to photobleaching, and are visible by EM (25). However, labeling of nuclear structures with quantum dots has required either suboptimal fixatives such as acetone or methanol, or post-embedding, section-surface labeling (25). Another strategy uses the photo-oxidative properties of fluorochromes to polymerize diaminobenzidine, which is readily visible by EM upon reaction with osmium fixative (26). This strategy has been used in an *in vivo* system to detect a recombinant protein carrying a small tetracysteine peptide motif that binds fluorescent biarsenical derivatives of fluorescein, FlAsH and ReAsH (27). These can then be used to photo-oxidize diaminobenzidine after fixation to permit visualization by EM.

Regarding the strategy for transfer of the specimen from LM to EM, we present here two possible strategies for cryosections (also see 22). One uses cryosections supported on glass coverslips which are subsequently embedded in resin before resectioning (see *Protocol 5*), whereas the other starts with cryosections directly on EM grids (see *Protocol 6*). The use of glass coverslips ensures high-quality LM imaging and provides easy handling of the specimen. However, transfer to EM is time-consuming and highly skilled in that it involves flat embedding with resin, sectioning, and collection of the first resin section containing the thin cryosection (6). Starting with cryosections directly on grids avoids resin embedding, thus shortening experimentation time, but renders handling more difficult and LM imaging suboptimal. Grids are mounted between a glass slide and coverslip for imaging by LM, dismounted, washed, and immediately processed for EM (24). If grids are treated at high temperatures (for example, for DNA FISH), the Formvar film becomes fragile and prone to detachment, requiring careful handling during the dismounting step (see *Protocol 6*). Nevertheless, this nonembedding approach is generally simpler and adequate for most studies, in particular if the purpose of the LM imaging is solely to locate the structures of interest. The resin embedding approach is advisable in studies where high-quality LM imaging is required, as in the analysis of transcription sites (6; see *Fig. 4*). In this study, correlation between EM and LM showed that 83% of the sites seen by EM were resolvable by LM, allowing a high-throughput analysis of site density by the latter. Furthermore, EM analysis allowed the measurement of site diameter, complementing the information obtained by LM.

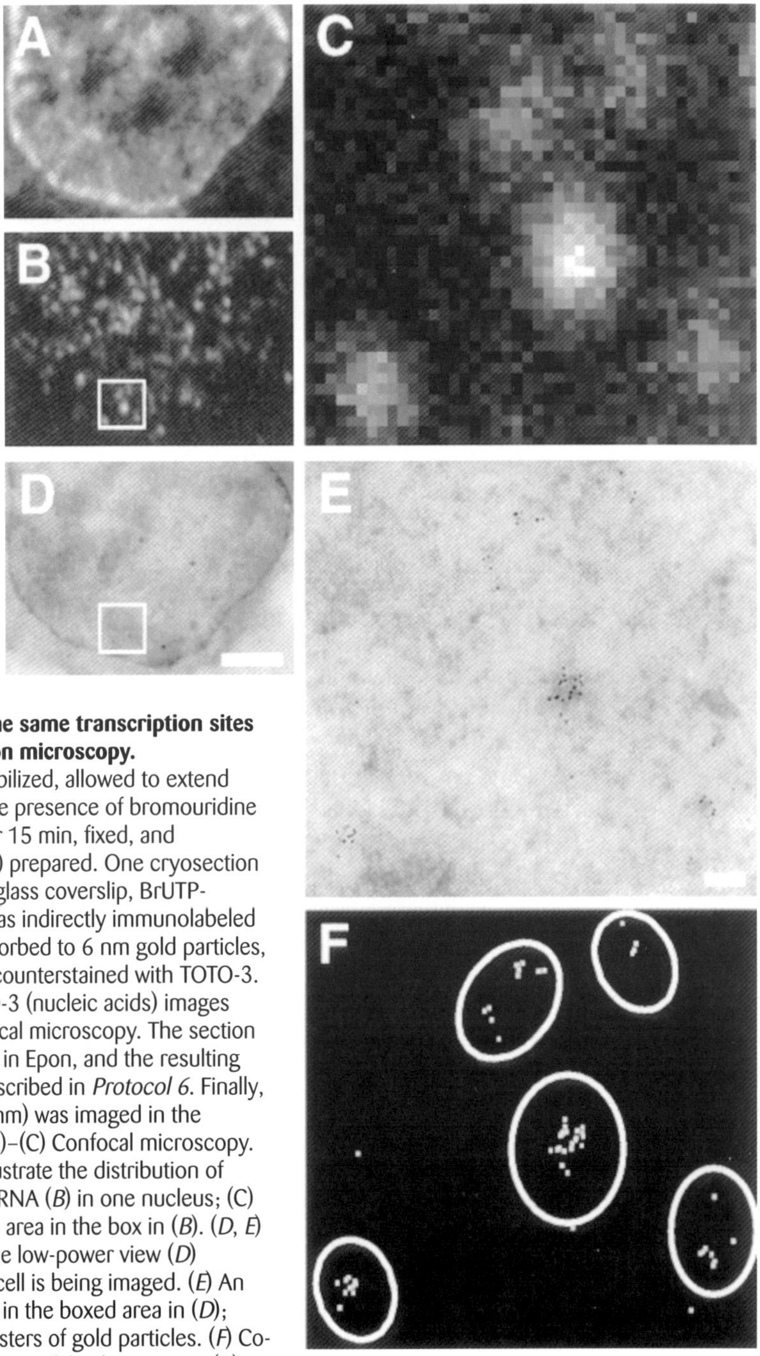

Figure 4. Imaging of the same transcription sites by confocal and electron microscopy.
HeLa cells were permeabilized, allowed to extend nascent transcripts in the presence of bromouridine triphosphate (BrUTP) for 15 min, fixed, and cryosections (~150 nm) prepared. One cryosection was transferred on to a glass coverslip, BrUTP-labeled RNA (BrRNA) was indirectly immunolabeled with FITC–Protein A adsorbed to 6 nm gold particles, and nucleic acids were counterstained with TOTO-3. FITC (BrRNA) and TOTO-3 (nucleic acids) images were collected by confocal microscopy. The section was then flat embedded in Epon, and the resulting block resectioned as described in *Protocol 6*. Finally, the end section (~150 nm) was imaged in the electron microscope. (A)–(C) Confocal microscopy. The low-power views illustrate the distribution of nucleic acids (*A*) and BrRNA (*B*) in one nucleus; (C) is an enlargement of the area in the box in (*B*). (*D, E*) Electron microscopy. The low-power view (*D*) confirms that the same cell is being imaged. (*E*) An enlargement of the area in the boxed area in (*D*); BrRNA is marked by clusters of gold particles. (*F*) Co-localization. The boundaries of the foci seen in (*C*) overlie the gold particles (shown in white) seen in (*D*), thus illustrating their co-localization. All sites were seen under both microscopes. Bars, 2 µm (*A, B, D*) and 100 nm (*C, E, F*). Reproduced, with permission from The Histochemical Society, from (6).

3. RECOMMENDED PROTOCOLS

Protocol 1

Cell fixation, embedding, and cryosectioning

Equipment and Reagents

- Cells
- Freshly depolymerized paraformaldehyde (PFA; 16% stock solution[a])
- 500 mM HEPES, pH 7.6
- 2.1 M sucrose (in phosphate-buffered saline (PBS))
- Cryo-ultramicrotome (Leica ULTRACUT UCT with Leica EM FCS cryo unit)
- Glass coverslips (no. 1.5; Chance Propper), or metal EM grids (Polysciences) coated with 0.3–0.5% Formvar (TAAB Laboratories) in chloroform

Method

1. Rinse and fix the cells in 4% PFA in 250 mM HEPES (pH 7.6) for 10 min at 4°C. For tissues, cut 1–2 mm pieces in this fixative and fix for 30 min at 4°C.
2. Fix the cells or tissue in 8% PFA in 250 mM HEPES (pH 7.6) for 1–2 h at 4°C.
3. For cells, during step 2, scrape the Petri dish and centrifuge at 200 *g* for 5 min, followed by 400 *g* for 2 min, 1200 *g* for 2 min, 2500 *g* for 2 min, and 5000 *g* for 2 min. Allow fixation to continue in the pellet for 20 min.
4. Dislodge the pellet with a sharpened wooden toothpick.
5. Transfer the cell pellet or tissue pieces through three drops of 2.1 M sucrose over 2 h and then on to a copper block.
6. Remove excess sucrose and shape the pellet into a cone appropriate for cryosectioning.
7. Freeze the pellet by immersion in liquid nitrogen[b] with continuous shaking until the fizzing stops.
8. Cut cryosections on a cryo-ultramicrotome; section thickness can be deduced from interference color.
9. Transfer the cryosections with drops of 2.1 M sucrose on a small-diameter metal bacteriological loop on to glass coverslips or metal EM grids[c].

Notes

[a]The solution should be freshly depolymerized in a water bath at 60°C for 1–2 h, with a stirrer. Add a few drops of 1 M NaOH until the solution is clear. Let the solution cool down and adjust the pH to ~7.5 before filtering.
[b]Keep blocks in liquid nitrogen for long-term storage.
[c]For long-term storage of cryosections, keep drops of sucrose on glass coverslips and store at –20°C. Grids can be floated off on the sucrose solution in a Petri dish and kept frozen for 1–2 weeks.

Protocol 2

Immunolabeling for LM

Equipment and Reagents

- Cryosections supported on glass coverslips
- 20 mM glycine (in PBS)
- 0.1% Triton X-100 (in PBS)
- Bovine serum albumin (BSA; 10% stock solution in PBS)
- Casein (0.5% stock solution in PBS)
- Fish skin gelatin (10% stock solution in PBS)
- PBS+ (PBS containing 1% BSA; 0.1% casein; 0.1% gelatin; pH 8.0; British Biocell International)
- Antibodies
- 0.1% Tween 20 (in PBS)

Method

1. Wash sucrose off with PBS.
2. If an aldehyde-containing fixative was used, quench the free aldehyde groups with 20 mM glycine for 20 min. Wash in PBS.
3. Incubate the coverslips with 0.1% Triton X-100 in PBS for 10 min[a]. Wash three times in PBS.
4. Block for 1 h with PBS+.
5. Incubate sections with the primary antibody (diluted in PBS+) for 2 h.
6. Wash the sections four times in PBS+ for >1 h.
7. Incubate with the secondary antibody (diluted in PBS+) for 1 h.
8. Wash the sections three to four times in PBS+ for >30 min, wash three times in PBS, and then wash in 0.1% Tween 20 in PBS.
9. Counterstain the nuclei with a DNA stain (e.g. DAPI) diluted in 0.1% Tween 20 in PBS.
10. Wash three times in PBS (can be stored in PBS at 4°C for <1 week).
11. Mount in VectaShield or equivalent mounting medium before imaging.

Notes

[a]Triton X-100 treatment should be omitted or made shorter (1 min) when cell membranes need to be recognized. In these conditions, antibody accessibility throughout sections may be compromised.

Protocol 3

Immunolabeling for EM

Equipment and Reagents

- Cryosections supported on EM grids or glass coverslips
- PBS+ (as in *Protocol 2*)[a]
- Antibodies
- Glutaraldehyde (25% stock solution; TAAB Laboratories)
- Methylcellulose[b]
- Uranyl acetate (3% stock solution; TAAB Laboratories)

Method

1. Proceed as in *Protocol 2* (immunolabeling for LM) until the point of incubation with the secondary antibody (step 7). Drops of Triton X-100 should be <10 µl to avoid the grid sinking.
2. Incubate the grids with a gold-conjugated antibody[c] (diluted in PBS+) for at least 4 h (often an overnight incubation gives the most efficient penetration and labeling).
3. Wash the grids three to four times in PBS+ for >3 h.
4. Wash six times in PBS.
5. Fix the antibodies with 0.5% glutaraldehyde in PBS for 10 min.
6. Wash six times in water.
7. Incubate the grids in methylcellulose with or without uranyl acetate[d] for 10 min at 4°C.
8. Collect the grids with metal loops, blot excess methylcellulose with hardened filter paper, and leave to dry.
9. Collect the grids from the loops.

Notes

[a]To avoid background labeling from protein aggregates visible by EM, clear stock solutions for PBS+ by ultracentrifugation (40 000 *g* for 1 h), except for casein, which should be centrifuged at 2000 *g* for 30 min.

[b]To prepare the methylcellulose solution, heat 150 ml of H_2O to 95°C, add 3 g of methylcellulose, and reheat to 95°C. Cool on ice, with stirring, add 150 µl of 20% sodium azide, and then leave at 4°C for 3–4 days, with stirring. Ultracentrifuge at ~300 000 *g* for 1.5 h. Store at 4°C, without disturbing the solution, and take aliquots for use from the top.

[c]The dilutions for gold-conjugated antibodies can be tested by LM, as described in section 2.2.

[d]The inclusion and final concentration of uranyl acetate depends on the amount of staining required and varies with cell type. The normal concentration range can go up to 0.3%. Dilute uranyl acetate stock solution in the appropriate amount of water to mix with methylcellulose in a 1:9 proportion.

Protocol 4

RNA *in situ* hybridization in cryosections

Equipment and Reagents[a]

- Cryosections supported on glass coverslips
- 2 mM vanadyl ribonuclease complex (in PBS)
- 20 mM glycine (in PBS)
- 0.1% Triton X-100 (in PBS)
- FITC-labeled poly(dT) probe (Sigma); final concentration 1 µM in hybridization mixture (Sigma or prepared in the laboratory[b])
- HybriSlips (Molecular Probes)
- Rubber cement (Qbiogene)
- Casein blocking solution (Vector Laboratories)

Methods

1. Wash sucrose off the sections with PBS.
2. Incubate sections with 2 mM vanadyl ribonucleoside complex in PBS for 15 min.
3. Wash in PBS (three 5 min washes).
4. Incubate with 20 mM glycine in PBS for 15 min.
5. Permeabilize the sections with 0.1% Triton X-100 in PBS for 10 min.
6. Wash in PBS (three 10 min washes). Rinse in 2× SSC (four 5 min washes). Rinse in water.
7. Dehydrate in ethanol series: 50, 70, 95, and 100% (5 min each at room temperature) and air dry at 37°C.
8. Denature the hybridization mixture (6 µl) containing FITC–poly(dT) at 70°C for 10 min.
9. Apply probe on to HybriSlips before overlaying coverslips containing sections. Seal with rubber cement and carry out hybridization in a humid chamber at 42°C for 24–60 h[c].
10. Wash coverslips in 2× SSC (three 30 min washes), and then in 1× SSC and 0.5× SSC (three 10 min washes each).
11. Amplify the probe signal with Alexa Fluor 488-labelled rabbit anti-FITC, followed by rabbit anti-Alexa Fluor 488 and Alexa Fluor 488-labelled goat anti-rabbit. Use the immunolabeling procedure in *Protocol 2*, but replace PBS+ with casein blocking solution[d].

Notes

[a]For all stages up until the amplification step, molecular biology-grade reagents (RNase- and DNase-free) should be used. Other aqueous solutions containing reagents that may not be RNase-free should be treated with diethyl pyrocarbonate.

[b]Our laboratory-prepared hybridization mixture contains 50% deionized formamide, 10% dextran sulfate (from an autoclaved 50% solution), 4× SSC, 2 mM EDTA, 20 mM Tris-HCl (pH 7.2), 10 µg/ml denatured salmon sperm DNA, and 10 µg/ml yeast tRNA.

[c]Control experiments include mock hybridization in the absence of poly(dT) probe in the hybridization mixture or pre-treatment of sections with 250 µg/ml RNase A in 2× SSC (1–2 h at 37°C).

[d]To 1× casein blocking solution (Vector Laboratories), add 2.6% NaCl, 0.2% BSA, and 0.1% fish skin gelatin (pH 7.5–8.0).

Protocol 5

Flat embedding for correlative analysis of cryosections supported on glass coverslips

Equipment and Reagents

- Cryosections supported on CELLocate glass coverslips (Eppendorf)
- Reduced osmium tetroxide (1% osmium tetroxide (Agar Scientific) in 1.5% potassium ferricyanide)
- 0.5% uranyl magnesium acetate (TAAB Laboratories)
- Reynolds' lead citrate[a]
- Glycid ether 100 resin[b] (Serva)
- Resin block molds (TAAB Laboratories)
- Trimmer (Leica EM TRIM)
- Ultramicrotome (Leica ULTRACUT UCT)

Method

1. After immunolabeling and imaging by LM, dismount coverslips, wash in PBS, and fix in 4% PFA in 250 mM HEPES for 1 h at room temperature.
2. Refix sections with reduced osmium tetroxide[c] for 1 h at room temperature. Wash six times in water.
3. Stain with 0.5% uranyl magnesium acetate[c] overnight at 4°C. Wash six times in water.
4. Stain with Reynolds' lead citrate[d] for 30 s. Wash six times in water.
5. Dehydrate through a graded ethanol series of 70, 90, and 100% (10 min each at room temperature).
6. Infiltrate sections with ethanol:glycid ether 100 resin[d] mixtures as follows: 3:1 for 30 min; 1:1 for 2 h; 1:3 overnight.
7. Remove the ethanol:glycid ether 100 resin mixture by leaving Petri dishes tilted over tissues.
8. Add degassed, freshly prepared, glycid ether 100 resin to block molds.
9. Place inverted coverslip over the mold opening, making sure that no air bubbles are in contact with the coverslip.
10. Polymerize for 2–3 days at 56°C.
11. Remove the resin block by cutting the plastic mold.
12. Detach the coverslip from the resin block by immersing in liquid nitrogen for short periods, followed by drying with compressed air. Repeat as many times as required.
13. Trim the resin block for section cutting using a mechanical trimmer.
14. Cut the first section (~150 nm), containing the sample, and transfer on to metal EM grids.

Notes

[a]Weigh 1.33 g of lead nitrate and 1.76 g of sodium citrate for a total of 50 ml of solution. Add 30 ml of freshly boiled, but cooled, H_2O and shake for 30 min. Add 8 ml of 1 M NaOH, fill up to 50 ml with freshly boiled, but cooled, H_2O, filter the solution, and aliquot into small, air-tight vials with silicone cap seals. Store at 4°C. Remove aliquots for use with a syringe. Do not use the solution if a white precipitate is visible.

[b]Weigh 21.2 g glycid ether 100, 9.2 g methyl nadic anhydride, and 14.8 g dodecenyl succinic anhydride, and mix well. The solution can be stored at –20°C. Before use, bring to room temperature, add activator (0.8 g Tris(dimethylaminomethyl)phenol), and mix well for >30 min. Unused resin can be stored at –20°C to be used for ethanol:glycid ether 100 resin mixtures in subsequent experiments.
[c]Make fresh solutions. These substances are extremely hazardous. Read the safety instructions on the vial and always work in the fume hood.
[d]Hazardous substances. Work in the fume hood. After polymerization, glycid ether 100 resin can be handled safely.

Protocol 6

Mounting and post-LM image processing of cryosections supported on metal grids

Equipment and Reagents

- Cryosections supported on finder grids (Pyser-SGI) coated with Formvar
- Glass slides
- Glass coverslips

Method

1. After immunolabeling, touch a drop of PBS or mounting medium[a] with both sides of the grid.
2. Lay the grid on a glass coverslip, with the side with the sections facing down.
3. Lower the glass slide on to the coverslip without dropping or pressing it. After mounting, do not move the coverslip sideways. Do not seal the coverslip.
4. Perform imaging by LM. Dismount the coverslip by adding excess PBS to the sides of the coverslip, allowing it to dislodge itself without mechanical aid.
5. Discard the coverslip. Thoroughly wash the grid in several drops of PBS and then water, on both sides. Air dry at 37°C after ethanol dehydration, or touch the side of the grid with a piece of hardened filter paper[b].
6. Depending on the labeling strategy, either proceed with immunogold labeling, or directly incubate grids in methylcellulose for EM imaging (see *Protocol 3*).

Notes

[a]If grids are processed for DNA FISH, Formvar films will be more fragile, making grids hard to dislodge from preparations using mounting medium without damaging or even losing the film. For such preparations, we recommend the use of PBS. Although certain fluorochromes such as FITC are more bleaching sensitive in such conditions, the addition of antioxidants (e.g. DABCO) helps to reduce bleaching.
[b]Air drying the grids is safer for preservation of the Formvar film than blotting and is thus recommended for DNA FISH-processed grids.

Acknowledgements

We thank Michael Hollinshead for help and advice, and the Medical Research Council (UK) and Fundação para a Ciência e Tecnologia (Portugal) for funding.

4. REFERENCES

1. **Jackson DA, Hassan AB, Errington RJ & Cook PR** (1993) *EMBO J.* **12**, 1059–1065.

★ 2. **Pombo A, Jackson DA, Hollinshead M, Wang Z, Roeder RG & Cook PR** (1999) *EMBO J.* **18**, 2241–2253. – *Optimised method for in situ labeling and visualisation of nascent transcripts with best preservation of transcriptional activity and cellular structure.*

3. **Fay FS, Taneja KL, Shenoy S, Lifshitz L & Singer RH** (1997) *Exp. Cell Res.* **231**, 27–37.

4. **Tokuyasu KT** (1980) *Histochem. J.* **12**, 381–403.

★ 5. **Tokuyasu KT** (1973) *J. Cell Biol.* **57**, 551–565. – *Ultramicrotomy using sucrose-embedded samples.*

★ 6. **Pombo A, Hollinshead M & Cook PR** (1999) *J. Histochem. Cytochem.* **47**, 471–480. – *High resolution correlative confocal and transmission electron microscopy using Tokuyasu cryosections.*

7. **Guillot PV, Xie SQ, Hollinshead M & Pombo A** (2004) *Exp. Cell Res.* **295**, 460–468.

8. **Weibel ER** (1979) *Stereological Methods,* Vol. 1. *Practical Methods for Biological Morphometry.* Academic Press, London.

9. **Stierhof YD & Schwarz H** (1989) *Scanning Microsc. Suppl.* **3**, 35–46.

10. **Grande MA, van der Kraan I, de Jong L & van Driel R** (1997) *J. Cell Sci.* **110**, 1781–1791.

11. **Pombo A, Cuello P, Schul W, *et al.*** (1998) *EMBO J.* **17**, 1768–1778.

12. **Green NM** (1969) *Adv. Immunol.* **11**, 1–30.

★★★ 13. **Griffiths G** (1993) *Fine Structure Immunocytochemistry.* Springer-Verlag, Heidelberg. – *Comprehensive information about Tokuyasu cryosectioning.*

14. **Griffiths G, Lucocq JM & Mayhew TM** (2001) *Cell. Microbiol.* **3**, 659–668.

15. **Doyle O, Corden JL, Murphy C & Gall JG** (2002) *J. Struct. Biol.* **140**, 154–166.

16. **Chow CM, Georgiou A, Szutorisz H, *et al.*** (2005) *EMBO Rep.* **6**, 354–360.

17. **Lafarga M, Berciano MT & Carmo-Fonseca M** (1997) *J. Neurosci. Methods,* **75**, 137–145.

18. **de Krom M, van de Corput M, von Lindern M, Grosveld F & Strouboulis J** (2002) *Mol. Cell,* **9**, 1319–1326.

19. **Custodio N, Carmo-Fonseca M, Geraghty F, Pereira HS, Grosveld F & Antoniou M** (1999) *EMBO J.* **18**, 2855–2866.

★★★ 20. **Levsky JM, Shenoy SM, Pezo RC & Singer RH** (2002) *Science,* **297**, 836–840. – *Single cell analysis of gene expression.*

21. **Koberna K, Malinsky J, Pliss A, *et al.*** (2002) *J. Cell Biol.* **157**, 743–748.

22. **Robinson JM, Takizawa T, Pombo A & Cook PR** (2001) *J. Histochem. Cytochem.* **49**, 803–808.

23. **Ren Y, Kruhlak MJ & Bazett-Jones DP** (2003) *J. Histochem. Cytochem.* **51**, 605–612.

24. **Takizawa T & Robinson JM** (2003) *J. Histochem. Cytochem.* **51**, 707–714.

★ 25. **Nisman R, Dellaire G, Ren Y, Li R & Bazett-Jones DP** (2004) *J. Histochem. Cytochem.* **52**, 13–8. – *Use of quantum dots in correlative microscopy.*

26. **Deerinck TJ, Martone ME, Lev-Ram V, *et al.*** (1994) *J Cell Biol,* **126**, 901–910.

★ 27. **Sosinsky GE, Gaietta GM, Hand G, *et al.*** (2003) *Cell Commun. Adhes.* **10**, 181–186. – *Use of ReAsH for correlative live-cell and transmission electron microscopy.*

28. **Spector DL, Goldman RD & Leinwand LA** (1998) *Cells: a Laboratory Manual.* Cold Spring Harbor Laboratory Press, Cold Spring Harbor, New York.

CHAPTER 11

Fluorescence lifetime imaging

Klaus Suhling

1. INTRODUCTION

Fluorescence lifetime imaging (FLIM) has emerged as a key technique for imaging the environment and interaction of specific proteins in living cells. The fluorescence lifetime is the average time a fluorophore remains in the excited state after excitation, typically nanoseconds, and is a function of the biophysical environment of the fluorophore. FLIM measures the fluorescence lifetime in each pixel of an image and thus provides image contrast according to this fluorescence lifetime. An image of the biophysical environment of a fluorescent probe in a cell can therefore be obtained without compromising the cell by biochemical assays. The increasing popularity is facilitated by commercial availability of FLIM add-on units to conventional microscopes: a chief advantage of wide-field time-gating and frequency-domain methods is the speed of acquisition, whereas for time-correlated single-photon counting (TCSPC)-based confocal or multi-photon excitation scanning it is sensitivity and accuracy in the fluorescence decay. FLIM has also been demonstrated in spectrally and polarization-resolved mode. Most biological applications of FLIM concern the detection of Förster/fluorescence resonance energy transfer (FRET) upon protein interaction, but FLIM has also been used to probe the local environment of fluorophores, reporting, for example, on the local viscosity, pH, refractive index, and ion and oxygen concentrations.

1.1. A very brief history of fluorescence, lifetime, and imaging

The phenomenon of luminescence had been known for a long time, in particular bioluminescence, but it was not until 1565 that Nicolas Monardes observed and gave an account of a blueish glow, which we now know to be fluorescence, from a wood extract used for medicinal purposes, lignum nephriticum (1). A similar observation was made in 1646 by Athanasius Kircher, and also later by others, e.g. Brewster, but none of them explained this phenomenon. In 1845, Herschel described the phenomenon of fluorescence of quinine and distinguished it from scattered light, and, in 1852, Stokes reinvestigated it and finally explained that

Cell Imaging: ***Methods Express*** (D. Stephens, ed.)

the emitted light was of a longer wavelength than the absorbed light – an effect now known as the Stokes shift. Above all, Stokes coined the term fluorescence in 1853 (1). The first reports on measuring nanosecond fluorescence lifetimes appeared in the mid-1920s (see *Fig. 1*) (2, 3).

The development of optical microscopy dates back to the end of the 16th century when the Dutch spectacle makers Zacharias Janssen and his son Hans noticed that two lenses in a tube allowed the magnification of small objects. In the 17th century, Robert Hooke started using compound microscopes to study small objects systematically (around the same time Galileo began using telescopes to study large and distant objects). Hooke's *Micrographia or Some Physiological Descriptions of Minute Bodies Made by Magnifying Glasses with Observations and Inquiries Thereupon*, published in 1665 and containing many images of small insects, fleas, seeds, plants, etc., appears to have inspired many others and was the scientific bestseller of its day (see Chapter 1 and *Fig. 1*). Indeed, to describe the compartmental structures he observed in cork, Robert Hooke coined the term 'cell' (but note that he was not describing what is now defined as a cell) (4).

In the early 1900s, fluorescence microscopy developed out of August Köhler's UV microscopy, when UV-excited autofluorescence was observed. However, it was not until the 1930s that Max Haitinger and others began systematically staining samples with fluorescent dyes, initially simply to make weakly autofluorescent biological samples visible. In the early 1950s, Albert Coons developed the technique of labeling antibodies with fluorescent dyes (5). This technique was extended to genetically encoded fluorescent labels, i.e. green fluorescent protein (GFP), in the 1990s (6). Single-point fluorescence decays of samples under a microscope were measured in the 1980s when the use of time-resolved fluorescence spectroscopy techniques expanded (e.g. 7), and the first FLIM images appeared at the end of the 1980s and in the early 1990s (8–10). Over the last decade, the FLIM technique has been improved and developed further to allow spectrally resolved and polarization-resolved FLIM and applied to an increasing number of studies in cell biology (11–13).

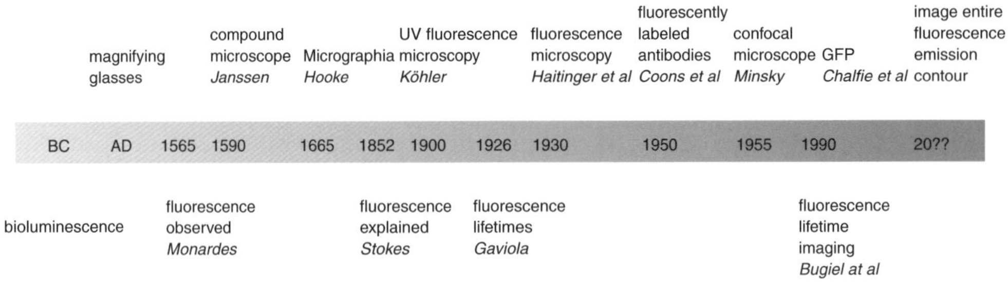

Figure 1. A brief history of fluorescence, microscopy, and their combination.
Despite hundreds of years of history, development, and research, there is still scope for improvement. An ideal FLIM system would acquire polarization-resolved and spectrally resolved fluorescence lifetimes, combined with high spatial resolution in 3D and high sensitivity in a minimum acquisition time. Of course, there is presently no FLIM technology that provides this unique combination of features.

1.2. Fluorescence microscopy

Although it has a modest spatial resolution compared with electron or X-ray microscopy, optical microscopy, and in particular fluorescence microscopy, remains a powerful and effective tool in the biological and biomedical sciences today, because it is minimally invasive and can be applied to living cells and tissues. Despite advanced light microscopy techniques such as phase contrast (a 1934 invention by Frits Zernike, which gained him the Nobel Prize in 1953), differential interference contrast (Nomarski), or modulation contrast (Hoffman and Gross), these transmitted light techniques only allow the visualization of small objects or organelles by their optical density or refractive index difference. Thus, fluorescent staining techniques were developed – due to the Stokes shift of fluorescence emission, the exciting light can be eliminated from the image so that only fluorescence on a dark background is detected, leading to a high contrast (14). Confocal microscopy, the principle of which is a 1955 invention by Marvin Minsky, is based on a spot of focused light scanned across the sample and detecting the fluorescence through a pinhole, which allows out-of-focus fluorescence emanating from above and below the focal plane to be eliminated from the image. The invention of the laser and computers with image-processing capacity, and the demonstration of the advantages of this type of microscopy for cell biology, have led to the widespread use of confocal microscopy over the last decade or so (15). (However, nonfluorescence-based optical microscopy techniques continue to be developed, such as, for example, coherent anti-Stokes Raman scattering and second-harmonic generation and third-harmonic generation imaging (14).)

Fluorescence microscopy can be performed by labeling a biological specimen with fluorescent dyes, quantum dots, or genetically encoded fluorescent proteins, and also by imaging endogenous fluorescence (autofluorescence), e.g. tryptophan, collagen, elastin, NADH, or, in the case of plants, chlorophyll (16, 17). An advantage of fluorescence imaging over bio- or chemiluminescence imaging is that a single fluorescent dye or protein can be excited repeatedly (around 10^5 times and emitting as many fluorescence photons), unlike bio- or chemiluminescence, which irreversibly produces only a single photon per chemical reaction of interacting molecules.

Conventional fluorescence microscopy relies on contrast according to the fluorescence intensity, with the fluorescence detection sensitivity extending down to the single molecule level. Two- and three-dimensional fluorescence images to locate, for example, labeled proteins can be recorded using wide-field or one- or two-photon confocal scanning techniques. Time-lapse imaging allows the temporal evolution of the system to be studied, and fluorescence bleaching techniques can provide information about the diffusion of the fluorescent probe. Imaging of FRET between a fluorescent donor and acceptor, enables the determination of protein interaction and conformational changes (18). In addition, recent developments in point spread function engineering techniques such as stimulated emission depletion, combined with 4Pi microscopy using two opposing objectives, have demonstrated fluorescence imaging of the microtubular network in mammalian cells with an axial resolution of 50 nm (19). The versatility of fluorescence microscopy can be extended even further to obtain more information about the environment of the fluorescent probe by imaging the fluorescence lifetime, spectrum, and polarization. To appreciate the benefits of this approach, some background and context of the phenomenon of fluorescence is essential (see *Table 1*).

Table 1. Advantages and disadvantages of FLIM

Advantages
- The fluorescence lifetime is a molecular property generally independent of variations in fluorophore concentration, illumination intensity, light path length, scatter, or photobleaching (as long as photoproducts do not fluoresce)
- FLIM can robustly and quantitatively probe the fluorophore's local environment directly, e.g. FRET, viscosity, pH, ions, etc., without the need to compromise the cell with biochemical assays
- FLIM of FRET by imaging the fluorescence decay of the donor is more robust than fluorescence intensity-based FRET and allows distinction between effects due to FRET efficiency and those due to probe concentration. Quantitative FRET studies between a spectrally similar donor and acceptor, e.g. GFP and YFP, are also possible (21)
- FLIM can experimentally distinguish spectrally similar probes (if their fluorescence lifetimes are different) with the same detector

Disadvantages
- Photophysical and time-resolved fluorescence spectroscopy expertise is required for data interpretation (e.g. fluorescent proteins usually have complex fluorescent decays)
- Interpretation of specific changes in fluorescence lifetime in terms of the underlying cell biochemistry may not always be straightforward

1.3. What is fluorescence?

Upon excitation into an excited state, a fluorescent molecule (fluorophore) can return to its ground state either radiatively by emitting a fluorescence photon, or nonradiatively, for example, by dissipating the excited state energy as heat (2, 3). This depends on the de-excitation pathways available. Fluorescence is the radiative deactivation of the lowest vibrational energy level of the first electronically excited singlet state, S_1 (see *Fig. 2*). The fluorescence lifetime, τ, is the average time a fluorophore remains in S_1 after excitation, and τ is defined as the inverse of the sum of the rate parameters for all depopulation processes:

$$\tau = \frac{1}{k_r + k_{nr}} \qquad \text{(Equation 1)}$$

where k_r is the radiative rate constant and the nonradiative rate constant k_{nr} is the sum of the rate constant for internal conversion, k_{ic}, and the rate constant for intersystem crossing to the triplet state, k_{isc}, so that $k_{nr} = k_{ic} + k_{isc}$ (see *Fig. 2*).

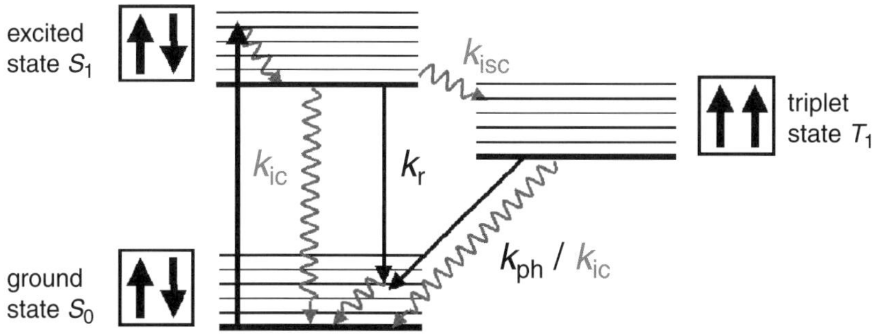

Figure 2. Jablonski energy level diagram of a fluorophore.
Schematic showing the ground state S_0, the first electronically excited singlet state S_1 (both with antiparallel electron spins), the triplet state T_1 (parallel electron spins), and the transitions between them (black arrows, radiative; gray wavy arrows, nonradiative). The thin lines represent vibrational energy levels. Fluorescence is the radiative deactivation of the first electronically excited singlet state, S_1. The radiative rate constant, k_r, is given by intrinsic properties of the molecule and the refractive index of the medium (87). The rate constants for intersystem crossing to the triplet state T_1 and internal conversion, k_{isc} and k_{ic}, represent nonradiative de-excitation paths. The transition to the triplet state involves a spin flip, and the energy level of the triplet state is always lower than the corresponding singlet state. The radiative triplet decay, with rate constant k_{ph}, is a forbidden transition, resulting in long lifetime phosphorescence (2, 3). In microscopy, fluorescence can be characterized not only by its intensity and position, but also by its fluorescence lifetime, polarization, and wavelength. Each of these parameters of the multi-dimensional fluorescence emission contour provides an additional spectroscopic dimension, which contains information about the biophysical environment of the fluorescent probe.

$\tau_0 = k_r^{-1}$ is the natural or radiative lifetime, which is related to the fluorescence lifetime τ via the fluorescence quantum yield, ϕ:

$$\phi = \frac{\tau}{\tau_0} = \frac{k_r}{k_r + k_{nr}} \qquad \text{(Equation 2)}$$

The fluorescence quantum yield can be thought of as the ratio of the number of fluorescence photons emitted to the number of photons absorbed (regardless of their energy) and is always less than one.

1.3.1. Fluorescence quenching

The previous definition of k_{nr} assumes no quenching processes. If there is quenching by another molecule present, Q, then this represents another pathway for the molecule to return to its ground state (with a rate constant k_Q) and in addition to the terms in the denominator there will appear the term $k_Q[Q]$ so that $k_{nr} = k_{ic} + k_{isc} + k_Q[Q]$. Quenching thus reduces ϕ and shortens τ.

While the fluorescence lifetime depends on the intrinsic characteristics of the fluorophore itself, it also depends in a measurable way upon the local environment (see *Fig. 3A*). In general, the local viscosity, pH, or refractive index, as well as interactions with other molecules, e.g. by collision or energy transfer, can all affect the fluorescence lifetime. In particular, FRET is a bimolecular fluorescence quenching process where the excited state energy of a donor fluorophore is nonradiatively transferred to an acceptor molecule by a dipole–dipole coupling process (see *Fig. 3B*) (2, 3, 18). The FRET efficiency varies with the inverse 6th power of the distance between donor and acceptor, and is usually negligible beyond 10 nm. FRET can therefore be used as a 'spectroscopic ruler' to probe intermolecular distances on the scale of the dimensions of the proteins themselves. This is a significant advantage over co-localization of two fluorophores, which is limited to the optical resolution limit (≈200 nm laterally, 500 nm axially). For FRET to occur, the donor emission spectrum and the acceptor absorption spectrum must overlap, and the transition dipole moments of the donor and acceptor must be oriented favorably (the orientation factor, κ^2, a function of the angles between the transition dipole moments of the donor and acceptor and their line of separation, must not be zero, $\kappa^2 \neq 0$) (2, 3, 18).

1.3.2. Decay of the excited state

After excitation, N fluorophores will leave the excited state S_1 according to the following rate equation:

$$dN = (k_r + k_{nr})N(t)dt \qquad \text{(Equation 3)}$$

Figure 3. Schematic representation of the photophysical phenomena that can be studied by FLIM and their effect on fluorescence decay.
(*A*) Fluorescence decay is a function of the environment of the fluorophore, schematically represented by the cylindrical structure (GFP). In general, it can be sensitive to pH, ions, oxygen, etc. The light gray molecules represent quenchers, and as their concentration increases, the quenching increases and the fluorescence lifetime, τ, decreases (see Equations 1 and 2). This is schematically represented on the right, where the light gray fluorescence decay is faster (shorter fluorescence lifetime) than the dark gray fluorescence decay (longer fluorescence lifetime). The black pulse represents the exciting light. In the specific case of GFP, the fluorescence lifetime is a function of the refractive index of its environment, and as the refractive index of the GFP environment increases, the GFP fluorescence lifetime decreases as described by the Strickler–Berg equation (88). (*B*) Interaction between two proteins can be identified with FRET by labeling one protein with a donor fluorophore (GFP) and the other with an acceptor fluorophore (dye). FRET occurs when a suitable donor and acceptor fluorophore are in close proximity, usually below 10 nm. Upon protein interaction, excited state energy from the GFP donor is transferred to the acceptor, and this FRET process can be identified by a shortened GFP fluorescence decay. Thus, imaging of FRET can demonstrate the proximity of fluorescent or fluorophore-tagged proteins with up to 100 times the resolution limit of far-field optical microscopy (18). (*C*) Polarization-resolved FLIM to perform TR–FAIM reveals the rotational mobility of a fluorophore (22, 25–27). This is affected by the viscosity of its surroundings, or by binding or conformational changes, and is characterized by the rotational correlation time, θ. The anisotropy can be calculated from the difference between the polarization-resolved fluorescence decays $I_{\parallel}$ and $I_{\perp}$ (Equation 5). A fast rotational motion leads to a rapid depolarization. Furthermore, fluorescence anisotropy may be used to probe homo-FRET (resonance energy transfer between the same type of fluorophore) (23).

In each of these applications, the fluorescence lifetime, τ (or in *C*, the rotational correlation time, θ), is used to generate contrast in the image (see *Fig. 5B* in color section).

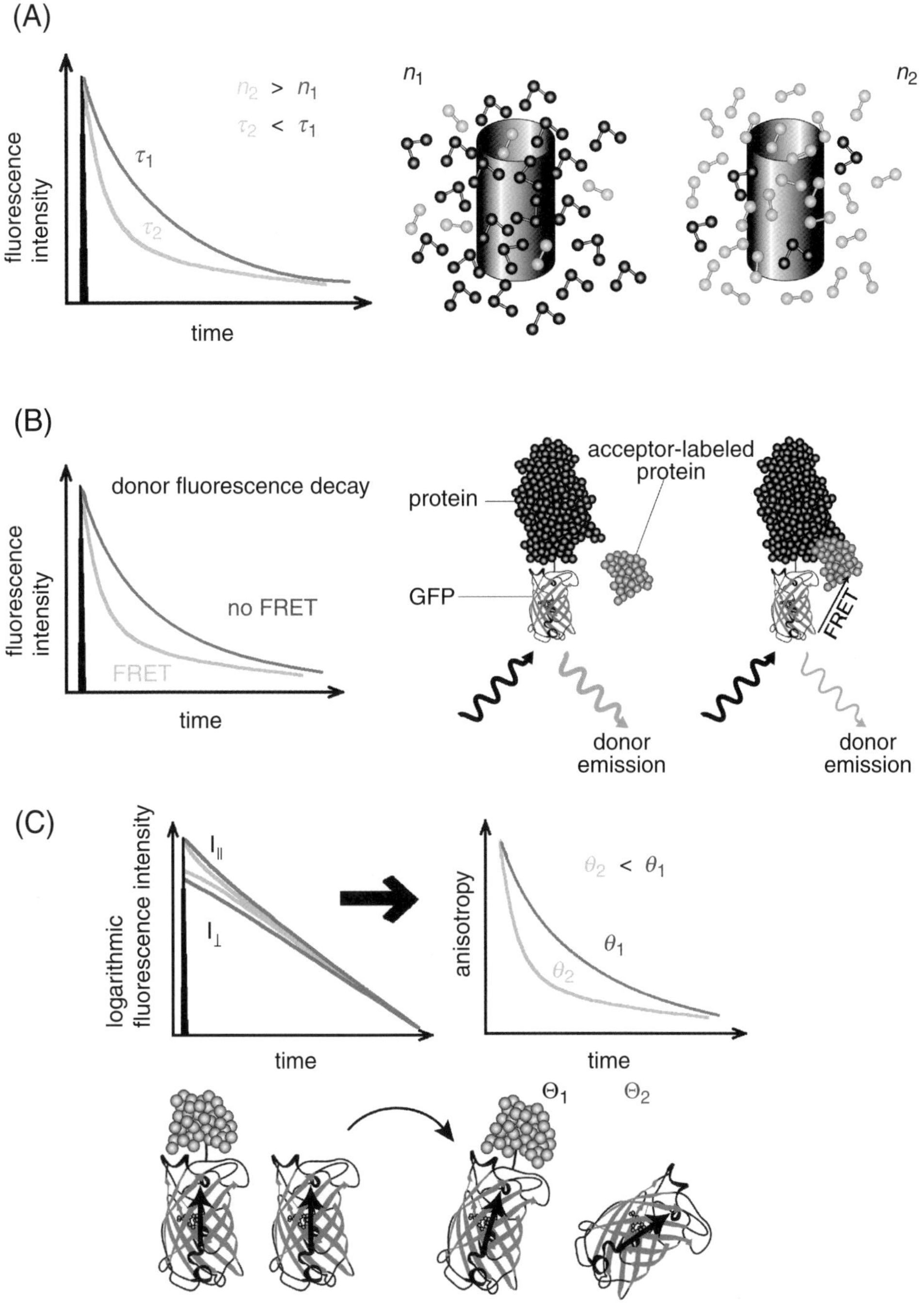
(A)
fluorescence intensity
n_2 > n_1
τ_2 < τ_1
τ_1
τ_2
time
n_1
n_2
(B)
fluorescence intensity
donor fluorescence decay
no FRET
FRET
time
acceptor-labeled protein
protein
GFP
FRET
donor emission
donor emission
(C)
logarithmic fluorescence intensity
$I_{\parallel}$
$I_{\perp}$
time
anisotropy
θ_2 < θ_1
θ_1
θ_2
time
Θ_1
Θ_2

where t is the time. Integration, using Equation 1, and taking into account that the fluorescence intensity $I(t)$ is proportional to the excited state population $N(t)$ yields:

$$I(t) = I_0 e^{-t/\tau} \qquad \text{(Equation 4)}$$

where I_0 represents the fluorescence intensity at $t = 0$. The decay of the fluorescence intensity thus follows an exponential decay law (2, 3), schematically shown in *Fig. 4(C)*.

1.4. What is FLIM and why is it used?

FLIM measures the fluorescence decay in each pixel of an image. Appropriate decay analysis extracts τ, which is then used to generate contrast. From the point of view of a fluorescence spectroscopist, FLIM is fluorescence lifetime spectroscopy in 2D, and from the point of view of a practitioner of imaging, FLIM utilizes the fluorescence lifetime as a contrast parameter. This is reflected in FLIM instrumentation development, but, either way, FLIM is a robust imaging method and has emerged as a key technique for imaging the environment and interaction of specific proteins in living cells (11–13, 16, 20).

Note that photophysical effects that are designed to occur in a sample, e.g. FRET, can be observed by FLIM or intensity-based fluorescence imaging. FLIM is particularly good at detecting these, since the fluorescence intensity is a time-averaged property that is dependent upon a variety of instrumental and sample influences, such as variations in the excitation intensity, fluorophore concentration, photobleaching, light path length, scatter, etc. It can thus be difficult to disentangle, interpret, or indeed quantify factors affecting the fluorescence intensity, whereas the fluorescence lifetime is a molecular property generally independent of variations in these factors. In addition, FLIM can be used to distinguish spectrally similar fluorophores if their fluorescence lifetimes are different (21) (see *Table 1*).

The fluorescence emission can be characterized not only by its fluorescence lifetime, intensity, and position, by also by its polarization and wavelength. Each of these parameters provides an additional spectroscopic dimension that contains information about the biophysical environment of the fluorescence probe. Thus, by imaging polarization-resolved or spectrally resolved fluorescence lifetimes, the multi-dimensional fluorescence emission contour can be accessed for more information about the biophysical environment of the fluorescent probe.

1.5. Polarization-resolved FLIM and spectrally resolved FLIM

1.5.1. Time-resolved fluorescence anisotropy imaging (TR–FAIM)

Fluorescence anisotropy involves measuring fluorescence at polarizations parallel and perpendicular to that of the exciting light. Steady-state, i.e. nontime-resolved, anisotropy imaging can be used to detect energy migration or homo-

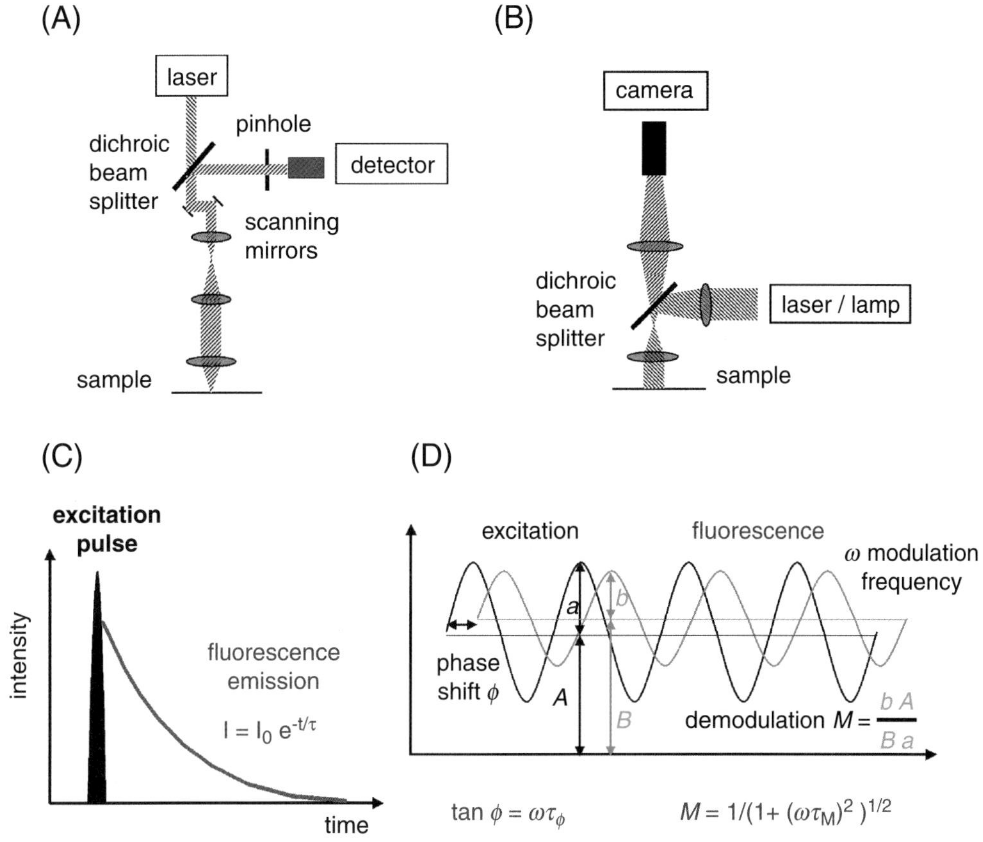

Figure 4. Measurement of fluorescence decay by FLIM.
Fluorescence imaging can be performed by confocal scanning (*A*) or wide-field imaging techniques (*B*) (34). For multi-photon excitation, there is no need to descan and use a pinhole, so the detector can be placed as close to the sample as possible for maximum sensitivity. Fluorescence decays can be measured in the time domain (*C*) or in the frequency domain (*D*) (37, 51), and FLIM has been performed using all combinations of the above. Each implementation has its characteristic features, strengths, and weaknesses, as summarized in *Table 2*.

FRET (resonance energy transfer between the same type of fluorophore), since it leads to a depolarization of the emitted fluorescence (22, 23). This technique has been used to study the proximity of isoforms of the glycosyl phosphatidylinositol-anchored folate receptor bound to a fluorescent analogue of folic acid to detect lipid rafts (24).

TR–FAIM, also known as polarization-resolved FLIM, can image the mobility of fluorophores in living cells (22, 25–27). This approach was recently extended to include linear dichroism imaging to study the orientation of BODIPY and DiO probes in the cell membrane, showing that cholesterol depletion disrupts molecular orientation in cell membranes (28).

Upon excitation with linearly polarized light, rotational diffusion of the fluorophore in its excited state results in a depolarization of the fluorescence emission. TR-FAIM measures fluorescence decays at polarizations parallel and perpendicular to that of the exciting light. The time-resolved fluorescence anisotropy $r(t)$ can be defined as:

$$r(t)=\frac{I_{\parallel}(t)-GI_{\perp}(t)}{I_{\parallel}(t)+2GI_{\perp}(t)} \quad \text{(Equation 5)}$$

where $I_{\parallel}(t)$ and $I_{\perp}(t)$ are the fluorescence intensity decays parallel and perpendicular to the polarization of the exciting light (2, 3). G accounts for different transmission and detection efficiencies of the imaging system at parallel and perpendicular polarization, and, if necessary, an appropriate background has to be subtracted (27). For a spherical molecule, $r(t)$ decays as a single exponential and is related to the rotational correlation time, θ, according to:

$$r(t)=(r_0-r_{\infty})\exp\left(-\frac{t}{\theta}\right)+r_{\infty} \quad \text{(Equation 6)}$$

where r_0 is the initial anisotropy and r_{∞} is the limiting anisotropy, which accounts for a restricted rotational mobility. For a spherical molecule in an isotropic medium, θ is directly proportional to the viscosity, η, of the solvent and the volume, V, of the rotating molecule:

$$\theta=\frac{\eta V}{kT} \quad \text{(Equation 7)}$$

where k is the Boltzmann constant and T the absolute temperature. Thus, imaging θ with TR-FAIM can measure the rotational mobility of a fluorophore in its environment (see *Fig. 3C*) and has been used to image the viscosity in living cells (27). Moreover, as the rotational diffusion can be slowed down by binding, TR-FAIM has the potential to visualize the binding of ligands and receptors in the cell. In addition, a hindered rotation of the fluorophore due to geometrical restrictions, e.g. in the cell membrane, can also be detected with this method.

1.5.2. Spectrally resolved FLIM

Spectrally resolved FLIM allows the fluorescence lifetime and spectra of two or more fluorophores to be observed simultaneously. This is advantageous in FRET studies, where the donor fluorescence lifetime can be monitored in one spectral channel and the acceptor fluorescence in another. A shortening of the average fluorescence lifetime of the donor cyan fluorescent protein (CFP) due to FRET to the acceptor yellow fluorescent protein (YFP), both linked by a short amino acid chain, was accompanied by an initial rise in the YFP fluorescence lifetime in the acceptor channel (acceptor in-growth) due to sensitized emission (29). Other spectrally resolved FLIM applications concern studies where the fluorescence lifetime of fluorophores emitting in different spectral regions is monitored simultaneously (30), including single-molecule studies (31). The spectral resolution in these cases is really a spectral separation, namely between the two spectral regions of fluorescence emission. However, true spectrally resolved FLIM with 10 nm bandwidth over a wide spectral range has been reported, both in the frequency domain (32) and the time domain (using a 16-anode photomultiplier) (33), allowing sophisticated analysis of multiple fluorophores, FRET, and possibly multiple donor–acceptor FRET pairs.

2. METHODS AND APPROACHES

2.1. Fluorescence imaging concepts

Fluorescence microscopy techniques can broadly be divided into two categories, mainly based on how the fluorescence is detected (34):

1. Confocal or multi-photon excitation scanning, where a focused beam is scanned over the sample (or by moving the sample through the focal point of an objective) and the image is acquired pixel by pixel using a nonimaging detector behind a pinhole, e.g. a photomultiplier or a photodiode (see *Fig. 4A*). (For multi-photon excitation imaging, nondescanned detection without a pinhole can be used.)
2. Wide-field camera-based imaging, where a position-sensitive detector such as a charge-coupled device (CCD) camera acquires the fluorescence in each pixel of the image simultaneously, i.e. records the entire field of view at the same time (see *Fig. 4B*).

In general, the parallel acquisition in wide-field imaging enables a much faster frame rate than sequential scanning in confocal imaging, but wide-field imaging suffers from out-of-focus blur and has a larger point spread function than confocal imaging. The main advantage of confocal scanning or multi-photon excitation microscopy is that it provides intrinsic optical sectioning, facilitating 3D imaging. Hybrid methods to combine wide-field imaging with optical sectioning exist, such as multi-beam multi-photon microscopy (28, 35), or spinning disk microscopy. Another way of overcoming the limited depth discrimination in wide-field microscopy is optical sectioning by structured illumination (36).

2.2. FLIM implementation

The nanosecond time resolution required to measure fluorescence lifetimes can be obtained either in the time domain by exciting the sample with a short optical pulse and directly observing the decay of the fluorescence intensity (see *Fig. 4C*), or in the frequency domain by modulating the excitation source and/or the detector gain (see *Fig. 4D*) (37).

2.2.1. Scanning FLIM

Confocal scanning or multi-photon excitation microscopes provide inherent optical sectioning, and here FLIM is essentially a series of single-channel fluorescence lifetime measurements where the fluorescence decay is acquired in each pixel of the image by TCSPC (8, 29, 31, 38–40), gated photon counting (41,

42), streak cameras (10, 43) (see *Fig. 4C*) or by phase modulation (30, 44, 45) (see *Fig. 4D*). Using dedicated electronics, in TCSPC the arrival time of single-fluorescence photons is recorded to picosecond accuracy after exciting the sample with a short laser pulse: a constant fraction discriminator (CFD) for the shaping of the detector pulse to enable precise timing, a time-to-amplitude converter (TAC) for the timing, and an analogue-to-digital converter (ADC) and multi-channel analyzer (MCA) to allocate the count to its time channel. By exciting the sample at MHz rates, i.e. millions of times per second, and recording the arrival time of many fluorescence photons, a probability distribution histogram of fluorescence photon arrival times is built up in the MCA. This is, in fact, the fluorescence decay curve (46, 47).

As excitation sources, tunable dye lasers can be used, but their small tuning range of ≈100 nm and cumbersome operation is a disadvantage. Tunable mode-locked solid-state lasers such as Ti:Sapphire lasers are much more user-friendly and provide subnanosecond or picosecond pulses over a wider tuning range (≈680–1050 nm). They have an average power of hundreds of milliwatts, a fixed repetition rate of about 80 MHz, which corresponds to 12.5 ns between pulses (roundtrip time of a pulse in the laser cavity), and are often used as excitation sources for TCSPC FLIM, in particular for two-photon excitation FLIM, but also frequency doubled for single-photon FLIM (≈340–525 nm). The repetition rate can be reduced by pulse pickers or cavity dumpers, which employ acousto-optical devices to select only a specified fraction of the pulses in the pulse train, or by long-cavity lasers (48). Recently, small and inexpensive low average power (<1 mW) picosecond diode lasers at fixed wavelengths with variable repetition rates have also been employed for TCSPC FLIM (39), and their variable repetition rate is particularly suited to measuring long fluorescence decays, e.g. those of most types of quantum dots. At the time of writing, the latest models released include a pulsed 470 nm diode laser, which is suitable for exciting GFP effectively.

A novel and innovative development is the use of a photonic crystal fiber as a tunable supercontinuum excitation source for FLIM (49). Ti:Sapphire laser pulses at 790 nm were coupled into a 30 cm long, 2 μm diameter, microstructured photonic crystal fiber to produce a continuum of pulses from 435 to 1150 nm. Appropriate spectral selection allowed the excitation of GFP and autofluorescence in confocal TCSPC and wide-field time-gated FLIM (49). The ease and simplicity with which the tunability is achieved over such a large range is a distinct advantage of this approach.

Photomultipliers in the photon-counting mode are the most frequently used detectors. They are small, reliable, and relatively inexpensive, but can be damaged by excessive signals and have a transit time spread of ≈150 ps, which is longer than a typical optical excitation pulse. Microchannel plates have the best time resolution, down to ≈30 ps, but they are expensive and can also be easily damaged by too high a light level. An alternative detector is an avalanche photodiode (or single-photon avalanche diode), which is inexpensive, has a high light detection sensitivity (quantum efficiency), and is not damaged by high light levels, but only has a small active area and a count-rate-dependent response. Devices with a transit time spread of only ≈40 ps have recently become available, but the light has to be focused very well on to its 20 μm diameter active area (a typical FLIM photomultiplier or microchannel plate cathode diameter is 8 mm).

TCSPC is a mature and reliable digital technique based on whether a photon is detected or not. The ease of reproducibility of measurements is due to the unique combination of advantages such as single-photon sensitivity, a high temporal resolution (picoseconds), linear recording characteristics independent of excitation intensity fluctuations, detector gain variations and photobleaching, well-defined Poisson statistics, wide dynamic range (in practice up to five orders of magnitude), and an excellent signal-to-noise ratio. To minimize acquisition time, it is normally run in time reverse mode, where the fluorescence photon triggers the detection electronics, rather than the laser excitation pulse (50). Drawbacks of TCSPC are that, as each photon is timed individually in each pixel of the image, the collection of many photons for a high statistical accuracy can be time-consuming (38, 51). The maximum photon flux that can be timed using a single channel (one detector, TAC, and ADC) is limited by photon pile-up and the dead time of the electronics, currently $\approx 10^6$ photons/s. Spatially resolved wide-field TCSPC with quadrant anode detectors has been demonstrated where not only the arrival time, but also the position of each photon is determined (52). However, as each photon still has to be timed individually, there is no advantage in throughput. TCSPC is also not suitable for measuring very long decay times such as lanthanide luminescence decays in the microsecond range (53). As an example of commercial availability, Becker & Hickl, PicoQuant, and Edinburgh Instruments specialize in PC-based TCSPC boards, which require a trigger signal from the excitation source, a signal from the detector, and, importantly, a scan synchronization signal (see *Table 2*).

A similar but rather faster approach is to bin all incoming photons within pre-set time windows after excitation (41, 42). This time-binning method is significantly faster than TCSPC because it is not necessary to reduce the fluorescence signal to the level of single-photon timing. It is commercially available from Nikon. However, it is less accurate than TCSPC and time resolution is lower (see *Table 2*). The use of multi-photon excitation streak-camera-based FLIM has also been reported (43). This technique allows rapid data acquisition, even for a large number of pixels, has a high temporal resolution, and has been commercialized (Hamamatsu). However, potential non-linearities and saturation may pose a problem and the laser repetition rate is only 1 Mhz (43).

Frequency-domain fluorescence lifetime measurements have also been combined with confocal/multi-photon laser scanning microscopy (30, 44, 45). The advantage here is that this approach is fast and can be inexpensive (44).

2.2.2. Wide-field FLIM

In the time domain, a fluorescence decay curve can be directly acquired after excitation of the sample, usually using a sampling technique (54, 55). After exciting the sample with a short laser pulse, time-gated snapshots of the fluorescence emission are taken at various nanosecond delays using high-speed gated image-intensified cameras. These stroboscopic approaches are fast, since all

Table 2. Summary of some of the advantages or strengths and disadvantages or weaknesses of various implementations of FLIM

Note that in wide-field FLIM, optical sectioning to remove out-of-focus blur can be achieved with structured illumination (36) or multiple beam scanning techniques (28, 35).

	Advantages/strengths	Disadvantages/weaknesses	Examples of commercial availability
Time-gated wide-field time domain	• Fast • All pixels acquired in parallel • Suitable for long-lifetime probes such as lanthanides – short-lived fluorescence can be gated away	• Low sensitivity; strong signal needed • Consecutive acquisition of time gates is vulnerable to artifacts, e.g. by photobleaching and sample movement • Out-of-focus blur and lower spatial resolution than confocal FLIM • Pulsed laser needed • Limited dynamic range	Kentech; Photek; Hamamatsu; LaVision Biotech; Photonic Research Systems
Wide-field frequency domain	• Fast • All pixels acquired in parallel • Easy to modulate continuous wave laser • No deconvolution (temporal) of instrumental response and fluorescence decay necessary	• Vulnerable to artifacts, e.g. by photobleaching and sample movement • Complex data and error analysis • Out-of-focus blur and lower spatial resolution than confocal FLIM	Lambert Instruments
Confocal/ multi-photon scanning with TCSPC	• Single photon sensitivity • Wide dynamic range • Linear recording characteristics independent of excitation intensity fluctuations, detector gain variations, or photobleaching • Easy visualization of fluorescence decays and well-defined Poisson statistics • Inherent optical sectioning • Best signal-to-noise ratio • High temporal resolution	• Slow; each photon has to be timed individually • Pulsed laser needed • Unsuitable for very long-lifetime probes, e.g. lanthanides • Saturates at high count rates	Becker & Hickl; PicoQuant; Edinburgh Instruments
Confocal/ multi-photon scanning with time-binning photon detection	• Fastest scanning technique • Inherent optical sectioning • Single-photon sensitivity	• Pulsed laser needed • Slower than wide-field imaging • Less accurate than TCSPC	Nikon
Streak camera FLIM	• Potential for very high temporal resolution • Fast • Easy visualization of fluorescence decays	• Expensive • Limited dynamic range • Pulsed laser needed • Potential non-linearities and saturation	Hamamatsu

pixels are acquired in parallel – a 100 Hz FLIM frame rate has been reported using two time gates and an optical delay (56) – but they lack single-photon sensitivity and accuracy, and their temporal resolution is limited to ≈80 ps (54) (see *Table 2*). A recent development is the use a segmented gated image-intensified camera, which allows the acquisition of four time gates simultaneously (57). Directly gated CCD cameras have also been developed (58), but their time resolution is lower than gated image-intensified cameras and they are more suited to imaging long-lifetime probes. Gated CCDs are commercially available from Photonic Research Systems. An advantage of time-gated FLIM is that it facilitates the imaging of long lifetimes, e.g. those of lanthanides (53, 59), without any contributions from short-lived fluorescence, e.g. autofluorescence (see *Table 2*). High-speed gated image-intensified cameras are available from Kentech Instruments, both the conventional and segmented variety, and also from Photek, Hamamatsu, and LaVision BioTec.

In frequency-domain wide-field FLIM, a sinusoidally modulated excitation beam and detector may be used to measure the phase shift and demodulation of fluorescence signals with respect to their excitation signals using modulated intensified cameras (9, 60–64). With this approach, a fluorescence lifetime may be calculated from both the phase shift and demodulation (at several modulation frequencies if necessary, e.g. for multi-exponential fluorescence decays; 65). For a simple monoexponential fluorescence decay, both calculations should yield the same value. For more complex decays, e.g. in the case of some fluorescent proteins such as CFP, the phase-shift lifetime is shorter than the demodulation lifetime (60). Commercial wide-field frequency domain FLIM systems are available from Lambert Instruments.

The main advantage of wide-field FLIM is the potential for rapid refresh rates due to the parallel acquisition of all pixels. This is important for real-time FLIM (66) for biomedical applications, e.g. endoscopy (67, 68).

There is a lively debate as to the relative merits of time- and frequency-domain approaches to FLIM, as reviewed recently (37). In theory, the two approaches are related by a Fourier transformation and, using a hybrid TCSPC and multi-frequency phase fluorometer, they have experimentally been demonstrated to be equivalent (69). To nonspecialists, the easy visualization of fluorescence decays in the time domain may be an advantage over the frequency domain, where the analysis of complex fluorescence decay profiles, such as stretched exponential decay profiles, is less tractable (70). However, for some applications the frequency-domain instrumentation is considered easier to implement since pulsed laser sources are not required, especially for longer lifetimes – although practitioners are increasingly using mode-locked lasers for frequency-domain measurements, particularly in multi-photon microscopes (45). Frequency-domain techniques are slightly more photon efficient than time-gating techniques (but this does not necessarily translate into more accurate fluorescence lifetimes) and require no temporal deconvolution of the instrumental response and the fluorescence decay. The signal-to-noise ratio is higher for TCSPC than for frequency-domain measurements (71), particularly at low intensities, but TCSPC saturates at high fluorescence intensities (51) (see *Table 2*).

One potential pitfall of the time-domain approach is that there should be sufficient time ($\approx 5\tau$) between excitation pulses for the sample fluorescence to decay completely in order to obtain accurate fluorescence lifetime values. In practice, this implies using mode-locked lasers with pulse pickers, cavity dumpers, lower-repetition-rate pulsed diode lasers (39, 72), long-cavity lasers (48), or appropriate fitting procedures to take residual fluorescence into account (73). This is not an issue for the frequency-domain approach.

The FLIM techniques continue to be improved, particularly by the reduction of acquisition times (42). The relative merits of the various FLIM implementations are summarized in *Table 2*, and the choice of system depends on the samples to be studied and the practitioner's preference for fast data acquisition and high temporal or spatial resolution.

3. RECOMMENDED PROTOCOLS

The protocols for FLIM sample preparation do not differ from those for confocal or wide-field intensity-based fluorescence microscopy. The data acquisition is followed by the main task of data analysis, i.e. extracting the fluorescence lifetimes from the raw data. Once these have been obtained, data interpretation in the context of models or hypotheses helps to verify or falsify different scenarios.

Protocol 1

FLIM of GFP-tagged proteins in cells

Equipment and Reagents

- Sample with GFP
- Hardware: inverted Leica TCS SP2 confocal scanning microscope (Leica Microsystems)
- Coherent Mira 900 Ti:Sapphire femtosecond laser with a Verdi V6 pump laser or Hamamatsu PLP-10 470 picosecond pulsed diode laser excitation sources
- Becker & Hickl SPC 830 board in 3 GHz, Pentium IV, 1 GB RAM computer with Windows XP
- Cooled Becker & Hickl PMC100-01 detector head based on Hamamatsu H5773P-01 photomultipliers, mounted on the microscope's X1 port
- DCC 100 detector control module
- Software: SPCIMAGE 2.8 by Becker & Hickl

Method

1. Place the sample on the microscope stage and obtain a transmission and fluorescence image to identify fluorescent cells and verify that the fluorescence emanates from the locations expected (e.g. cell membrane, cytoplasm). Obtain a fluorescence emission spectrum and verify that it is that of GFP. As a negative control, image a nonlabeled sample and verify that it does not fluoresce. Although this step is not essential specifically for FLIM, it is good practice in general and does help to verify that the sample is what you think it is.

2. Switch to FLIM mode – this is easily accomplished by moving a mirror out of the fluorescence detection beam path ('External Detector' button on 'Beam Path Setting' panel on the Leica TCS SP2 acquisition control software). An appropriate fluorescence emission filter to block any exciting light from reaching the detector must be in the fluorescence detection beam path.
3. Scan the sample and check, on the computer controlling the FLIM acquisition, that the detector count rate (black bar labeled CFD on acquisition control software of the Becker & Hickl SPC 830 board) is no more than about 1% of the laser repetition rate (green bar labeled SYNC acquisition control software) (see pile-up in section 5). If it is, reduce the laser excitation intensity, e.g. by placing a neutral density filter in the laser beam path.
4. Acquire a FLIM image, typically for 3–5 min. Stop scanning and save the raw data (a 3D data 'cube' consisting of spatial coordinates *x* and *y*, and time).
5. Open the raw data in the fluorescence decay analysis software package to display the fluorescence intensity image. This is simply the integrated fluorescence decay, i.e. the area under the fluorescence decay curve (see *Fig. 4C*), in each pixel.
6. Select a typical pixel by placing the cursor on it and inspect the fluorescence decay in that pixel. If the peak count is below 100, use spatial binning of pixels. The counts of adjacent pixels (e.g. 3×3 or 5×5) are added into the central pixel, so that a higher peak count is obtained there. This degrades the spatial resolution of the image, but provides a higher statistical accuracy for the next step. Alternatively, the measurement could be repeated for a longer acquisition time (step 4). For 30–50 min, an approximately tenfold higher peak count (and total counts) is obtained, but this is far too long an acquisition time for most biological samples because of the danger of artifacts due to sample movement, microscope drift, phototoxicity, and photobleaching.
7. Select a global pixel threshold value (above which the decay in a pixel is fitted) and apply a single exponential decay fit to the image. The result yields a fluorescence lifetime for each pixel above the threshold, which is then encoded in color. Each pixel is colored with the result of the fit, and a FLIM map is obtained. Check the reduced χ^2 values for various pixels – around 1 (and up to 1.3) indicates a good fit. Inspect the corresponding residuals, which should be randomly distributed around zero.
8. The fluorescence lifetime histogram plots how often certain fluorescence lifetimes occur vs. the fluorescence lifetime itself. This plot should have the shape of a Gaussian distribution and its mean value should broadly agree with many other FLIM studies of GFP in cells. Adjust the color range such that the Gaussian fluorescence lifetime distribution fits into the color range (see *Fig. 5C* in color section).

Protocol 2

FLIM of FRET with GFP and monomeric red fluorescent protein (mRFP)-tagged proteins in cells

Equipment and Reagents

- Sample with GFP and mRFP
- Equipment as in *Protocol 1*

Method

1. Carry out steps 1–4 in *Protocol 1*, but with a narrow-band GFP emission filter in step 2 to exclude any acceptor emission. The acquisition time may need to be increased accordingly.
2. Photobleach the acceptor by directly exciting it with a high intensity, for example with a mercury lamp, the attenuated Ti:Sapphire pump laser, or with a photonic crystal fiber as a tunable supercontinuum excitation source. Verify that the mRFP acceptor has been bleached.
3. Repeat step 4 of *Protocol 1*.
4. For a quick qualitative analysis, select a pixel threshold and fit a single exponential decay to the donor image before and after bleaching. As single exponential fits are usually much faster than double exponential fits, this allows you to obtain a rapid indication of GFP fluorescence lifetime contrast. After bleaching the acceptor, the fluorescence lifetime of the GFP donor should increase to about the value obtained in step 8 of *Protocol 1* due to the abrogation of FRET. Proceed to the more accurate double exponential fit for quantitative analysis (step 5 of this protocol). However, if the GFP fluorescence lifetimes before and after acceptor bleaching are the same, and similar to the value obtained in step 8 of *Protocol 1*, then it appears that no FRET is taking place.
5. Fit a double exponential decay to the donor image. The donor image should show two fluorescence lifetimes – one is the unquenched donor, which does not take part in FRET and should be similar to the value obtained in step 8 of *Protocol 1*. The second (shorter) fluorescence lifetime is the GFP donor undergoing FRET. The fit will also yield the pre-exponential factors or amplitudes that give an indication of the relative amount of donors undergoing FRET. After acceptor photobleaching (step 2 of this protocol), the second shorter component should disappear, or its pre-exponential factor (amplitude) be greatly diminished. It can be useful to fix the donor lifetime of the free component to the value measured in a control experiment where FRET does not occur (see step 8 of *Protocol 1*).
6. The results for the fluorescence lifetimes, pre-exponential factors, and the lifetime ratio and the pre-exponential factor ratio for each pixel can then be encoded in color. Each pixel is colored according to its value, and contrast due to lifetimes, pre-exponential factors, and their ratios is obtained. Check the reduced χ^2 values (which can also be encoded in color and displayed as an image) – a value of around 1 (and up to 1.3) indicates a good fit. Inspect the residuals, which should be randomly distributed around zero.
7. The interpretation of the images is that the long donor fluorescence lifetime is due to GFP not undergoing FRET, and its corresponding pre-exponential factor shows the relative amount. The short fluorescence lifetime is due to GFP acting as a donor undergoing FRET. The shorter the fluorescence lifetime, the more efficient the FRET. Its corresponding pre-exponential factor again indicates the relative amount of GFP undergoing FRET. The GFP fluorescence lifetime ratio indicates the FRET efficiency, whereas the pre-exponential factor ratio indicates the proportion of GFP undergoing FRET.

8. Fluorescence lifetime histogram plots should accompany the images for easy visualization of average fluorescence lifetime values and the fluorescence lifetime distribution (see *Fig. 5C* in color section).

4. DATA ANALYSIS

4.1. Frequency domain

In most cases, the FLIM experiment is carried out and the fluorescence decay data in each pixel are saved for subsequent analysis. In the frequency domain, the fluorescence lifetime can be determined from the phase-shift and demodulation data, possibly at several different modulation frequencies (65), according to:

$$\tau_\varphi = \frac{\tan\varphi}{\omega} \qquad \text{(Equation 8)}$$

and

$$\tau_m = \frac{1}{\omega}\left(\frac{1}{M^2} - 1\right)^{1/2} \qquad \text{(Equation 9)}$$

where ω is the modulation frequency, φ the phase shift, and M the demodulation (modulation depth ratio (AC/DC ratio) of fluorescence and excitation) as defined in *Fig. 4D*, τ_m the modulation lifetime, and τ_φ the phase-shift lifetime. Several comprehensive publications on frequency-domain FLIM data analysis exist (e.g. 74, 75). τ_m and τ_φ are identical and independent of ω in the case of a single exponential decay ($\tau_\varphi = \tau_m$). τ_m and τ_φ are different and a function of ω for nonsingle exponential decays ($\tau_\varphi \neq \tau_m$). Practitioners of frequency-domain FLIM often plot τ_m vs. τ_φ to look for deviations from strictly monoexponential behavior, for example as a qualitative indication of FRET (21, 61, 62). A refined method (AB plots) has recently been devised that allows fluorescence lifetimes in two-component mixtures to be determined using a single modulation frequency (76). A temporal deconvolution of the instrumental response and the fluorescence decay, as in TCSPC FLIM, is not necessary in frequency-domain FLIM.

4.2. Time domain

In time-domain FLIM, the fluorescence lifetime is calculated from an exponential fit to the decay data, and the details of such a fitting procedure depend on the technique used. In wide-field time gating, for example, only the part of the fluorescence decay after the excitation pulse is measured. For the special case where only two gates are used, and no deconvolution is necessary, the decay time can be calculated analytically from:

$$\tau = \frac{t_2 - t_1}{\ln \frac{I_1}{I_2}} \qquad \text{(Equation 10)}$$

where t_1 is the start of the first time window and t_2 is the start of the second time window (of equal duration), and I_1 and I_2 are their respective intensities (55, 56). This enables a rapid determination of the fluorescence lifetime and can therefore be used for fast video-rate (or faster) FLIM, e.g. for endoscopy FLIM (67). In practice, however, a third time window should be added so that the background can also be calculated (57).

In TCSPC, the time window usually includes the moment of the excitation, and the fluorescence decay includes the rising edge (the excitation light itself is blocked by a filter) (46, 47). A measurement of the excitation pulse using a scattering sample yields the instrumental response function (IRF) with a full width at half maximum Δt. This is the response of the detection system if no fluorescence is measured and comprises contributions from the width of the optical pulse Δt_{op}, the jitter in the electronics Δt_{jit} and, most importantly, the transit time spread of the detector Δt_{tts} (46):

$$\Delta t = \sqrt{\Delta t_{op}^{\,2} + \Delta t_{tts}^{\,2} + \Delta t_{jit}^{\,2}} \qquad \text{(Equation 11)}$$

The finite width of the IRF has to be taken into account when calculating the fluorescence lifetime from the decay data. In single-point measurements, this is routinely done by measuring the IRF using a scattering solution and deconvolving it when fitting the fluorescence decay. In FLIM, measurement of the IRF can be cumbersome and not so straightforward, particularly for two-photon excitation FLIM (77). However, one solution is to estimate the IRF from the rising edge of the fluorescence decay (commercially available SPCIMAGE software; Becker & Hickl). This works remarkably well, but does not take into account any IRF afterpeaks. For a detailed and accurate fluorescence decay analysis, afterpeaks cannot be disregarded (46, 47).

Another point to bear in mind for TCSPC-based FLIM decay fitting is the statistical accuracy of the decay data, i.e. the number of counts in the peak (or the number of counts collected – the integrated decay). The standard deviation is given by the square root of the number of counts because counting experiments have Poisson statistics. For 100 counts in the peak, the standard deviation in the peak count is 10%, whereas for 10 000 counts in the peak, the standard deviation is only 1% (but the decay takes at least 100 times longer to accumulate) (46, 47). In order to obtain good statistics, the number of time channels used in imaging is much less than in conventional cuvette-based single decay measurements: usually only 64 time channels are used for FLIM, whereas up to 4000 can be used for single decay curves. In addition, the spatial binning of pixels allows a higher peak count to be obtained at the expense of spatial resolution. However, low counts are still frequently encountered, for example in dim regions of the image. The problem of fitting fluorescence decays with a low number of counts has been investigated and it was found that conventional least squares fitting algorithms yield an erroneous lifetime for low counts (78, 79). This fitting artifact has to be taken into account when interpreting the data. The best way to avoid it is to set a threshold for the peak count in the entire image below which the decay is not fitted, or use a different fitting algorithm, e.g. maximum likelihood estimation (78, 79).

4.3. Decay models

There is a variety of decay fitting models, and the resulting values for the fitting parameters have to be interpreted in the context of the experimental situation. A single exponential decay was earlier described in Equation 4. A bi-exponential fluorescence decay model has the form:

$$I(t) = B + A_1 e^{-t/\tau_1} + A_2 e^{-t/\tau_2} \quad \text{(Equation 12)}$$

where $I(t)$ is the fluorescence intensity at time t, B is the background, A_1 and A_2 are the pre-exponential factors (amplitudes) and τ_1 and τ_2 are the fluorescence lifetimes. For $A_2 = 0$, a single exponential decay is obtained, which is an appropriate model for a single fluorophore with a single emitting state. The lifetime may vary with changes in the environment; typically k_{nr} varies, which consequently changes τ (2, 3, 46, 47).

The fitting works by choosing a mathematical decay model and then minimizing the square of the distance between each data point and a point on the curve representing the fitting model for all data points in the decay. The goodness of fit is judged by the distribution of real data points above and below the curve of best fit, which should be random (flat residuals) and the normalized χ^2, which for Poisson noise data should be around 1 (46, 47).

Bi-exponential fluorescence decay would be an appropriate model for two fluorophores with a single emitting state each. The pre-exponential factors would be a measure of the relative concentration of each fluorophore, whereas the lifetimes would report on the environment via k_{nr}. A bi-exponential decay is also often used as a simple model for FRET analysis of the donor fluorescence decay. In this case, the pre-exponential factors A_1 and A_2 yield information about the relative unquenched (no FRET) and quenched (FRET) population, whereas the fluorescence lifetimes τ_1 and τ_2 correspond to the unquenched lifetime and an averaged quenched lifetime (see *Protocol 2*). This latter approximation averages over all orientations and distances, which is a fair assumption to make as a first approximation.

A different model is the stretched exponential decay:

$$I(t) = B + I_0 e^{-(\frac{t}{\tau})^{\frac{1}{h}}} \quad \text{(Equation 13)}$$

where h is the heterogeneity parameter ($h > 1$) (70). This model can be used to fit a continuous distribution of lifetimes, where h is related to the width of the lifetime distribution, or for FRET, where h is related to the dimensionality of the system (70). A stretched exponential decay model has, for example, been used to fit wide-field time gating FLIM data of complex, heterogeneous biological samples such as tissue autofluorescence. Mathematically speaking, it has one more fitting parameter than a single exponential (Equation 4, with background), but one less than a bi-exponential model (Equation 12). For only 10–20 time gates in wide-field time-gated FLIM and a poor single exponential fit, a bi-exponential model with five free-fitting parameters will always produce a statistically good fit, although it may not be physically relevant. The stretched exponential function represents a good compromise between a single and a bi-exponential fit model to generate contrast in an image (70).

4.4. Global analysis

Global analysis is a data analysis and decay fitting method that introduces constraints to large data sets that have certain features in common. It was originally introduced in fluorescence spectroscopy where data from different measurements under different conditions were analyzed simultaneously (80). This drastically reduces the number of independent fitting parameters and thus improves the ability to distinguish between different fitting models, or it improves the ability of the fitting model to recover the relevant parameters with greater accuracy. Its relevance for FLIM is that the fluorescence decays are not fitted as independent measurements in each pixel, but as, for example, two spatially invariant lifetimes, i.e. two lifetimes that are equal for all pixels in an image, one for FRET and one for no FRET (74, 81). In this case, only the pre-exponential factors, which then are a measure of the relative population of each species, vary in each pixel. Global analysis has been implemented for both frequency-domain (74) and time-domain FLIM (81), and in the frequency domain this approach allows the determination of two lifetimes from single-frequency data (74).

4.5. Image display

To display a FLIM image, or a TR-FAIM image, it is customary to encode the fluorescence lifetime or the rotational correlation time image according to a continuous rainbow color code to generate contrast. For example, the shortest lifetime in all the pixels of an image is blue and the longest lifetime is red. All other lifetimes in between are represented by a color between blue and red. The intensity contrast can be retained by using the brightness of the color. FLIM images can, of course, also be represented in black and white – in this case the gray scale represents the lifetime and no information on the fluorescence intensity can be given in the same image.

5. TROUBLESHOOTING

In both confocal scanning and wide-field FLIM instruments, as well as in the data analysis, there may be artifacts that can complicate data interpretation (82). Instrumental artifacts include scattered light, which, in time-domain FLIM, will show up as a peak on top of the beginning of the fluorescence decay and may be confused with a short decay time, or as a small peak after the IRF, which may be caused by reflections inside the microscope. These scattered light artifacts can be identified as such because they can be distinguished with spectral discrimination – they are always at the same wavelength as the exciting light. Remembering that, in air, light travels 30 cm in 1 ns helps to pinpoint the origin of reflections.

Filter or glass fluorescence can also cause artifacts, especially at low sample fluorescence, but this can easily be identified by taking a measurement without the sample – if a decay is obtained under these circumstances, it is due to the instrument and has nothing to do with the sample! On the other hand, it should be noted that sample autofluorescence may also contribute to a fluorescence decay.

In TCSPC, TAC nonlinearities may cause poor fits, but can be identified by blocking the excitation and shining ambient light, e.g. from the transmitted light source, on to the sample and measuring the timing. A constant background, B, should be obtained in each pixel of the image, which can be thought of as $A_1 = A_2 = 0$ in Equation 12. Regions where deviations from a constant value for B occur will never yield a good fit, and these should be avoided for the measurement if they cannot be removed by adjusting the parameters for the TCSPC card.

One infamous artifact in TCSPC is photon pile-up, which is caused by too high a photon detection rate (46, 47). This leads to only the first photon being timed, with any subsequent photons being ignored, because the electronics are busy timing and processing the first photon. Pile-up leads to a shortening of the lifetime; the best way to avoid this is to keep the photon count rate at around 1% of the laser repetition rate (46, 47) (see step 3 of *Protocol 1*).

In wide-field intensified CCD-based FLIM, the phosphor decay of the image intensifier may cause artifacts. Depending on the composition of the phosphor, its decay time can vary from less than a microsecond to a millisecond, and the energy of the electrons impinging on it may also affect its decay time (a higher energy shortens the phosphor decay time). For time-gated applications, there should always be a delay before the time window is moved. It is practical to begin the acquisition with a long delay after the excitation, i.e. the tail of the fluorescence decay, and move the time window up to the peak to minimize any residual phosphor decay to be imaged in the subsequent time window (54).

Another artifact in wide-field FLIM can be caused by sample movement between successive frames, either in time-gating or for multi-frequency phase FLIM, and photobleaching can also cause artifacts (83–85). The problem of aliasing caused by the presence of higher harmonics in frequency-domain FLIM has also been addressed (83).

The design of experiments may also cause artifacts. For example, for high fluorescence probe concentrations on fluorescent beads, reabsorption of the emitted fluorescence has been observed (if the absorption and emission spectra overlap) (86), but for cells this is an unlikely artifact.

6. APPLICATIONS OF FLIM

The most frequent application of FLIM concerns the identification of FRET in cell biology, principally to study the interactions of proteins (20), but FLIM has also been used to probe the local environment of fluorophores, reporting, for example, Ca^{2+}, oxygen, or pH concentration, and to report on sensitivities in photodynamic therapy (11–13). FLIM of autofluorescence has been used to provide intrinsic contrast in unstained tissue, as reviewed previously (16, 17).

7. OUTLOOK

There are various implementations of FLIM, and, depending on the application, each has its advantages and drawbacks. The ideal fluorescence microscope would acquire the entire multi-dimensional fluorescence emission contour of intensity, position, lifetime, wavelength, and polarization in a single measurement, with single-photon sensitivity, maximum spatial resolution, and minimum acquisition time (see *Fig. 1*). Needless to say, there is presently no technology with this unique combination of features, and to build one remains a challenge for instrumentation developers. The application of new physical techniques to important problems in cell biology is often the path to unexpected discoveries (4), and there is clearly a tremendous way to go before we are close to saturating the capabilities of fluorescence imaging for cell biology. Indeed, imaging fluorescence parameters such as lifetime, spectrum, and polarization, as well as imaging more rapidly in 3D at higher spatial resolution, is certain to reveal exciting new aspects of cell biology.

Acknowledgements

I would like to gratefully acknowledge David Bacon from the Biological Sciences Department at Imperial College London for assistance with *Fig. 3*. I would also like to thank my colleagues for critical comments on the manuscript, especially Dan Elson from the Photonics Group at Imperial College London and Rainer Heintzmann from the Randall Division of Cell and Molecular Biophysics at King's College London. Financial assistance from the UK's Engineering and Physical Sciences Research Council (EPSRC), the Royal Society, and Leica Microsystems is also gratefully acknowledged.

Note added in proof

A new textbook which deals with many of the issues raised in this chapter has just been published (89).

8. REFERENCES

1. **Harvey EN** (1957) *A History of Luminescence.* American Philosophical Society, Philadelphia.
2. **Lakowicz JR** (1999) *Principles of Fluorescence Spectroscopy,* 2nd edn. Kluwer Academic/Plenum Publishers, New York.
3. **Valeur B** (2002) *Molecular Fluorescence.* Wiley-VCH, Weinheim, Germany.
4. **Amos B** (2000) *Nat. Cell Biol.* **2**, E151–E152.
5. **Rost FWD** (1995) *Fluorescence Microscopy.* Cambridge University Press, Cambridge, UK.
6. **Chalfie M, Tu Y, Euskirchen G, Ward WW & Prasher DC** (1994) *Science,* **263**, 802–805.
7. **Rodgers MAJ & Firey PA** (1985) *Photochem. Photobiol.* **42**, 613–616.
8. **Bugiel I, König K & Wabnitz H** (1989) *Lasers Life Sci.* **3**, 47–53.
9. **Morgan CG, Mitchell AC & Murray, JG** (1990) *Proc. R. Microscop. Soc.* **1**, 463–466.
10. **Minami T & Hirayama S** (1990) *J. Photochem. Photobiol. A,* **53**, 11–21.

★★★ 11. **Herman P, Lin H-J & Lakowicz JR** (2003) In: *Biomedical Photonics Handbook,* pp. 9–30. Edited by T Vo-Dinh. CRC Press, New York. – *A detailed, thorough, and comprehensive review mainly of FLIM instrumentation and techniques, including less-well-known methods.*

★★★ 12. **Clegg RM, Holub O & Gohlke C** (2003) In: *Methods in Enzymology: Biophotonics,* part A, pp. 509–542. Edited by G Marriot and I Parker. Academic Press, San Diego. – *A FLIM review focusing on the photophysical aspects and some applications of FLIM.*

13. **Suhling K, French PMW & Phillips D** (2005) *Photochem. Photobiol. Sci.* **4**, 13–22.
14. **Michalet X, Kapanidis AN, Laurence T, *et al.*** (2003) *Annu. Rev. Biophys. Biomol. Struct.* **32**, 161–182.
15. **Amos WB & White JG** (2003) *Biol. Cell,* **95**, 335–342.
16. **Elson DS, Requejo-Isidro J, Munro I, *et al.*** (2004) *Photochem. Photobiol. Sci.* **3**, 795–801.
17. **Urayama P & Mycek M-A** (2003) In: *Handbook of Biomedical Fluorescence,* pp. 211–236. Edited by M-A Mycek & BW Pogue. Marcel Dekker, New York.

★★★ 18. **Jares-Erijman EA & Jovin TM** (2003) *Nat. Biotechnol.* **21**, 1387–1395. – *A comprehensive review of the background theory and context of FRET and how to image it using different methods including FLIM.*

19. **Hell SW** (2003) *Nat. Biotechnol.* **21**, 1347–1355.

★★★ 20. **Wouters FS, Verveer PJ & Bastiaens PI** (2001) *Trends Cell Biol.* **11**, 203–211. – *A review focused on FLIM and FRET applications in biology.*

21. **Harpur AG, Wouters FS & Bastiaens PI** (2001) *Nat. Biotechnol.* **19**, 167–169.
22. **Lidke DS, Nagy P, Barisas, *et al.*** (2003) *Biochem. Soc. Trans.* **31**, 1020–1027.
23. **Squire A, Verveer PJ, Rocks O & Bastiaens PI** (2004) *J. Struct. Biol.* **147**, 62–69.
24. **Varma R & Mayor S** (1998) *Nature,* **394**, 798–801.
25. **Siegel J, Suhling K, Lévêque-Fort S, *et al.*** (2003) *Rev. Sci. Instrum.* **74**, 182–192.
26. **Clayton AHA, Hanley QS, Arndt-Jovin DJ, Subramaniam V & Jovin TM** (2002) *Biophys. J.* **83**, 1631–1649.
27. **Suhling K, Siegel J, Lanigan PMP, *et al.*** (2004) *Opt. Lett.* **29**, 584–586.
28. **Benninger RKP, Önfelt B, Neil MAA, Davis DM & French PMW** (2005) *Biophys. J.* **88**, 609–622.
29. **Becker W, Bergmann A, Biskup C, Zimmer T, Klöcker N & Benndorf K** (2002) *Proc. SPIE,* **4620**, 79–84.
30. **Carlsson K & Liljeborg A** (1998) *J. Microsc.* **191**, 119–127.
31. **Tinnefeld P, Herten DP & Sauer M** (2001) *J. Phys. Chem. A,* **105**, 7989–8003.
32. **Hanley QS, Arndt-Jovin DJ & Jovin TM** (2002) *Appl. Spectrosc.* **56**, 155–166.
33. **Bird DK, Eliceiri KW, Fan CH & White JG** (2004) *Appl. Opt.* **43**, 5173–5182.

34. **Amos WB** (2000) In: *Protein Localization By Fluorescence Microscopy, A Practical Approach*, pp. 67–108. Oxford University Press, Oxford, UK.
35. **Straub M & Hell SW** (1998) *Bioimaging*, **6**, 177–185.
36. **Neil MAA, Juskaitis R & Wilson T** (1997) *Opt. Lett.* **22**, 1905–1907.
37. **Valeur B** (2005) In: *Fluorescence Spectroscopy in Biology: Advanced Methods and Their Applications To Membranes, Proteins, DNA, and Cells*, pp. 30–45. Springer-Verlag, Berlin/Heidelberg/New York.
★★ 38. **Becker W, Bergmann A, Hink MA, König K, Benndorf K & Biskup C** (2004) *Microsc. Res. Tech.* **63**, 58–66. – *A very good review of TCSPC-based FLIM.*
39. **Kress M, Meier T, Steiner R, Dolp F, Erdmann R, Ortmann U & Rück A** (2003) *J. Biomed. Opt.* **8**, 26–32.
40. **Treanor B, Lanigan PM, Suhling K, *et al.*** (2005) *J. Microsc.* **217**, 36–43.
41. **Buurman EP, Sanders R, Draaijer A, *et al.*** (1992) *Scanning*, **14**, 155–159.
42. **Gerritsen HC, Asselbergs NAH, Agronskaia AV & van Sark WGJHM** (2002) *J. Microsc.* **206**, 218–224.
43. **Krishnan RV, Masuda A, Centonze VE & Herman B** (2003) *J. Biomed. Opt.* **8**, 362–367.
44. **Booth MJ & Wilson T** (2004) *J. Microsc.* **214**, 36–42.
45. **French T, So PTC, Weaver DJ, *et al.*** (1997) *J. Microsc.* **185**, 339–353.
46. **Birch DJS & Imhof RE** (1991) In: *Topics in Fluorescence Spectroscopy: Techniques*, Vol. 1: *Techniques*, pp.1–95. Plenum Press, New York.
47. **O'Connor DV & Phillips D** (1984) *Time-correlated Single-photon Counting.* Academic Press, New York.
48. **Lévêque-Fort S, Papadopoulos DN, Forget S, Balembois F & Georges P** (2005) *Opt. Lett.* **30**, 168–170.
49. **Dunsby C, Lanigan PMP, McGinty J, *et al.*** (2004) *J. Phys. D Appl. Phys.* **37**, 3296–3303.
50. **Haugen GR, Wallin BW & Lytle FE** (1979) *Rev. Sci. Instrum.* **50**, 64–72.
★★★ 51. **Gratton E, Breusegem S, Sutin J, Ruan Q & Barry N** (2003) *J. Biomed. Opt.* **8**, 381–390. – *An excellent comparison of different approaches to FLIM.*
52. **Emiliani V, Sanvitto D, Tramier M, *et al.*** (2003) *Appl. Phys. Lett.* **83**, 2471–2473.
53. **Vereb G, Jares-Erijman E, Selvin PR & Jovin TM** (1998) *Biophys. J.* **74**, 2210–2222.
54. **Dowling K, Dayel MJ, Lever MJ, French PMW, Hares JD & Dymoke-Bradshaw AKL** (1998) *Opt. Lett.* **23**, 810–812.
55. **Wang XF, Periasamy A, Herman B & Coleman DM** (1992) *Crit. Rev. Anal. Chem.* **23**, 369–395.
56. **Agronskaia AV, Tertoolen L & Gerritsen HC** (2003) *J. Phys. D Appl. Phys.* **36**, 1655–1662.
57. **Elson DS, Munro I, Requejo-Isidro J, *et al.*** (2004) *New J. Phys.* **6**, 180.
58. **Mitchell AC, Wall JE, Murray JG & Morgan CG** (2002) *J. Microsc.* **206**, 233–238.
59. **Marriott G, Clegg RM, Arndt-Jovin D & Jovin TM** (1991) *Biophys. J.* **60**, 1374–1387.
60. **Pepperkok R, Squire A, Geley S & Bastiaens PIH** (1999) *Curr. Biol.* **9**, 269–272.
61. **Ng T, Squire A, Hansra G, *et al.*** (1999) *Science*, **283**, 2085–2089.
★★ 62. **Gadella TWJ, Jovin TM & Clegg RM** (1993) *Biophys. Chem.* **48**, 221–239. – *An early FLIM paper with a good overview of the topic.*
★★ 63. **Lakowicz JR, Szmacinski H, Nowaczyk K, Berndt KW & Johnson M** (1992) *Anal. Biochem.* **202**, 316–330. – *An early FLIM paper with a good overview of the topic.*
64. **Morgan CG, Mitchell AC & Murray JG** (1992) *Trends Anal. Chem.* **11**, 32–41.
65. **Squire A, Verveer PJ & Bastiaens PIH** (2000) *J. Microsc.* **197**, 136–149.
66. **Schneider PC & Clegg RM** (1997) *Rev. Sci. Instrum.* **68**, 4107–4119.
67. **Requejo-Isidro J, McGinty J, Munro I, *et al.*** (2004) *Opt. Lett.* **29**, 2249–2251.
68. **Mizeret J, Stepinac T, Hansroul M, Studzinski A, van den Bergh H & Wagnieres G** (1999) *Rev. Sci. Instrum.* **70**, 4689–4701.
69. **Hedstrom J, Sedarous S & Prendergast FG** (1988) *Biochemistry*, **27**, 6203–6208.
70. **Lee KCB, Siegel J, Webb SED, *et al.*** (2001) *Biophys. J.* **81**, 1265–1274.
71. **Philip J & Carlsson K** (2003) *J. Opt. Soc. Am. A*, **20**, 368–379.
72. **Elson DS, Siegel, J, Webb, SED, *et al.*** (2002) *Opt. Lett.* **27**, 1409–1411.
73. **Sakai Y & Hirayama S** (1988) *J. Lumines.* **39**, 145–151.
74. **Verveer PJ, Squire A & Bastiaens PIH** (2000) *Biophys. J.* **78**, 2127–2137.

75. **Gadella TWJ, Clegg RM & Jovin TM** (1994) *Bioimaging,* **2**, 139–159.
76. **Hanley QS & Clayton AH** (2005) *J. Microsc.* **218**, 62–67.
77. **Habenicht A, Hjelm J, Mukhtar E, Bergstrom F & Johansson LB** (2002) *Chem. Phys. Lett.* **354**, 367–375.
78. **Tellinghuisen J & Wilkerson CW** (1993) *Anal. Chem.* **65**, 1240–1246.
79. **Nishimura G & Tamura M** (2005) *Phys. Med. Biol.* **50**, 1327–1342.
80. **Knutson JR, Beechem JM & Brand L** (1983) *Chem. Phys. Lett.* **102**, 501–507.
81. **Barber PR, Ameer-Beg SM, Gilbey J, Edens RJ, Ezike I & Vojnovic B** (2005) *Proc. SPIE,* **5700**, 171–181.
82. **vandeVen M, Ameloot M, Valeur B & Boens NI** (2005) *J. Fluores.* **15**, 377–413.
83. **van Munster EB & Gadella TWJ** (2004) *J. Microsc.* **213**, 29–38.
84. **van Munster EB & Gadella TWJ** (2004) *Cytom. Part A,* **58A**, 185–194.
85. **Hanley QS, Subramaniam V, Arndt-Jovin DJ & Jovin TM** (2001) *Cytometry,* **43**, 248–260.
86. **Siegel J, Elson DS, Webb SED, *et al.*** (2001) *Opt. Lett.* **26**, 1338–1340.
87. **Toptygin D** (2003) *J. Fluores.* **13**, 201–219.
88. **Suhling K, Siegel J, Phillips D, *et al.*** (2002) *Biophys. J.* **83**, 3589–3595.
★★★ 89. **Becker W** (2005) *Advanced Time-Correlated Single Photon Counting Techniques.* Springer, Berlin. – *A comprehensive and up-to-date review of state-of-the-art TCSPC instrumentation and applications including multidimensional TCSPC and FLIM, with many illustrations and references.*

CHAPTER 12

Homo-FRET measurements to investigate molecular-scale organization of proteins in living cells

Rajat Varma and Satyajit Mayor

1. INTRODUCTION

Cell biological processes are controlled by signal transduction pathways that involve protein–protein or protein–lipid interactions. These interactions occur on the length scale of a few nanometers. Understanding the mechanism of signal transduction at molecular-length scales in living cells is a challenging question. Several fluorescence-based techniques have been developed in the past couple of decades that can be used to examine protein–protein interactions in living cells. One such technique is based on a phenomenon known as Förster's or fluorescence resonance energy transfer (FRET) between two fluorophores (1). FRET involves transfer of energy from one fluorescent molecule in its excited state to another and occurs in a distance-dependent manner, generally in the length scale of 1–10 nm (1, 2). Hence, this technique has been termed a molecular ruler. Imaging FRET processes in a cell expands the resolution of light microscopy for detection of proximity. It can provide nanometer-scale information about interactions between molecules that are only resolved at optical resolution (~250 nm).

There are many ways that FRET can be measured. In this chapter, we focus on one of the ways of measuring FRET that occurs between like fluorophores called homo-FRET. This is based on determining the polarization state of emitted fluorescence in the steady state. We will present the scope of the technique and the range of questions it can address, the design of the experiment including the choice of fluorophores, the modification of the microscope and data collection, and finally data analysis and interpretation.

Cell Imaging: ***Methods Express*** (D. Stephens, ed.)

1.1. Theoretical concepts

Fluorescence is a phenomenon exhibited by certain molecules, which when excited near their absorption maxima emit photons of longer wavelengths (see *Fig. 1A*). While the molecules lose some energy in nonradiative processes, the molecules spend some time in their excited state before emitting a photon. The lifetime of the excited state in fluorescence is in the range of nanoseconds to

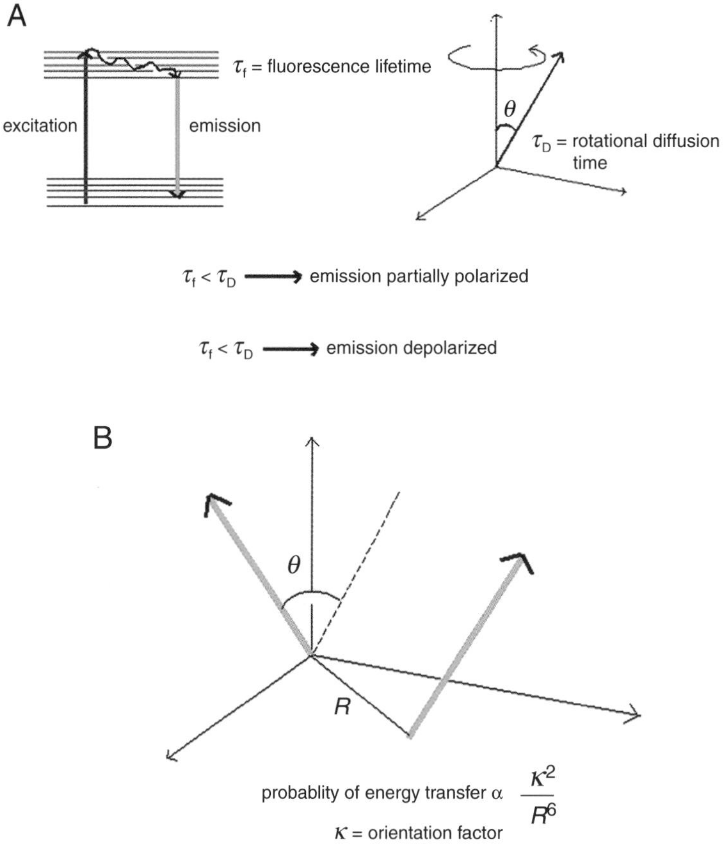

Figure 1. Theoretical concepts.
(*A*) Fluorophores emit photons of longer wavelengths than those they are excited by. The time fluorophores spend in their excited state is called the lifetime of the excited state. Rotational diffusion directly influences anisotropy of fluorescence emission. If the lifetime of the fluorophore is less than the rotational diffusion time, the emission is partially polarized when excited with plane-polarized light; however, if the lifetime is more than the rotational diffusion time, the emission is depolarized, e.g. a fluorophore in water compared with a fluorophore in glycerol. (*B*) FRET is a phenomenon that depends on the orientation factor, κ, between the participating dipoles or fluorophores and on the inverse sixth power of the distance between the dipoles. The dependence on orientation factor leads to a depolarization of acceptor emission when donors are excited with plane-polarized light.

microseconds and reflects the probability of transition from the excited state to ground state.

Simply put, FRET is a phenomenon that occurs between fluorophores as a result of dipolar interactions. Due to the dipolar nature of this interaction, it is inversely related to the sixth power of distance between the fluorophores and is directly proportional to the orientation factor of the two dipoles. The orientation factor is a measure of the angle between the two fluorophores. FRET is also proportional to the spectral overlap between the donor emission and acceptor absorption spectra, bringing the molecular specificity into the equation.

At a more technical level, FRET is a quantum mechanical property of a fluorophore resulting in nonradiative energy transfer between the excited state of the donor fluorophore and a suitable acceptor fluorophore via dipole–dipole interactions. This is a process that was well described in the early 1950s by Förster (1) who proposed a theory for understanding this phenomenon. This theory, now called Förster's theory of resonance transfer, shows how the probability of resonance depends on the local configuration of fluorophores. As mentioned above, consequences of fluorophore interactions (including the range and orientation dependence) may be traced to this dipole–induced dipole interaction. Therefore, energy transfer efficiency depends on the relative orientation and separation between the two transition dipoles, as well as on the overlap between donor emission and acceptor absorption spectra as given by the equations below.

Energy transfer efficiency (E) varies inversely as the sixth power of the distance between the donor and acceptor (2):

$$E = 1/\left[1 + (r/R_0)^6\right] \quad \text{(Equation 1)}$$

where r is the distance of separation between the donor and the acceptor fluorophore. R_0 ('Förster distance') encapsulates molecular properties of the fluorophores that sets the length scale for the transfer process, and is defined as the separation for which the energy transfer efficiency is 50 %. It is calculated using the following expression:

$$R_0 = [8.8 \,.\, 10^{-23} \,.\, n^{-4} \,.\, Q \,.\, \kappa^2 \,.\, J(\lambda)]^{1/6} \quad \text{(Equation 2)}$$

where n is the refractive index of the medium in the range of overlap, Q is the quantum yield of the donor in the absence of acceptor, and $J(\lambda)$ is the spectral overlap. κ^2 is the orientation factor, which depends on the relative orientation of the two dipoles and is defined by:

$$\kappa^2 = (\text{Cos}\theta_T - 3\text{Cos}\theta_A \,.\, \text{Cos}\theta_D) \quad \text{(Equation 3)}$$

It can be quickly estimated that this orientation factor can vary from 0 to 4, but it is usually taken to be 2/3, a value corresponding to a random orientation of the donors and acceptors over a spherical surface. Unless explicitly determined by measurements of fluorescence anisotropy (3), it is often erroneous to assume a value for κ^2, since this may result in significant errors in the measurement of distances (4). Typically, R_0 varies between 1 and 10 nm for various pairs of fluorophores (2). Using the expression for the energy transfer efficiency and a statistical distribution of fluorophores, the probability of nonradiative transfer between any pair in an assembly of fluorophores can be explicitly calculated.

2. METHODS AND APPROACHES

2.1. Experimental approaches to measuring molecular proximity using FRET

There are several consequences of a FRET event. FRET leads to: (i) quenching of donor fluorescence; (ii) sensitized emission of acceptor fluorophore; (iii) reduction in donor lifetime and hence a direct effect on rates of donor photobleaching; and (iv) depolarization of acceptor fluorescence. In each case, the efficiency of a FRET event is described by the equations above. Numerous reviews and detailed handbooks are available to measure FRET due to consequences (i) to (iii) above (5–7), so we will just briefly summarize these methods below.

The design of a FRET experiment depends on which of the consequences is being monitored (6, 8). For methods that primarily quantify steady-state fluorescence emission (i, ii and iv, above) for imaging purposes, this translates into collecting an image of donor fluorescence and a separate image of acceptor fluorescence (i, ii), or anisotropy images of donor or acceptor fluorescence (iv). The ratio of donor fluorescence to acceptor fluorescence is then compared with the ratio of donor fluorescence to acceptor fluorescence collected under conditions where there is no likelihood of FRET between donor and acceptor. The use of ratio imaging is particularly important, since this will generally take care of local variations in donor and acceptor fluorescence (9, 10).

2.1.1. FRET measurements based on quenching of donor fluorescence

As mentioned above, FRET leads to a quenching of donor fluorescence. The simplest form of a FRET measurement is based on this phenomenon. This involves collecting donor and acceptor fluorescence images, and calculating ratios of donor and acceptor fluorescence in every pixel under conditions where there is FRET and compare it with a situation where there is no FRET. This methodology takes care of local variations in donor and acceptor fluorescence, but is prone to errors due to quenching of donor fluorescence by phenomena other than FRET. The extent of donor quenching may be taken as a good measure of FRET efficiency (E) and this can be calculated from the relative fluorescence yield in the presence (F_{AD}) and in the absence (F_D) of the acceptor:

$$E = 1 - (F_{AD}/F_D) \qquad \text{(Equation 4)}$$

Another experimental determination of donor quenching is by measuring the extent of dequenching upon acceptor photobleaching (7, 11). This is a very useful technique, as it directly yields energy transfer efficiencies and is generally unaffected by the environmental factors mentioned above. The extent of increase in donor fluorescence post-bleaching is used to calculate energy transfer efficiencies given by Equation 4, where F_{AD} is now the fluorescence yield of the donor in the presence of and F_D after photobleaching of the acceptor.

2.1.2. FRET measurements based on sensitized emission of acceptor fluorescence

FRET can also be measured based on sensitized emission of acceptor fluorescence (5, 12). The experiment involves exciting the donor only and collecting data (sensitized emission image) in the acceptor channel. Though a straightforward concept, the experiment becomes error prone, since the donor excitation is likely to excite the acceptor directly and the donor fluorescence is also likely to cross over into the acceptor channel. The choice of excitation and emission wavelength bandwidths is critical. This is because the sensitized emission signal collected is a composite of (i) fluorescence due to the direct excitation of the acceptor at the donor excitation wavelength; (ii) spillover fluorescence from the donor into the acceptor fluorescence channel; (iii) autofluorescence; and finally (iv) a contribution from sensitized emission signal. 'Crosstalk' corrections can be difficult to implement and if not done appropriately might mask the energy transfer signal completely (13, 14). A useful rule of thumb to remember is that the energy transfer signal should be at least 10–15% above the total signal observed in the acceptor channel and be relatively free of cellular autofluorescence, which can tend to be nonhomogeneous.

2.1.3. FRET measurements based on reduction in donor lifetime

A consequence of FRET between spectrally distinct donors and acceptors is that donor species are depleted from the excited state by the FRET process and thus the fluorescence lifetime of the donor species is reduced. Energy transfer efficiency (E) may also be directly calculated from the fluorescence lifetime of the donor in the presence (τ_{AD}) or absence (τ_D) of the acceptor as:

$$E = 1 - (\tau_{AD}/\tau_D) \qquad \text{(Equation 5)}$$

Donor lifetimes may be measured by two different methods: a time-domain or a frequency-domain method. This may be directly measured via a recently evolving and powerful methodology called fluorescence lifetime imaging (FLIM), described in Chapter 11 and refs (15–17). The main advantage of the FLIM technique is that the FRET signal depends only on the excited rate reactions and not on the donor concentration or light path length. In the time domain, fluorescence decays are directly measured after exciting with a short pulse of light; the most common technique used is time-correlated single-photon counting (18). In the frequency domain, the sample is excited with a light wave whose intensity oscillates sinusoidally with a range of frequencies in the region of the reciprocal of the lifetime that is being measured. The intensity of fluorescence emitted will also vary sinusoidally with the same frequency, but with a different phase and amplitude, which may be used to calculate phase τ_φ and modulation τ_M lifetimes. However, both these methods require involved instrumentation (15–17).

Another implementation of the time-dependent method exploits the change in donor excited state lifetimes resulting in a reduction in the number of fluorophores in the excited state. This is a much simpler method and does not require complex instrumentation. The reduction in donor excited state populations in turn is expected to reduce the rate of photobleaching of the donor fluorophores in the presence of acceptor fluorophores. This method was developed by Jovin and colleagues (19–21) and is termed photobleaching FRET. It involves measuring the rate of donor photobleaching over a much slower timescale (milliseconds to seconds). However, extracting quantitative information about the energy transfer efficiency by this process is not straightforward, since photobleaching is a complex multi-exponential process (19–21). FRET may differentially affect some, but not all, of these processes, making the analyses complicated.

2.1.4. FRET measurements based on depolarization of acceptor fluorescence

In this chapter, we describe specific implementation of the homo-FRET microscopy method, taking into account the consequence of depolarization of acceptor fluorescence due to FRET (see (iv) in section 2.1). For this purpose, we first need to understand the term fluorescence polarization or anisotropy of emission.

The probability of absorption of photons depends on the angle between the transition dipole of the fluorophore and the polarization of incident light. Light emitted from such molecules can also be polarized. The plane of polarization of the emitted photons need not be the same as that of the incident photons and it depends on several factors.

The polarization of fluorescence or fluorescence emission anisotropy is measured by exciting the fluorophores by plane-polarized light and recording fluorescence intensities in the two perpendicular directions through polarizers aligned parallel and perpendicular with respect to exciting light. Fluorescence anisotropy is defined as:

$$r = \frac{I_{\parallel} - I_{\perp}}{I_{\parallel} + 2I_{\perp}} \qquad \text{(Equation 6)}$$

where $I_{\parallel}$ is the intensity observed in the parallel direction and $I_{\perp}$ is the intensity observed in the perpendicular direction (22). On the other hand, polarization (P) is defined as:

$$P = \frac{I_{\parallel} - I_{\perp}}{I_{\parallel} + I_{\perp}} \qquad \text{(Equation 7)}$$

where P and r are related to each other by the following equations:

$$P = \frac{3r}{2+r} \text{ and } r = \frac{2P}{3-P} \qquad \text{(Equation 8)}$$

Fluorescence anisotropy is the more convenient form for representing polarization of fluorescence, since it is an additive term and its denominator represents the total intensity. Hence, the fluorescence anisotropy of a solution can be considered as a sum of fluorescence anisotropies of single fluorophores.

At low concentrations of fluorophores, the rotational diffusion of the fluorophores dictates the polarization of emitted light. If the rotational diffusion times are less than the lifetime of the fluorophore, the fluorescence will be depolarized, but if the rotational diffusion times are larger than the lifetime of the fluorophore, the fluorescence is partially polarized (see *Fig. 1A*). Viscosity of a solution or medium governs the rotational diffusion times of fluorophores; hence, such measurements can be used to determine viscosities of media. The equation that governs the relationship between rotational diffusion and fluorescence anisotropy is given by the Perin equation:

$$\frac{1}{r}=\frac{1}{r_0}+\frac{\tau RT}{r_0\eta V} \quad \text{(Equation 9)}$$

where T is the temperature, V is the volume, η is the viscosity, τ is the rotational correlation time, and R is the gas constant (22). Here, r_0 is defined as the value of fluorescence anisotropy of a sample in the absence of any depolarizing factors such as rotational movement or energy transfer. Consider a sample that is frozen such that each transition dipole is not undergoing any rotational diffusion. If all the dipoles were aligned, then r_0 would solely depend on the angle between the excitation and emission dipole moments. Since in a real sample all dipoles need not be oriented, r_0 also depends on the distribution of transition dipoles in the sample relative to the polarization of exciting light.

The dependence of FRET on the orientation factor term leads to depolarization of fluorescence emission resulting from a FRET event. Hence, if one excites the donor in a system of fluorophores capable of undergoing FRET with plane-polarized light, then the fluorescence emitted from the acceptor molecule is depolarized (see *Fig. 1B*). This system of fluorophores could be two spectrally distinct fluorophores and hence may undergo hetero-FRET, or could be the same fluorophore, thus undergoing homo-FRET. Fluorescence anisotropy or polarization measurements thus allow us a means of monitoring homo-FRET events.

3. RECOMMENDED PROTOCOLS

3.1. Fluorophore choice for homo-FRET microscopy

Since fluorescence anisotropy is an intrinsic property of fluorescence emission, it is independent of the light path and other environmental parameters that affect fluorescence intensity measurements. A requirement of the homo-FRET method is that the donor fluorophore must have a nonzero value of anisotropy to begin with and that the neighboring 'acceptor' species must have a relatively random orientation and/or some rotational freedom to register sufficient depolarization of fluorescence emission (23). It should be noted that fluorescence emission anisotropy is also sensitive to the viscosity of the environment and the mass attached to the fluorescent probe (22), since these factors affect the rotational rates. Thus, the choice of fluorophore and the method for tagging to the specific protein of interest are important. It is important that the fluorophore is photostable in its environment. It should satisfy the criterion that there be one fluorophore per protein molecule. If there is more than one fluorophore per protein molecule, intramolecular energy transfer (intramolecular homo-FRET) occurs. This would lead to inherent depolarization and make it difficult to estimate energy transfer due to protein–protein interactions. Green fluorescent protein (GFP) has been used successfully for these measurements (24, 25). This is a good choice for genetically tagging proteins to a fluorescent molecule and satisfies all the criteria mentioned above. FRET between different spectral variants of GFP fluorophores provides a molecular scale in the range of 2–6 nm (see *Table 1*) (26).

Table 1. R_0 values for different fluorescent protein pairs

	Acceptor (enhanced GFP variant or DsRed)				
Donor	**Blue**	**Cyan**	**Green**	**Yellow**	**Red**
Blue	2.61 ± 0.05	3.77 ± 0.08	4.14 ± 0.08	3.82 ± 0.08	3.17 ± 0.06
Cyan	–	3.28 ± 0.07	4.82 ± 0.10	4.92 ± 0.10	4.17 ± 0.08
Green	–	1.93 ± 0.04	4.65 ± 0.09	5.64 ± 0.11	4.73 ± 0.09
Yellow	–	1.00 ± 0.02	3.25 ± 0.07	5.11 ± 0.10	4.94 ± 0.10
Red	–	1.40 ± 0.03	2.84 ± 0.06	3.14 ± 0.06	3.54 ± 0.07

R_0 values are given in nanometers. Uncertainties indicated for these quantities are estimated standard deviations of R_0. Reproduced from (26) with permission from Elsevier.

3.2. Experimental design of homo-FRET measurements

The homo-FRET technique can be used to address several biological questions. This technique can be used to determine the oligomerization state of proteins in living cells, as well as to determine quantitatively how the oligomerization state of one protein may be influenced by overexpression of another protein. Instrumentation required for steady-state anisotropy measurements can be easily implemented in a conventional microscope with the proper placement and alignment of excitation and emission polarizers (27) (see below). The homo-FRET method is particularly advantageous while probing organizations such as small clusters at membrane surfaces, in the cytoplasm, or in solution (23, 25, 27, 28). A wide variety of fluorophores are capable of undergoing homo-FRET, thereby allowing the measurement of homo-FRET with different Förster's radii suitable for uncovering distances in the 2–6 nm range (26). GFP and its spectral variants have recently been shown to be suitable probes for homo-FRET (24, 25), providing a useful tool to study the organization of many GFP-tagged proteins inside cells at FRET-scale resolution.

3.3. Modifications of wide-field fluorescence microscopy for anisotropy measurements

Anisotropy measurements of fluorescently labeled proteins or lipids in living cells can be performed on a fluorescence microscope equipped with a stable light source, excitation and emission shutters and filter wheels, and a cooled charge-coupled device (CCD) digital camera as a detector, with minor modifications. It is also important that the shutters, filter wheels, and camera are controlled electronically via a computer and acquisition software (for example, METAMORPH or IPLAB). Although the measurement is straightforward, it is important to understand how each component of the microscope affects the polarization of emitted light.

3.3.1. Alignment of polarizers

The first step in setting up a fluorescence microscope for anisotropy measurements is to align polarizers in filter sliders and fit them into the microscope as shown in *Fig. 2*. A set consisting of three polarizers can be obtained from Chroma Corporation. These polarizers have two white dots placed on them. These dots indicate the axis of the polarizers. One of these is placed in the excitation slider so that its axis is along the P-plane of the dichroic. It is important to note that dichroic films are anisotropic and can introduce rotation of the plane of polarization of light (29). Hence, the excitation polarizer should be aligned in such a way that its plane of polarization is the same as the P-plane of the dichroic. If the excitation polarizer is aligned along the S-plane of the dichroic, a depolarization of excited light is observed.

The next step is to align the emission polarizers parallel and perpendicular with respect to this excitation polarizer. The polarizers can be aligned using reflected light from the base of a coverslip as the polarization of reflected light is maintained. This is achieved by measuring the average intensity of reflected light in the perpendicular mode and slowly rotating the emission slider in such a way as to minimize the intensity in the perpendicular mode. The same procedure is repeated for the parallel mode and the polarizer is rotated so that maximum intensity is recorded. A polarization ratio $I_{\perp}/I_{\parallel}$ of 0.01 is easily achievable.

3.3.2. Illumination source

Mercury arc lamps are probably the most commonly used light sources for fluorescence microscopy because they are very bright and have a spectrum covering the entire UV-visible range. Unfortunately arc lamps are inherently unstable as light sources. It is difficult to achieve a light output variation of less than 1%; most commonly 2–3% variation is observed. These variations in light intensity are detrimental to anisotropy measurements because a 1% variation in light intensity can introduce an error of 0.01 in anisotropy units. Arc wandering is the most common cause of the light intensity variation and cannot be overcome using a stabilized power supply. Tungsten halogen lamps are best suited for such measurements because they have a light intensity variation better than 0.001% if a stabilized power supply is used (see *Fig. 3A*). The other option is to use stabilized

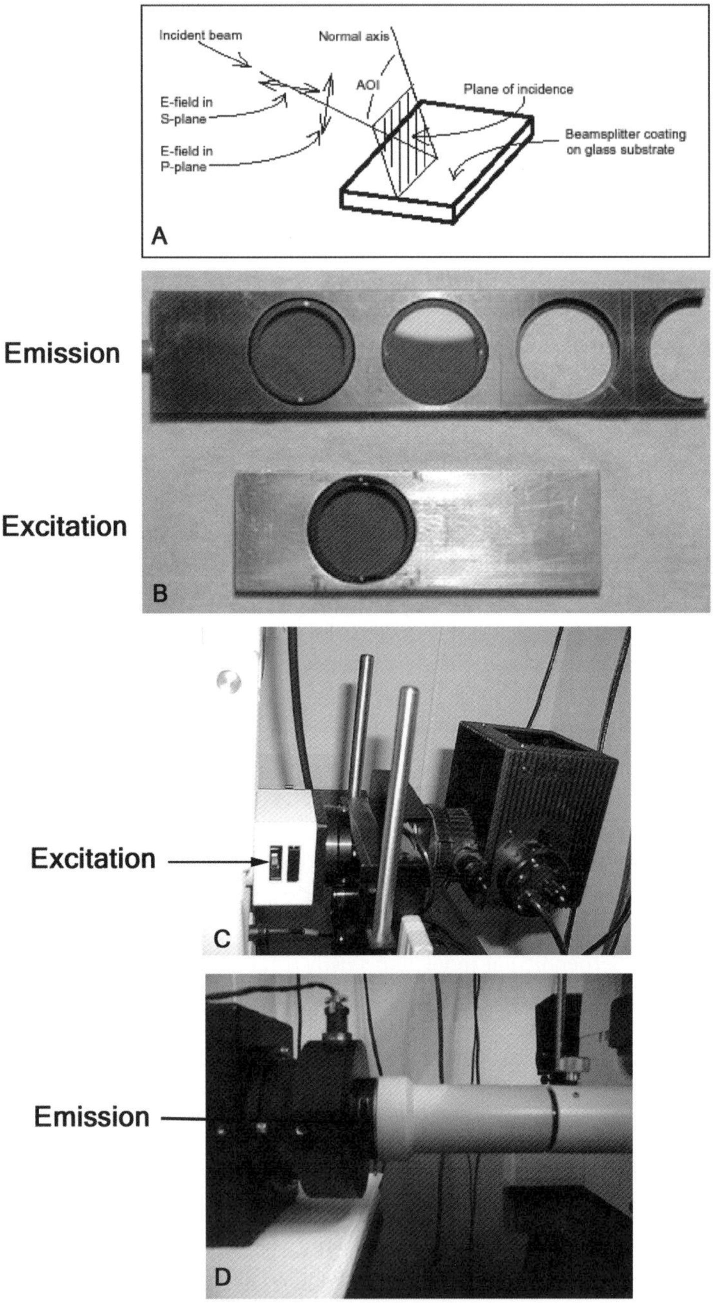

Figure 2. Setting up the microscope for anisotropy measurements.
(*A*) Schematic demonstrating the S- and P-planes of a dichroic. Dichroics are anisotropic thin films and have different transmittance properties in S- and P-planes. Excitation polarizers should be aligned so that their axis is the same as the P-plane of the dichroic. (*B*) Polarizers installed in excitation and emission sliders. The two white dots indicate the axes of the polarizers. (*C*) Position indicating where the excitation polarizer is installed in the microscope between the light source and the dichroic. (*D*) Position indicating where the emission polarizers are installed just before the CCD camera.

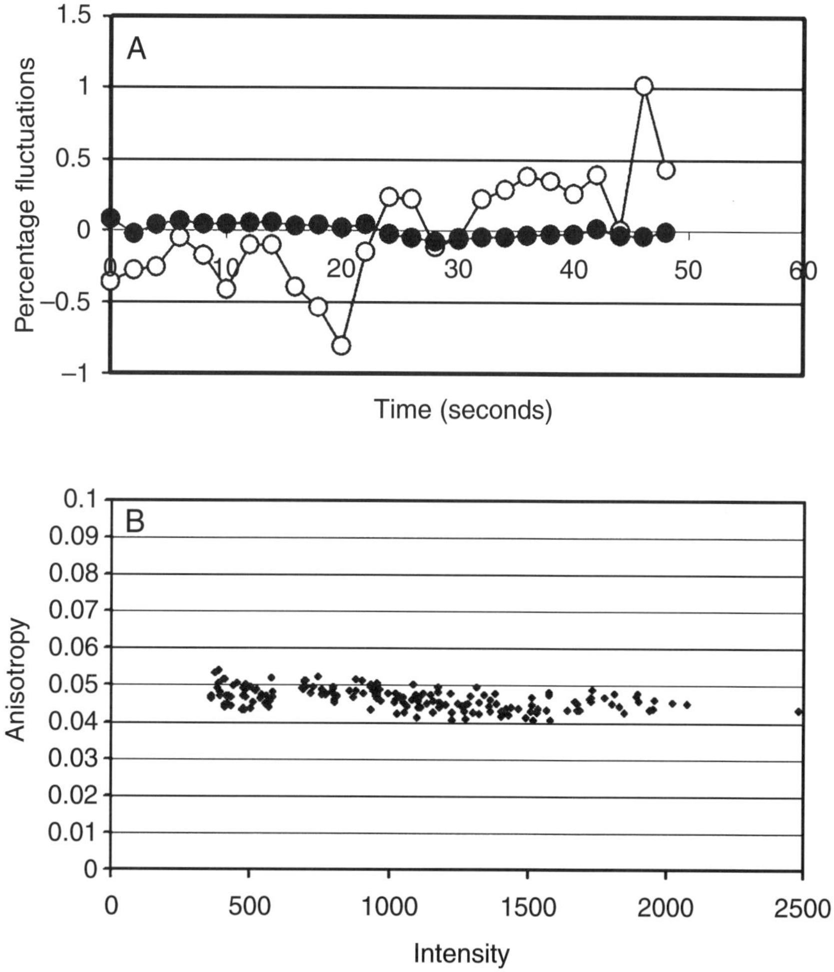

Figure 3. Light source stability.
A stable light source is necessary for anisotropy measurements. (*A*) Graph showing percentage fluctuation in intensity for a mercury arc lamp vs. a halogen lamp. Light reflected off the base of a coverslip was recorded using neutral-density filters. Exposure time was 1 s and images were collected every 2 s for 25 times. The average intensity of each field was determined. The difference between the individual field averages and the global average was divided by the global average and multiplied by 100 to obtain the percentage fluctuation. Halogen lamps (closed circles) are more stable than arc lamps (open circles) and hence are best suited for anisotropy measurements. (*B*) Scatter plot of anisotropy vs. intensity for rhodamine 6G in water. The anisotropy value is not G-corrected. The concentration independence of this value is a good indicator of the system working properly.

lasers with appropriate beam expanders. Any light source capable of giving light output stability better than 0.1% is suitable for anisotropy measurements.

3.3.3. Emission path and image collection

Images of cells are collected using a cooled CCD digital camera using emission polarizers aligned parallel and perpendicular with respect to the polarization of the exciting light. This alignment has already been described. Anisotropy measurements are prone to errors if the response of the detector is nonlinear. Hence, it is extremely important to characterize and understand the properties of the cooled CCD camera that one is using for measurements. After start-up, the detector and its dark currents should be allowed to stabilize before starting measurements. During the course of the measurement, the dark currents should remain constant. A small increase in the value of the dark current may be observed over several hours of operation. The linearity of the detector can be measured in the following way. Images of a dilute solution of rhodamine 6G or any other fluorophore that does not bleach are collected using a range of exposure times from 1 to 10 s. Appropriate backgrounds are subtracted from each of the images and a ratio of the background-subtracted intensity and the corresponding exposure time is determined. This ratio is plotted against the exposure time. This should give rise to a single ratio value for each exposure time. If the variation in this ratio is less than 1%, it indicates that the response of the detector is linear.

Another manifestation of the anisotropic property of dichroics is the introduction of a *g* value in the measurement due to different transmittivities along the P- and S-planes (29). The *g* value can be experimentally obtained by exciting the sample with depolarized light and taking the ratio of images collected using the two aligned emission polarizers. While collecting parallel and perpendicular images sequentially, photobleaching of the sample during collection of the first image should be avoided.

One can calibrate the system using anisotropy of rhodamine 6G in water at different concentrations. This will not only ensure that the response of the system is linear, but also that there are no other parameters influencing the anisotropy of a standard. *Fig. 3(B)* shows the anisotropy of rhodamine 6G in water at different concentrations.

3.4. Spatial resolution, pixel shifts, and precision

The resolution of a microscope objective can be understood only by using principles of wave optics and considering light emitting from a point source as a train of waves (30). A point source of light, like a fluorescent subresolution bead, when imaged by an objective generates a 3D diffraction pattern, which when sectioned in the 2D focal plane is known as the airy disc. The radius (*r*) of the airy disc is governed by the numerical aperture (NA) of the objective:

$$R = \frac{0.61\lambda}{\text{NA}} \qquad \text{(Equation 10)}$$

The airy disc for each objective can be imaged using subresolution fluorescent beads and the digital camera. The resulting intensity distribution of such a point source of light is known as the point spread function (PSF) for that objective. The full width at half maximum of the PSF gives the diameter of the airy disc and can be used as a measure of the resolution of the objective. The shape of the PSF also gives information regarding the quality of the imaging system. If the PSF is not radially symmetrical, this indicates aberrations in the imaging system. Chromatic aberrations in the imaging system are determined by obtaining the PSF for all the wavelengths. If the PSFs for all the wavelengths overlap and have the same shape, this indicates the absence of chromatic aberrations.

One lacuna in anisotropy measurements is the inability to make the measurements at high resolution. High-resolution and high-NA objectives (e.g. 60× 1.4 NA) depolarize the exciting light and thus the measurements become difficult. The cause for this depolarization has been discussed (31). From electromagnetic focusing theory, it is known that when focusing linear (*x*-polarized) light at high apertures, the focal field also exhibits significant perpendicular and longitudinal components. The depolarization is due to the curvature of the spherical wave front of the focused field and is predicted to be stronger at high apertures. Such data can be corrected, but a prior knowledge of the orientation of the transition dipoles must be obtained, which is difficult to obtain for fluorophores in biological samples. Such corrections are described in (31). Hence, we make all measurements with a 20× objective with 0.75 NA to avoid these depolarization effects.

Every imaging system is characterized by imaging subresolution fluorescent beads. We have found that the PSF corresponding to the parallel image may not necessarily overlap the PSF corresponding to the perpendicular image. This results in an inability to perform anisotropy measurements for point objects. One can overcome such errors by binning the images, so that both the PSFs with their distortions are accommodated in the same pixel, but this leads to a loss in resolution. *Fig. 4* shows images of beads representing PSFs under two conditions, unbinned (see *Fig. 4D–F*, also available in color section) and binned images (see *Fig. 4A–C*, also available in color section). The two PSFs do not overlap because of asymmetric distortion of either. These distortions are caused by one or many of the following factors – spherical aberrations in objectives and projection lenses, and/or refractive effects in emission polarizers. These effects give rise to errors in anisotropy measurements of point objects or small objects covering a few pixels, such as endosomes in cells.

Anisotropy measurements can be made for a given area or a set of pixels. The minimum area for which one obtains accurate measurement can be determined by doing repeated measurements on a single field of a slide containing a dilute solution of rhodamine in 70% glycerol. A significant fluctuation in the intensity is observed from pixel to pixel for a homogeneous sample such as rhodamine in glycerol. Since exposure times in these measurements are sufficiently long (in the order of seconds), these fluctuations are not inherent to the sample and hence probably arise solely due to the read noise of the CCD. Due to these fluctuations, it is not possible to measure the anisotropy of a single pixel accurately. Different

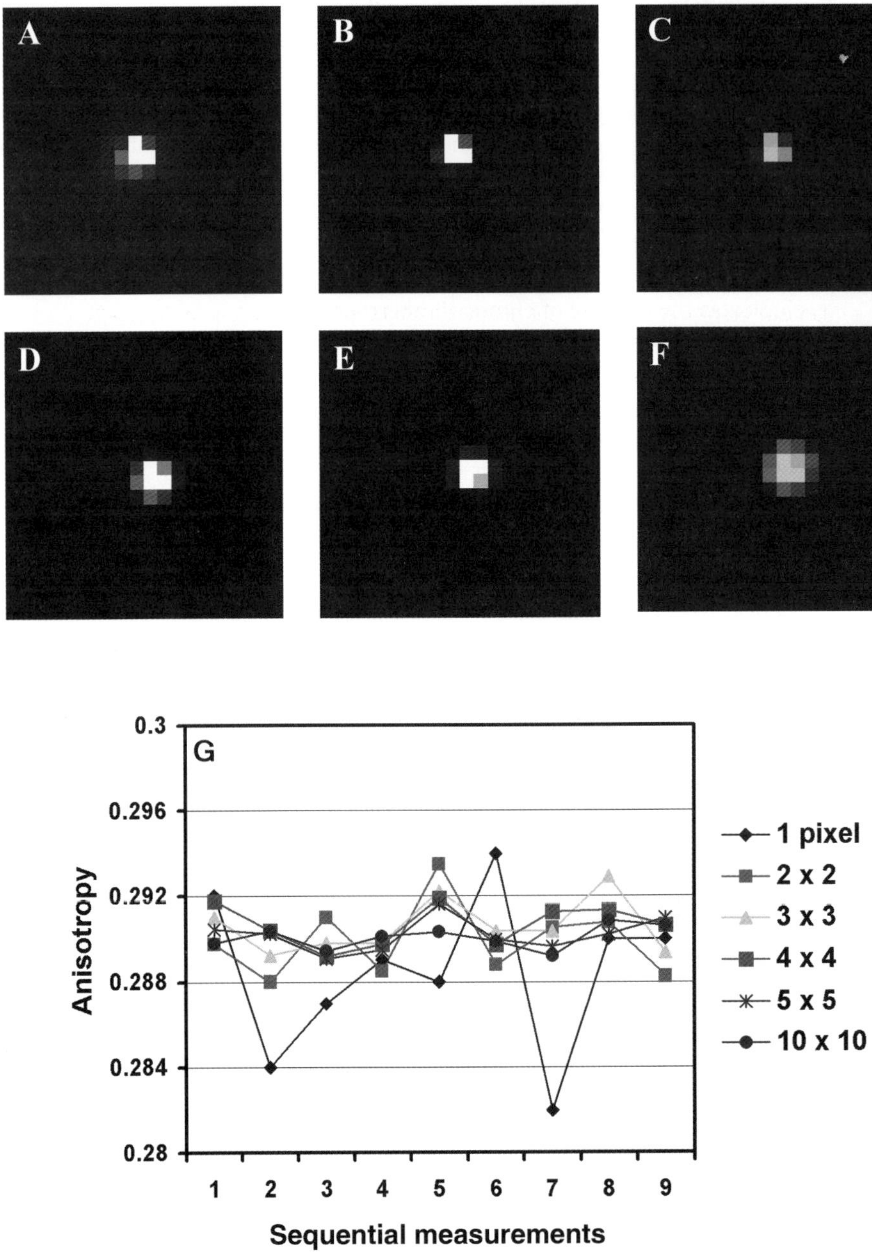

Figure 4. Spatial resolution in anisotropy measurements (see page xxvii for color version). (*A*)–(*C*) Images of subresolution beads under 3 × 3 binning showing the overlap of the PSFs collected in the parallel (*A*) and perpendicular (*B*) directions. (*D*)–(*F*) Images of subresolution beads taken at no binning showing the nonoverlap of the PSFs for parallel (*D*) and perpendicular (*E*) directions. (*C*) and (*F*) are fluorescence overlays of parallel and perpendicular directions by pseudo-coloring the parallel and perpendicular images red and green. (*G*) Effect on the measurement of anisotropy of averaging over a larger number of pixels. Fluctuations in anisotropy of rhodamine in glycerol for areas on the images ranging from 1 pixel to 100 pixels are shown. Anisotropy can be determined accurately for an area of 4 × 4 or larger.

pixel areas were tested for accuracy and it was found that accurate measurements were for pixel areas 4 × 4 onwards (see *Fig. 4G*, also available in color section). If the lamp intensity is stable, this depends solely on the properties of the detector system.

3.5. Analysis

Images collected through parallel and perpendicular filters are background subtracted. Background images are collected from 'blank' dishes containing the same medium and under the same illumination conditions as the experimental dishes. As anisotropy is an additive quantity, any fluorescent substance sticking to the base of the experimental dish will contribute to the data with its own anisotropy; therefore appropriate background subtraction is crucial for accurate determination of the anisotropy of a labeled sample. As an example, we encountered problems with determination of the anisotropy of GFP-tagged proteins expressed in Chinese hamster ovary cells due to nonspecific sticking of GFP on the coverslips contributing to a background that was uniform but had a different anisotropy to that of GFP on cells. We overcame this problem by growing our cells on noncoated coverslip dishes. As far as possible, the background should be reduced experimentally. Software methods such as median filtering do estimate nonuniform background, but have a drawback that they can subtract some of the signal as well and hence cannot be used for anisotropy measurements.

Once the images are background subtracted, anisotropy can be measured for a single pixel or an area. Two kinds of analysis may be used for different purposes – per pixel and per cell. For per cell analyses, average intensities of cells are measured in the parallel and perpendicular images by cutting out the regions that outline the cell boundaries. This is done using the METAMORPH software. Per pixel analysis is also done by logging pixel values via compatible software packages. Subsequent analysis is carried out by importing these values into EXCEL worksheets, where total intensity and anisotropy values are calculated using the intensities in parallel and perpendicular directions, using Equation 6.

3.6. Prototype measurements

We will describe two examples to illustrate the measurement and phenomenon of concentration-dependent depolarization of fluorescence.

In the first example, we took lissamine rhodamine in 70% glycerol at different concentrations, ranging from 0.5 μM to 12 mM, and measured its anisotropy by making a thin film of the solution between two coverslips without any spacers. *Fig. 5(A)* shows the dependence of fluorescence anisotropy of lissamine rhodamine in 70% glycerol as a function of concentration. This curve fitted the theoretically predicted curves using Förster's theory, as described in (32). The rationale behind using 70% glycerol was that mixing of the fluorophores at this viscosity is more easily achieved than at higher percentages of glycerol.

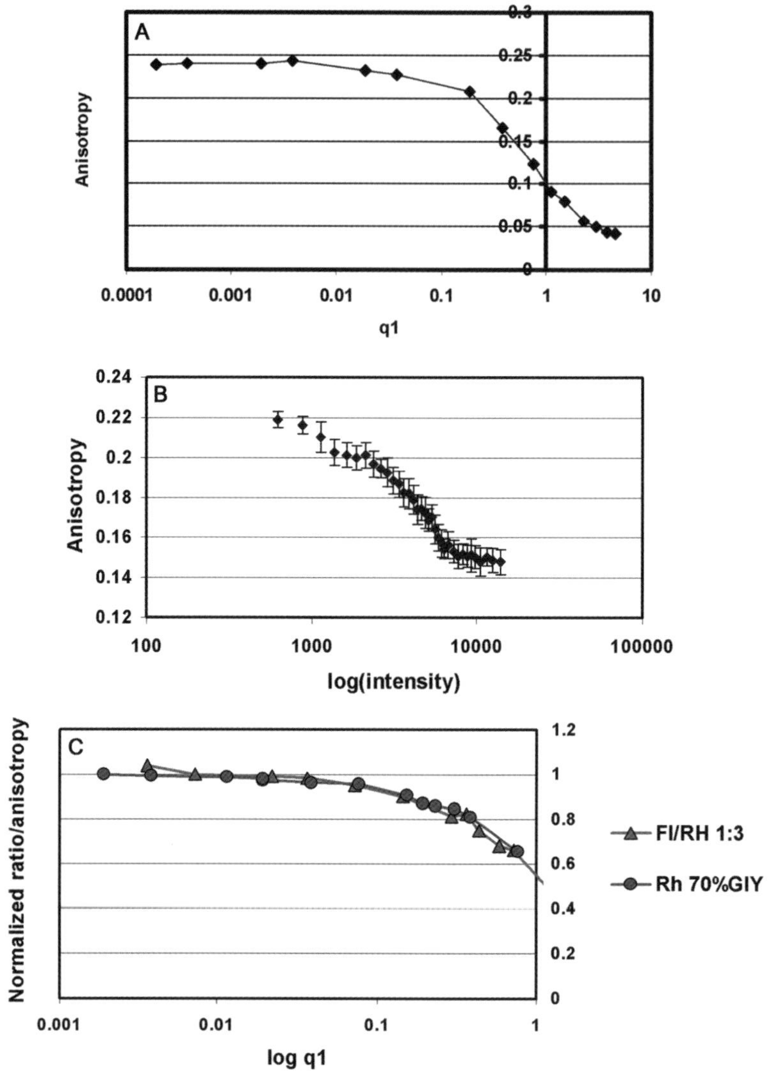

Figure 5. Prototype measurements.
(*A*) Anisotropy of lissamine rhodamine in 70% glycerol as a function of number of acceptors per unit volume defined by the Förster's radius. The data is well described by Förster's theory. (*B*) Anisotropy vs. intensity for C6-NBD-SM incorporated at the surface of mildly fixed Chinese hamster ovary cells. C6-NBD-SM was incorporated at different concentrations, the data were pooled based on total intensities, and the average anisotropy for different intensity ranges was obtained. (*C*) Comparison of the sensitivity of anisotropy measurements and ratio measurements for detection of FRET. For this experiment, we used a ratio of fluorescein-sodium, a water-soluble sodium salt of fluorescein, and lissamine rhodamine at a fixed ratio of donor to acceptor (1:3) and at different total concentrations in 70% glycerol. If the relative change in ratio of fluorescein to rhodamine in solution for the same change in total concentration is less or more than the relative change in anisotropy of rhodamine for the same change in total concentration, then one can make a statement regarding sensitivity of one technique with respect to another. We found that the relative change in anisotropy (circles) was the same as the relative change in the ratio (triangles) for the same change in total concentration.

Nonhomogeneous mixing gives rise to errors in anisotropy because of large changes in local concentrations.

The first example demonstrated fluorophores distributed in three dimensions. Next, we examined an analogous experiment in two dimensions – the surface of a cell. We incorporated fluorescent lipid, C6-NBD-sphingomyelin (C6-NBD-SM), on the surface of mildly fixed cells at different concentrations. C6-NBD-SM is a lipid probe that is expected to be distributed evenly on the surface of the cell. *Fig. 5*(*B*) shows the anisotropy of C6-NBD-SM as a function of total intensity. Data were collected over 12 dishes in which a range of concentrations from 1 to 5 μM was used. Finally, all data were pooled and local averages were taken in different intensity ranges, showing that anisotropy of a membrane-localized probe also changes upon increasing concentration. This may be explained by Förster's theory as described previously (see supplementary text in 24).

3.7. Analyzing co-localization at the nanometer scale

Homo-FRET offers an elegant way of analyzing co-localization between different proteins at the nanometer scale. Consider an experiment wherein the oligomerization state of protein A is being measured using homo-FRET. Protein B is proposed to interact with protein A and participate in oligomerization. At the nanometer scale, this would translate as an increase in the average distance among protein A species and hence reduced homo-FRET among them. Protein A and protein B can be tagged with fluorophores that are not FRET pairs (e.g. GFP and Cy5; see circles in *Fig. 6*, also available in color section), and hence protein B fluorescence can be used to estimate its concentration. The homo-FRET among protein A species can be compared with the ratio of expression levels between protein A and protein B. This relationship will allow us to determine the nature of co-localization or co-oligomerization at the nanometer scale. We have used this method to determine the co-clustering of multiple glycosylphosphatidylinositol (GPI)-anchored proteins (see *Fig. 5* in 24).

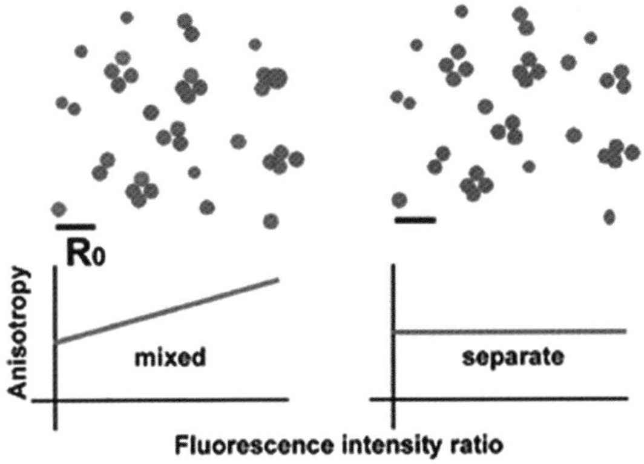

Figure 6. Co-localization at the nanometer scale using homo-FRET (see page xxviii for color version).
In the schematic, if two different proteins (red and green circles) occupy the same cluster (left panel), increasing expression of one protein (red circles) will lead to a decreasing number of homo-FRET events for the second protein (green circles). As a result, homo-FRET between the green species will decrease. Consequently, there will be an increase in emission anisotropy of the green protein species being monitored. Alternatively, if the different protein species are present in separate clusters (right panel), there will be no change in the anisotropy of the fluorescent species being monitored with increased expression of either of the proteins.

3.8. Quantitation of FRET efficiencies from the homo-FRET experiment

We have described applications where homo-FRET has been used to examine the nanometer-scale organization of proteins on cell surfaces. In most cases, a comparison of the data set is made with situations where there is no FRET. It would be convenient if we could use an expression for homo-FRET efficiency and carry out a quantitative comparison, as in other FRET techniques. The value of the measured anisotropy depends on the following factors: statistical distribution of the relative orientations of the fluorophore dipole moments with respect to the incident polarization and to each other, the rotational diffusion coefficient, and the relative separation between fluorophores. The anisotropy is highly sensitive to the relative orientation of the dipole moments, and thus even if the relative distance between fluorophores is slightly greater than R_0, there is an appreciable depolarization if the dipole moments of the two fluorophores are not parallel to one another. Under conditions where there is no energy transfer, it can be shown that the anisotropy (A) = 0.4 for donor dipoles distributed uniformly over a sphere (22). The homo-FRET efficiency expression based on measuring anisotropies r and r_0 of donor fluorescence in presence or absence of FRET conditions, respectively, is:

$$E = 1 - (r/r_0) \qquad \text{(Equation 11)}$$

It must be noted that the above equation is valid only under the simplest situations where the sole reason for the change in anisotropy may be attributable to nonradiative transfer to other donor species, where excitation after leaving the donor never returns to the same donor species, and where there is no change in the donor lifetimes (32).

3.9. Comparison of the sensitivities of homo- and hetero-FRET in bulk solution

We have so far described anisotropy measurements that can be used either to obtain information regarding rotational diffusion of a fluorophore or to monitor homo-FRET. In contrast, hetero-FRET is measured between two different fluorophores that have an overlap between the absorption and emission profiles of the acceptor and donor, respectively. The ratio of the fluorescence intensity of donor and acceptor is one such measure of conventional hetero-FRET. To compare differences in sensitivities of homo- and hetero-FRET methods, the relative changes in ratio vs. anisotropy for the same change in the number of acceptors per unit volume in solutions may be monitored. This was done by comparing the relative change in anisotropy of lissamine rhodamine in 70% glycerol to the relative change in ratio of fluorescein-sodium and lissamine rhodamine in 70% glycerol as a function of change in concentration. *Fig.* 5(*C*) shows normalized changes in ratio or anisotropy for the same change in total concentration, indicating that hetero-FRET measurements are as sensitive as homo-FRET measurements when detecting FRET events in bulk solution.

3.10. Comparison between homo-FRET and hetero-FRET measurements for small clusters

Although homo- and hetero-FRET techniques have equal sensitivity in the limit of bulk solution measurements, certain arrangements of molecules are only detected by homo-FRET methods. For example, if there is a small fraction of clusters in a sea of monomers, homo-FRET provides the only solution (24). We have extensively used the homo-FRET technique to examine the 'raft' association of GPI-anchored proteins at the surface of living cells. In this particular instance, only homo-FRET experiments provided measurement of interactions between GPI-anchored proteins, consistent with submicron-scale clustering of these proteins. Hetero-FRET experiments failed to detect significant clustering of these molecules. The detection of homo-FRET (24, 27), but not hetero-FRET (33–35), between GPI-anchored proteins requires a consistent explanation. Theoretical models based on a probabilistic approach towards calculating the extent of hetero-FRET observable from varying fractions of small clusters of molecules that ranged in size from two to seven molecules per cluster have provided clues to this discrepancy. Hetero-FRET efficiencies expected from some of these models are shown in the curves in *Fig. 3* in (24). Consistent with the size and fraction of clusters obtained from homo-FRET methods, theoretical models to predict hetero-FRET efficiencies have also shown that at the low fraction of clusters in the membrane and at the scale of the clusters (maximum of two to four species per cluster), it would be unrealistic to expect significant hetero-FRET above background fluctuations in FRET signals at low levels of protein expression in the membrane. Homo-FRET measurements obviously will report on all events of homo-oligomerization, thus providing a higher signal-to-noise ratio in some instances. This analysis underlines the value of performing both types of FRET experiments when the molecular organization to be probed is at the small-oligomer scale and at a low proportion.

4. FURTHER IMPROVEMENTS IN ANISOTROPY MEASUREMENTS

So far, all descriptions in this chapter have been limited to implementation of anisotropy measurements on a wide-field microscope to determine steady-state anisotropy. The measurements described have involved collecting two images sequentially. This leads to a loss in time resolution (4–5 s for each composite measurement), potentially missing events that are dynamic in the timescale of the measurement. It is possible to custom build a microscope system consisting of two identical cameras, so that parallel and perpendicular images can be collected simultaneously using a cube-type polarizing beam splitter to split the images into parallel and perpendicular components. It would be extremely useful if the measurements could be made on a confocal microscope, which would not only provide simultaneous acquisition, but also increased *z*-resolution. This type of application has now been implemented in Zeiss confocal microscopes with some degree of success (unpublished observations). These advances in technologies are likely to provide us with dynamic information regarding oligomeric organization of molecules in living cells.

FRET efficiency measured from steady-state experiments is a combination of two components – a configurational term regarding molecules undergoing FRET and the fraction of species undergoing FRET. For example, the measurement of FRET efficiency by steady-state methods will not be able to distinguish between the two steady-state situations where 50% of species are undergoing FRET at 100% transfer efficiency for the FRET process or 100% of molecules are undergoing FRET at 50% efficiency. Anisotropy measurements in the time domain can deconvolve this information and provide information about the contribution from rotational modes.

Time-domain measurements involve measuring the rate of anisotropy decays; measuring the rate of decay of fluorescence anisotropy directly measures FRET efficiency. This is related to average distances between fluorescent species by the following equation:

$$\omega = \frac{3}{2}\kappa^2 \left(\frac{R_0}{R}\right)^6 \tau_F^{-1} \qquad \text{(Equation 12)}$$

where the anisotropy decay rate due to homo-FRET

$$\tau_{r1} = \frac{1}{2\omega} \qquad \text{(Equation 13)}$$

and τ_F is average fluorescence lifetime and $\kappa^2 = 2/3$. In case there are multiple anisotropy decay components, the amplitude of the decay component due to FRET also indicates the fraction of molecules undergoing FRET (25).

However, measurement of anisotropy decays at the picosecond timescale requires expensive instrumentation; it is implemented with the use of a pulsed light source, flash lamp or laser (for single-photon or multi-photon excitation), and detection devices adapted either for time-correlated single-photon counting, or for frequency and phase modulation detection. The implementation of these techniques requires a separate treatment and will not be discussed in further detail here.

5. SUMMARY

In summary, we have described how to implement steady-state anisotropy measurements on a wide-field microscope. Keeping all the parameters discussed, in principle it should be possible to implement these measurements on scanning or spinning disc confocal systems. With the implementation of two detector-based fluorescence emission polarization measurements, time resolution may be enhanced several fold. High spatial resolution is still a limitation of this technique due to intrinsic problems of image registration. Homo-FRET measurement is the method of choice for certain types of molecular configurations where there is a small fraction of homo-oligomers in the presence of a larger fraction of monomers, a situation often encountered in biological processes.

6. REFERENCES

★ 1. **Förster T** (1948) *Ann. Phys.* **2**, 55–75. – *First quantitative description of FRET.*
2. **Stryer L** (1978) *Annu. Rev. Biochem.* **47**, 819–846.
3. **Dale RE, Eisinger J & Blumberg WE** (1979) *Biophys. J.* **26**, 161–193.
4. **Wu P & Brand L** (1992) *Biochemistry*, **31**, 7939–7947.
★★ 5. **Gordon GW, Berry G, Liang XH, Levine B & Herman B** (1998) *Biophys. J.* **74**, 2702-2713. – *A useful review on basic FRET measurements by conventional means.*
★★ 6. **Jovin TM & Arndt-Jovin DJ** (1989) In: *Cell Structure and Function by Microspectrofluorometry*, pp. 99–117. Edited by E Kohen & JG Hirschberg. Academic Press, San Diego, CA. – *A good book reviewing conventional basic FRET measurements.*
★★ 7. **Kenworthy AK & Edidin M** (1999) In: *Protein Lipidation Protocols*, pp. 37–49. Edited by MH Gelb. Humana Press, Totowa, NJ. – *Another useful review of basic FRET measurements.*
8. **Jovin TM & Arndt-Jovin DJ** (1989) *Annu. Rev. Biophys. Biophys. Chem.* **18**, 271–308.
9. **Bright GR, Fisher GW, Rogowska J & Taylor DL** (1989) *Methods Cell Biol.* **30**, 157–192.
10. **Dunn KW, Mayor S, Myers JN & Maxfield FR** (1994) *FASEB J.* **8**, 573–782.
11. **Bastiaens PI, Majoul IV, Verveer PJ, Soling HD & Jovin TM** (1996) *EMBO J.* **15**, 4246–4253.
12. **Herman B** (1989) *Methods Cell Biol.* **30**, 219–243.
13. **Billinton N & Knight AW** (2001) *Anal. Biochem.* **291**, 175–197.
14. **Nagy P, Vamosi G, Bodnar A, Lockett SJ & Szollosi J** (1998) *Eur. Biophys. J.* **27**, 377–389.
15. **Oida T, Sako Y & Kusumi A** (1993) *Biophys. J.* **64**, 676–685.
★★ 16. **Bastiaens PI & Squire A** (1999) *Trends Cell Biol.* **9**, 48–52. – *A good introduction to FLIM measurements and how FRET can be measured using FLIM.*
17. **Wallrabe H & Periasamy A** (2005) *Curr. Opin. Biotechnol.* **16**, 19–27.
★★ 18. **Srivastava A & Krishnamoorthy G** (1997) *Arch. Biochem. Biophys.* **340**, 159–167. – *A comprehensive description of time-resolved anisotropy measurements.*
19. **Kubitscheck U, Kircheis M, Schweitzer-Stenner R, Dreybrodt W, Jovin TM & Pecht I** (1991) *Biophys. J.* **60**, 307–318.
20. **Kubitscheck U, Schweitzer-Stenner R, Arndt-Jovin DJ, Jovin TM & Pecht I** (1993) *Biophys. J.* **64**, 110–120.
21. **Jurgens L, Arndt-Jovin D, Pecht I & Jovin TM** (1996) *Eur. J. Immunol.* **26**, 84–91.
22. **Lakowicz JR** (1999) *Principles of Fluorescence Spectroscopy*, 2nd edn. Plenum Press, New York.
23. **Runnels LW & Scarlata SF** (1995) *Biophys. J.* **69**, 1569–1583.
★★★ 24. **Sharma P, Varma R, Sarasij RC, *et al.*** (2004) *Cell*, **116**, 577–589. – *This paper uses a variety of homo- and hetero-FRET techniques in combination with theoretical modeling to understand the nanoscale organization of GPI-anchored proteins at the surface of living cells. A good reference to look at applications of FRET techniques is described in this paper. The supplementary text is a good starting point for theoretical aspects of FRET and its application for obtaining structural information about small clusters.*
★★★ 25. **Gautier I, Tramier M, Durieux C, *et al.*** (2001) *Biophys. J.* **80**, 3000–3008. – *This paper uses time-resolved anisotropy measurements to monitor homo-FRET between GFP-tagged proteins in the cytoplasm. Homo-FRET is used to understand the monomer–dimer transition of the GFP-tagged proteins.*
26. **Patterson GH, Piston DW & Barisas BG** (2000) *Anal. Biochem.* **284**, 438–440.
27. **Varma R & Mayor S** (1998) *Nature*, **394**, 798–801.
28. **Rocheleau JV, Edidin M & Piston DW** (2003) *Biophys. J.* **84**, 4078–4086.
29. **Reichman J** (2000) *Handbook of Optical Filters for Fluorescence Microscopy.* Chroma Technology Corporation, Brattleboro, VT.
30. **Inoue S** (1996) *Video Microscopy.* Plenum Press, New York.

★ 31. **Axelrod D** (1990) *Methods Cell Biol.* **30**, 333–352. – *This is an important reference that describes the implementation of anisotropy measurements on the microscope. Factors that affect anisotropy of fluorescence are discussed in detail and it is a very useful review.*

★★★ 32. **Agranovich VM & Galanin MD** (1982) *Electronic Excitation Energy Transfer in Condensed Matter.* North-Holland Publishing, Amsterdam. – *The first two chapters of this book discuss the theoretical aspects of FRET and fluorescence anisotropy.*

33. **Kenworthy AK & Edidin M** (1998) *J. Cell Biol.* **142**, 69–84.

34. **Kenworthy AK, Petranova N & Edidin M** (2000) *Mol. Biol. Cell,* **11**, 1645–1655.

35. **Glebov OO & Nichols BJ** (2004) *Nat. Cell Biol.* **6**, 238–243.

CHAPTER 13

High-content and high-throughput screening

Elizabeth P. Roquemore

1. INTRODUCTION

Fluorescence microscopy has long been recognized as a powerful tool for studying the physical and biochemical functions of cells. Detailed examination of cells at the microscopic level provides a means of probing complex signaling mechanisms, elucidating protein function, identifying drug candidates, and characterizing compound effects. The automation of fluorescence microscopy and image analysis, coupled with advances in optical probe technologies, has made high-throughput, information-rich cell assays more accessible for academic and pharmaceutical screening environments. Because microscopy has the potential to yield significantly more information than other detection methods, the term 'high content' has been adopted primarily to describe screens and assays that are detected and quantified using automated imaging platforms.

1.1. High-content, high-throughput screening for pre-clinical drug discovery

Increasingly, high-content assays are being applied throughout the pre-clinical drug discovery process, from target selection through to lead optimization (see *Fig. 1*). At the level of target selection and validation, a growing application area for high-content, cell-based assays is in phenotypic functional screening (1–8). Libraries of small molecules, small interfering RNAs (siRNAs), or cDNAs are screened to elucidate gene function using living cells. With the help of automated image analysis routines, a large number of cellular features can be assessed to monitor the effects of test agents on cell phenotype.

When used in high-throughput screens for lead identification, high-content assays not only provide a relevant biological context for assessing the effect of compounds on the target of interest, but can also simultaneously yield valuable

Cell Imaging: *Methods Express* (D. Stephens, ed.)

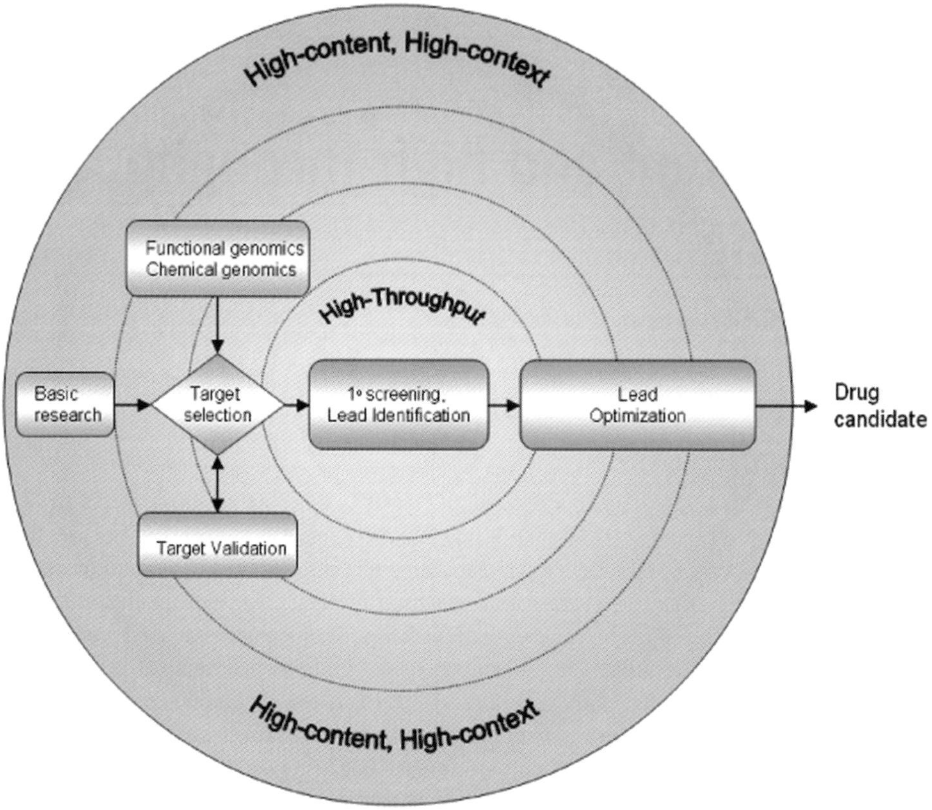

Figure 1. Application of high-content, high-throughput assays in pre-clinical drug discovery. A schematic diagram of the pre-clinical drug discovery process is superimposed on the imaginary space encompassed by high-content and high-throughput cell-based assays. Primary screening occupies the region of highest throughput, while activities upstream and downstream of lead identification extend into regions involving higher-content analysis.

information about compound availability and toxicity (9). Once lead compounds have been identified, high-content assays can be an effective means of characterizing activity, specificity, mechanism of action, potency, and off-target effects (9, 10).

High-content, cell-based assays are a natural complement to more-rapid screening assays. Pairing of relatively low-content primary screens with a panel of high-content secondary assays is becoming more common (8, 10). For example, Lundholt and colleagues (10) recently used a high-throughput fluorescence imaging plate reader to perform a primary cell-based screen for compounds that inhibit the Akt1 pro-survival pathway. They then used a confocal imaging system (IN Cell Analyzer 3000; GE Healthcare) to conduct a series of higher-content secondary assays to characterize the hits. With the microimaging platform, they were able to confirm hit compound activity by directly visualizing translocation of the target (Akt1) and then identify the mechanism of action of novel compounds by examining their effects on different components of the signaling pathway.

Still further downstream in the drug discovery pathway, high-content assays are being used to perform early toxicology studies on lead compounds. Compounds can be profiled against a range of general toxicity assays such as those for cell viability, proliferation, and mitotic index (11, 12). Genotoxicity testing is also facilitated by automated analysis of micronuclei formation (13, 14).

1.1.1. Automated fluorescence microscopy for high-content screening

Before the advent of commercial automated fluorescence microscopy systems, mainstream high-throughput, cell-based screening assays primarily employed macroscopic detection systems such as plate readers and flow cytometers to monitor gross changes in intensity of optical probes in whole cells or cell lysates. Such assays, while enabling relatively high throughput, tend to provide less information than can be gained by analyzing the cells on a microscopic level. Flow cytometry systems are sometimes described as high-content platforms (15), in that they can detect multiple probes associated with individual cells. Subpopulations of cells can be distinguished and characterized. Nevertheless, flow cytometers generally detect only gross changes in fluorescence intensity. In contrast to flow cytometers, most plate readers do not resolve varying response levels within heterogeneous cell populations, reporting instead population-averaged signals. While some macroimaging plate readers are capable of detecting individual cells, subcellular features are not resolved. With relatively few exceptions (see, for example, 10 and 16), spatial information from the probe is lost, making it difficult or impossible to assess subcellular localization, an important indicator of protein function. As a result, the pool of targets accessible through cell-based screens has generally been limited to a subset whose activity can be quantified by intensity measurements (e.g. cell-surface receptors, secreted proteins/peptides, transporters, ion channels, and enzymes) – to the exclusion of a large number of potential targets that exert their effects through spatial and morphological changes.

True high-content, cell-based screening assays are enabled by combining automated fluorescence microscopy, digital image capture, and automated image analysis. Such systems enable maximum information to be derived from high-resolution images at a speed that is amenable to the screening environment. Integrated analysis software enables automated quantification of the intensity and spatial and temporal characteristics of fluorescent probes. Characterization of compound effects can be based on analysis of multiple cell features and parameters. For example, in addition to measuring subcellular distribution of a reporter, high-content systems can simultaneously capture information about cell shape, organelle integrity, cell number, and overall fluorescence intensity of subcellular compartments, all of which can be used to score and filter individual cells and/or wells for compound toxicity. Identification and analysis of each individual cell in an image allows characterization of differential behaviors within heterogeneous cell populations. Some analysis packages allow implementation of classification methods that can be used to automate assignment of cells to defined categories (e.g. active/inactive/nonexpressing) based on a large number of parameters rather than a single indicator.

Microimaging systems also allow multiplexing of two or more assays in a single cell or well, enabling the behavior and interactions of multiple targets or pathway components to be assessed in the same environmental context. For example, within the same cell, behavior of the cell-cycle regulator Cdc25B could be correlated with a second marker for cell-cycle phase, such as cyclin B1, which starts to accumulate in late S phase (17). The versatility of automated fluorescence microscopy systems lends itself well to the increasing emphasis, in pre-clinical drug discovery, on gaining a breadth of physiologically relevant information from a wide range of cell types, assays, and probes. *Table 1* shows examples of the types of assay applications that can be configured for high-content, high-throughput analysis.

Table 1. Types of application suitable for high-content, high-throughput analysis

Assay application	What is quantified	Example references
Receptor function		
Ligand binding and receptor internalization	Ligand or other entities associated with plasma membrane receptors. Translocation of receptors from plasma membrane to pits/endosomes (probe associated with ligand or receptor)	23–25, 33, 36
Receptor activation/ desensitization/signaling	Translocation of reporter to sites of phosphorylated receptor (e.g. Transfluor), calcium flux, recruitment of signaling molecules (e.g. phospholipase C)	19, 12
Hormone receptors	Nuclear translocations	26
Signal transduction, pathway dissection		
Gene expression	Gene expression reporters – usually enzyme-based	12
Protein expression and protein–protein interactions	GFP fusion protein intensity, specific antigens, enzyme substrates, protein–protein interactions (via protein complementation, fluorescence resonance energy transfer, etc.)	30–32
Intracellular signaling	Redistribution of signaling molecules and specific probes to a variety of locations: plasma membrane, nuclei, endosomes, Golgi, mitochondria, membrane ruffles, cytoplasm, endoplasmic reticulum, cytoskeletal elements, etc.	9, 10, 28, 29
Gap junction function	Markers of gap junction communication, intercellular translocations	38
Cytotoxicity		
Cell proliferation	Cell count, BrdU incorporation, specific antigens (Ki-67, pRb)	11, 12
Cell viability	Esterase activity, cell permeability	12
Micronuclei formation	Micronuclei count (particularly that associated with binucleate cells)	13, 14
Cell-cycle disruption	Trafficking and expression of phase-specific reporters and antigens, DNA content, BrdU incorporation	7, 8, 3, 6

Assay application	What is quantified	Example references
Apoptosis and necrosis	Cell morphology (shrinking, swelling, blebbing), DNA condensation and fragmentation, Annexin binding, protease activities, mitochondrial integrity and function, etc.	34
Mitotic index	DNA content, cell count, phospho-histone H3	6
Cell processes		
Angiogenesis	Vessel formation and characteristics	39
Wound healing	Wound shape analysis	16
Neurite outgrowth	Neurite formation and characteristics	35
Phagocytosis, pinocytosis	Uptake of particles or probes	40
Cell adhesion and migration	Cell spreading, cell count, *x*-*y* migration, transwell migration	3, 16, 37
Cell-cycle progression	Trafficking and expression of phase-specific reporters and antigens, DNA content, BrdU incorporation	7, 8, 3, 6
Secretory trafficking	Trafficking of markers such as temperature-sensitive VSV–G variant	3, 27
Intracellular morphology changes	Changes in morphology and staining of the nucleus, endoplasmic reticulum, Golgi apparatus, endosomes, cytoskeleton	4

1.1.2. Components and features of high-content screening platforms

High-content screening systems are typically composed of an automated high-resolution fluorescence microscope, one or more high-speed cameras (usually charge-coupled devices (CCDs)), robotic plate- and sample-handling capability, and integrated software for automated image processing and analysis. Additional features important for live-cell screening include liquid handling systems for on-line addition of reagents, rapid autofocus and image acquisition capability, programmable scheduling software to allow kinetic data capture, and an environmentally controlled read chamber.

Throughput is a prime consideration when choosing a high-content platform for screening. For practical purposes, throughput can be defined as the number of samples that can be imaged *and analyzed* per unit of time. Throughput rates cited for commercial systems vary considerably, ranging anywhere from 5000 to 100 000 samples or wells per day. There is often not enough information available to assess the validity of some claims. For example, what is considered a 'day' in terms of screening may not be specified. In many cases, reported throughput rates do not include analysis time, which can vary considerably depending on the software and analysis method. The maximum throughput achievable for a given assay will depend not only on the mechanical speed of the system components, but also on factors that are specific to the assay, such as scheduling of liquid additions and kinetic time points, exposure times, the amount of resolution needed, the number of cells required to be imaged per treatment condition, and the analysis requirements.

Both confocal and nonconfocal systems are available for high-content screening. Confocality may be attractive for applications requiring analysis of probe co-localization, or accurate optical sectioning for 3D reconstruction. Examples of confocal imaging systems suitable for such applications are Opera (Evotec) and Pathway HT (BD Biosciences), both of which employ Nipkow spinning disk technology to achieve high-speed confocality. There is a widespread assumption that confocal systems result in better assay results than nonconfocal systems. While this is sometimes true, it may not always be the case, particularly when working with relatively dim samples. Because most confocal systems necessarily reject signal above and below the focal plane, fewer objects may be detected and the overall intensity of detected objects may be decreased. This can lead to decreased assay signal-to-noise ratios. If the confocal optical section is relatively thick and the excitation source is powerful, a strong potential advantage of confocality is the ability to reduce background signal emanating from such sources as unbound probe, floating cell debris, assay medium, and plastic culture dishes. The IN Cell Analyzer 3000 (GE Healthcare) is an example of a confocal system that is optimized for maximum background reduction with minimal loss of signal from the sample. Confocal systems can use either lasers or white light for illuminating the sample. Lasers may be preferable for exciting dim samples. However, the number of lasers that can practically be incorporated into commercial systems is limited, thus limiting the range of fluorescent probes that can be detected. By contrast, white light-based confocal systems allow excitation across the spectrum, limited only by the available filters and dichroics. Confocal systems tend to be more costly than their nonconfocal counterparts, particularly if a UV laser is incorporated. Hybrid systems – in which confocality is an optional mode or upgrade – offer the most versatility, but generally resolution and/or sensitivity are compromised.

Other factors to consider when choosing a microimaging system include number of cameras, accuracy and range of analysis solutions, environmental control options, and automation capability (18). The relative importance of various system features will depend on the volume of samples that needs to be processed, the range and type of potential applications, and the available budget. Features and benefits of some of the leading automated high-content systems are described in *Table 2*.

1.1.3. Image analysis software

Automation of image analysis has increased the speed at which information can be extracted from image data and has also minimized bias due to human subjectivity. The amount and quality of information obtained is dependent not only on image quality, but also on the power of the analysis software to distinguish features of interest from the surrounding background and to report meaningful measurements. Most of the leading high-content systems now offer a

Table 2. Leading automated high-content fluorescence microscopy systems

Nonconfocal systems		
Supplier	**System**	**Features/benefits**
Fisher Scientific (formerly Cellomics)	ArrayScan Certified, VTI, and KineticScan HCS readers	A range of readers employing Zeiss epifluorescence microscope with UV and mercury/xenon illumination sources and a single CCD camera. The kinetic reader has on-board liquid handling, scheduling software, and environmental control. Upgradeable with Zeiss ApoTome grid projection hardware and software for optical sectioning capability. Well-rounded package of image analysis tools and compatible reagent kits for high-content screening. Cellomics has set the industry standard with a high-content informatics solution for integrated data storage, mining, and analysis.
GE Healthcare	IN Cell Analyzer 1000 system	A modular bench-top system employing a Nikon epifluorescence microscope with a xenon arc lamp illumination source and a single CCD camera. Phase contrast and differential interference contrast options for transmitted light. Z-stack acquisition and analysis. Accepts slides as well as microtiter plates of well densities from six to 1536. On-board liquid dispensing and temperature control optional. Open architecture for robotic integration. Versatile suite of image analysis tools with supervised training facility for developing automated multi-parameter cell classification routines with learning capability. Fully compatible with IN CELL DEVELOPER TOOLBOX software for creating new analysis routines without knowledge of programming or scripting languages. Gateway software available from Cellomics to import images into CELLOMICS STORE and VHCS DISCOVERY TOOLBOX. Open architecture for robotic integration.
Molecular Devices	Discovery-1	A compact system with a mercury arc lamp illumination source and single CCD camera. Optional transmitted light source for bright-field imaging. Generic robotics interface. METAXPRESS image analysis application modules based on METAMORPH analysis software.
Molecular Devices	ImageXpress 5000A	Originally developed by Axon Instruments (now a part of Molecular Devices), ImageXpress employs Nikon optics and is designed for higher throughput than the Discovery-1. Upgradeable for live-cell imaging with full environmental control and optional single-channel liquid handling. Image analysis toolbox has a range of analysis tools, including automated object classification capability. Open software allows access to the scripting interface for both acquisition and analysis. A wizard facility allows scripting of analysis protocols. Integrated database for image storage, retrieval, and organization.
BD Biosciences (formerly Attobioscience)	Pathway HT	Nipkow spinning disk confocality with dual mercury arc lamp excitation and ability to switch between confocal and nonconfocal imaging modes. White-light LED for transmission microscopy. Environmentally controlled chamber and single-channel pipetting for on-line liquid addition. Analysis software for on-line object recognition and basic intensity measurements. For off-line analysis, some pre-set analysis protocols are available, as well as the capability to create automated classification protocols.

Supplier	System	Features/benefits
GE Healthcare	IN Cell Analyzer 3000	Laser-based confocal system purpose-built for high-resolution, high-speed screening. Employs proprietary line-scanning technology for sensitive, high-throughput confocality with three parallel excitation and detection channels and a range of filters for each camera. Line-scanning methodology leads to far greater signal capture and image contrast than other confocal methods, resulting in significantly increased sensitivity and higher assay signal-to-noise ratios. Dual-channel liquid addition and full environmental control are standard. Open architecture for robotic integration. Versatile suite of rapid, on-line analysis modules, most of which analyze images in less time than it takes to move from one well to the next (200 ms). Images can be imported by IN CELL DEVELOPER TOOLBOX. Gateway software available from Cellomics to import images into CELLOMICS STORE and VHCS DISCOVERY TOOLBOX.
Evotec	Opera	Laser-based confocal system employing Nipkow spinning disk technology for high-speed confocality with up to four parallel detection channels and six-position filter wheels for each CCD camera. Basic image analysis capability can be complemented with Acapella software to script analysis routines that can be run on line or off line. Open hardware for robotic integration.

range of pre-developed analysis routines that have been validated against specific applications and are integrated into the system for batch analysis of image stacks from one or more sample plates. These solutions can vary significantly in speed, flexibility, batch processing ability, and interactive options. Systems also vary as to whether they enable analysis to occur in parallel with image acquisition ('on line'), or must be run in sequence ('off line'), after the entire acquisition run has been completed.

In instances when pre-developed modules are not available for a particular application, a customized analysis solution may need to be developed. Some systems (e.g. ImageXpress 5000A, Molecular Devices; Pathway HT, BD Biosciences; Opera, Evotec) allow users to script and integrate their own analysis routines. Others (e.g. Discovery-1, Molecular Devices) have integrated third-party software for the same purpose, or have facilitated export of images in generic format (e.g. IN Cell Analyzer 1000, GE Healthcare; ArrayScan, Cellomics) so that they can be analyzed with third-party solutions such as METAMORPH (Universal Imaging Corporation), IMAGE-PRO PLUS (Media Cybernetics), and IMAGEJ (http://rsb.info.nih.gov/ij/). Most of these options require specialized knowledge of programming or scripting languages and may be difficult to implement for batch analysis of image stacks. GE Healthcare recently launched powerful analysis development software (IN CELL DEVELOPER TOOLBOX) that is image-stack friendly and allows users to design custom analysis routines without learning complex scripting languages.

The latest wave of image analysis products to hit the market includes those that allow development and incorporation of automated cell classification routines. Rather than second guessing which cell parameters are important for describing a particular response, these analysis tools allow users to assign cells to two or more phenotypic classes and then let the software indicate which combination of parameters is most important for distinguishing the different classes. The methods used to train the classification routine may vary from relatively simple quadratic discrimination of classes to more advanced 'machine-learning' algorithms that automatically improve their own performance through experience. A number of vendors (e.g. BD Biosciences, GE Healthcare, Molecular Devices) now offer image analysis routines with automated cell classification capabilities. An example of the use of automated classification to distinguish two populations of cells is shown in *Fig. 2* (also available in color section).

2. METHODS AND APPROACHES

A consequence of the versatility inherent in high-content screening is that there is a wide range of methods and factors to consider at every stage of the screen, from assay design and optimization through to image acquisition and analysis.

2.1. Assay development

2.1.1. Assay design

One of the first steps in assay design is to identify an appropriate cell type or model system for the assay. If the cells are engineered to express fluorescent or bioluminescent reporters, it may take some time to isolate and characterize clones, or to optimize transient transfection protocols.

When choosing the cell system, some consideration should be given to the characteristic cell morphology and growth habits. Most detection platforms are designed to image and analyze cells growing in monolayers. Cells that tend to cluster together or grow on top of each other may be difficult to resolve and analyze. Likewise, cells that adhere loosely and take on spherical morphologies are usually more difficult to image and analyze than those that adhere well to the plate and spread out flat across its surface. When cells are relatively thick or rounded up due to poor adherence, various structures can lie in different planes of focus. This can lead to imaging problems, since organelles lying outside of the chosen focal plane may appear blurry or faint, can obstruct the resolution of features within the focal plain, and may not be detected at all if the optical section is very thin. The method by which an imaging system finds optimal focus can have an effect on how well the system performs on samples of varying thickness. The majority of high-content systems now employ laser-based hardware autofocus mechanisms that locate the well bottom and then focus a specified offset distance above this location. Some systems offer additional software autofocus options that determine the offset automatically by finding the peak intensity of probes associated with the cells. This can be of benefit if the optimal plane of focus differs for various fluorescent probes, since it allows the

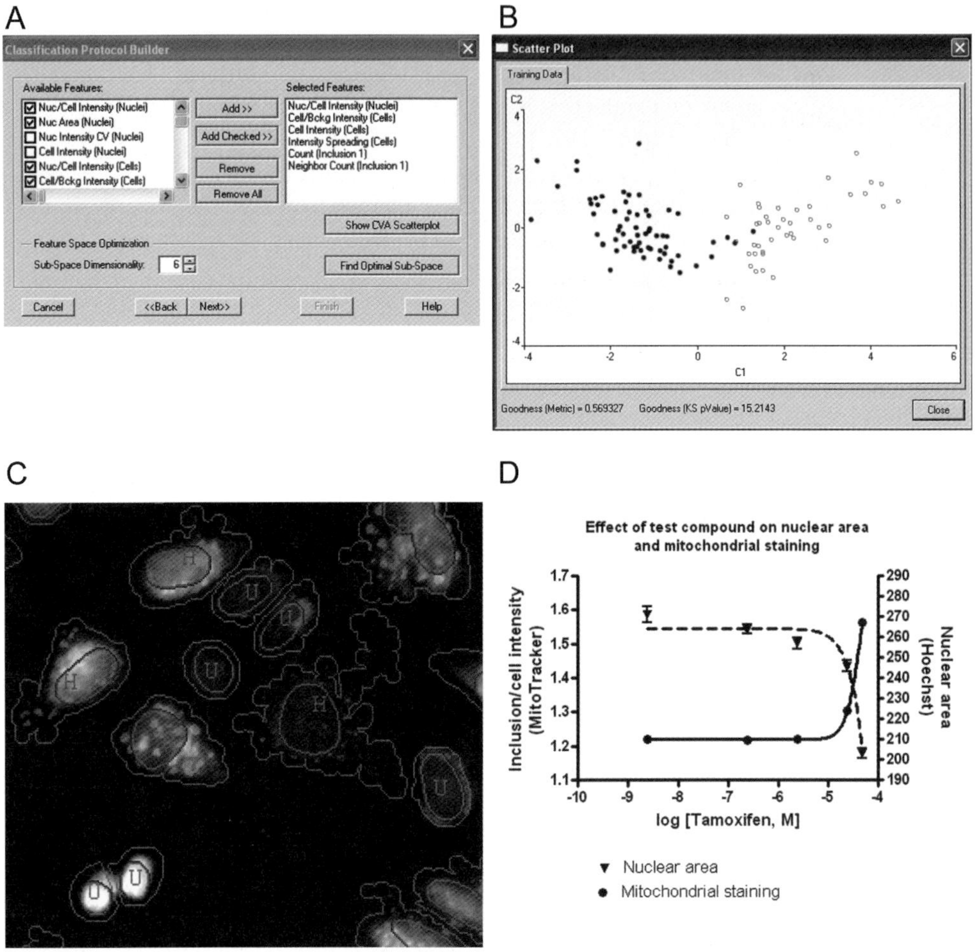

Figure 2. High-content analysis of the EGFP–FYVE domain profiling assay (see page xxix for color version).

IN Cell Analyzer 1000 analysis software was used to analyze the assay described in *Protocol 5*. (*A*) The software facilitates design of a classification protocol, allowing the protocol developer to choose from a list of 39 available measures for classification. The Sub-Space Dimensionality field is shown here being used to reduce the number of parameters to the six most influential. (*B*) Color-coded data visualization scatter plot (printed here in gray scale) allowing assessment of how well a classification routine separates two control cell populations used to train the classifier. Filled circles represent healthy control cells; open circles represent cells exposed to a toxic compound. (*C*) After classification, cell assignments appear labeled and color-coded in the Image View window. The developer has chosen H/green and U/magenta to label the healthy and unhealthy classes, respectively. Nuclei are labeled with Hoechst (blue), mitochondria are stained with MitoTracker (red), and endosomes are labeled with EGFP–FYVE domain reporter (green). (*D*) Analysis results showing a dose-dependent effect of tamoxifen on nuclear size and mitochondrial staining intensity. Results are shown as mean ± SEM, with $n = 12$ replicate images (wells) per data point.

optimal offset to be set independently for separate channels (wavelengths). However, contrast-based software autofocus can be problematic if the brightest objects in the image (for example, dead cells or dust) are not the objects of interest.

Once the model cell system is established, an assay method is devised. Concomitantly, an analysis method must be identified and developed (see section 2.3.1). Most cell-by-cell analysis approaches require the use of at least two optical probes – one to identify cells and a second to follow the reporter. Optical probes must be compatible with the excitation sources and optical filters of the detection system, and not introduce unwanted toxic side effects. For example, dyes that require UV excitation may need to be avoided if the target is involved in a stress-response pathway that is triggered by UV light. Adjustment of probe concentrations may also be necessary to maximize assay signal-to-noise ratios and minimize required exposure times.

During assay optimization, time-course experiments are performed to identify the optimal end point. Sensitivity of the assay to compound solvents is assessed. Dye concentrations, incubation times, and assay buffers are systematically varied to assess cell health, minimize variability, and maximize the assay dynamic range. Dose-response studies using known activators and inhibitors are typically conducted, and metrics such as 50% effective concentration (EC_{50}), 50% inhibitory concentration (IC_{50}) and Hill slope are compared with literature values and other in-house data to verify the biological relevance of the assay system.

Clearly, developing a robust screening assay can be a complex, time-consuming, and costly process. The purchase of validated commercial assays and reagents for cell-based screening may help to reduce development time and expense. Fortunately, the number of commercial assays and kits is on the increase. Cellomics has developed a range of bio-application kits that contain validated combinations of antibodies, fluorescent dyes, and other reagents. GE Healthcare offers a panel of cell lines stably expressing green fluorescent protein (GFP) reporter proteins that are involved in key cell-signaling pathways. The cell lines are well characterized and accompanied by validated assay protocols. In addition, the expression vectors used to establish the cell lines are provided so that customers can use them to genetically engineer cell lines of their choice. Xsira Pharmaceuticals have developed Transfluor bioassays that have been well validated for screening G protein-coupled receptor (GPCR) ligands and other compounds that regulate GPCRs (19).

2.1.2. Fixed- and live-cell assay formats

One of the decisions to be made during assay development is whether to image the cells live or to preserve them with fixative prior to imaging. For higher-throughput screens, fixed-cell formats have a number of advantages over live-cell protocols. Fixed assay plates can be transported easily to the screening facility. Prior to screening, samples can be accumulated and stored for several weeks or months, making it easier to schedule screening runs. There is no need for environmental control during the screen, and temporal variability is reduced, leading to better assay signal-to-noise ratios. After screening, fixed samples can be returned to storage and then retrieved at a later date for re-imaging if necessary.

Some applications may not be amenable to fixation, or may perform better in a live-cell format. For example, reporter fluorescence or localization may not be preserved upon fixation, the cellular response may occur too rapidly to enable fixation, or the fixation process may introduce artifacts. Imaging the cells while they are alive can overcome these problems, as well as simplify the logistics of kinetic assays. For live-cell screens, environmental control can be critical for maintaining cell health and minimizing temporal variability throughout the duration of the screen. Most mammalian cell cultures require a humidified environment with a constant temperature of 37°C and 5% CO_2.

2.1.3. Automating an assay for screening

Once an assay has been developed, it is scaled up and usually automated for screening. A typical screening workflow comprises seeding the cells to microtiter plates, pre-equilibrating with assay medium, adding reagents and test compounds, fixing (optional), and then reading the assay on the imaging platform. One or more washes and additional incubation steps may also be necessary. Depending on available resources, some or all of these steps may be automated.

Scale up and automation often introduce additional sources of variability that must be identified and minimized. Automated plate washers and liquid handling systems may be harsher than human operators, disrupting the cell monolayer or stimulating stress response mechanisms that have an unwanted impact on cell, target, or probe behavior. Adverse effects may also be encountered when cells are grown at higher well densities, since smaller volumes increase the likelihood of problems such as medium evaporation, nutrient depletion, and overcrowding. Variability is often associated with the edge wells of multi-well plates (20) and can be traced to a variety of sources such as leaching of adhesives used in plate manufacture, thermal gradients, and differential evaporation rates across the plate. As for any screen involving microtiter plates, plate-to-plate and batch-to-batch variability can also be expected. Controls included on each plate of the screen, or interspersed at regular intervals throughout the sample set, can be used to normalize for these variations. If the assay is imaged live, temporal variability is also likely and must be minimized.

The most commonly used metric to assess assay quality during optimization for screening is the Z′ coefficient, a dimensionless factor that takes into account the assay dynamic range and the data variation associated with positive and negative controls (21). A Z′ value of 1 represents an ideal assay with no variation or an infinitely large dynamic range. Assays with Z′ values greater than 0.5 are considered excellent for screening. In addition to Z′, other factors such as EC_{50} and Hill slope of control dose-response curves are monitored during the optimization process to ensure that the assay continues to display acceptable characteristics.

2.1.4. Cell handling

How the cells are handled during maintenance and sample preparation will have a direct impact on assay quality. Harsh conditions can apply selection pressure to cell cultures, resulting in population heterogeneity and changes in phenotype, including loss of reporter expression. In our experience, it is advisable to passage cells while they are in exponential-phase growth, avoid seeding the cells at densities that are too sparse or too confluent, and discard them after about 25 passages. When preparing samples for screening, seeding density and evenness of distribution can be critical for minimizing variability. The use of automated cell-culture and sample preparation systems can reduce the time, cost, and labor involved in cell handling, and can help improve the consistency and accuracy of the assay results.

2.2. Image acquisition

2.2.1. Controlling image quality

During acquisition set-up, a variety of parameters can significantly affect image quality. The amount of magnification (i.e. which objective is used) will determine whether key features are resolved. The combination of excitation filters, emission filters, and dichroics must be appropriate for the chosen fluorophores to ensure efficient probe excitation and signal capture, and to avoid bleed-through of signal into other channels. The system must be well aligned and regularly calibrated to ensure that the field of view is uniformly illuminated and that image data from multiple channels is in register. Most systems require flat-field correction to correct nonuniformities. Autofocus offsets must also be carefully chosen to ensure that features of interest will be in focus. Exposure times should be set to ensure sufficient capture of signal from the sample without exceeding the dynamic range of the detector.

Depending on the detection platforms available, the requirement for confocality may need to be assessed and the confocal aperture adjusted (if this is possible). If the primary purpose of confocality is noise reduction, a relatively thick optical section should be chosen. If, on the other hand, the aim is to gain resolution in two or three dimensions, then smaller confocal apertures may be favorable. In cases where signal fluorescence is weak, confocality may not be desirable at all, since significant signal loss may lead to poor identification of objects and lower assay signal-to-noise ratios.

2.2.2. Increasing acquisition rates

During optimization of the acquisition protocol, exposure times should be optimized to help maximize acquisition rates. In addition, consider how much resolution is required for the assay. Assays that involve detection of relatively large structures or compartments may not require the highest possible resolution that the system can deliver. For example, a nuclear translocation assay involves detection of relatively large subcellular compartments (cytoplasm and nucleus) that do not need to be well resolved in order to detect probe redistribution. In such cases, if the imaging system uses a CCD detector, pixel binning may be a good way of decreasing the assay read time. Binning is the process of combining the charges of neighboring CCD pixels during readout. The net result is a decrease in resolution of the final image, coupled with a significant increase in both sensitivity and the speed of image acquisition. Because binning increases the signal-to-noise ratio, exposure times can sometimes also be decreased, further increasing the acquisition rate. Almholt *et al.* (9) have reported the use of pixel binning to maximize acquisition rates for a MAPKAPK2 nuclear translocation assay. For assays requiring maximum resolution, binning may not be an option. For example, an assay such as Akt1 redistribution (10) is based on detection of reporter protein associated with fine ruffle-like structures at the cell periphery. High-resolution images may be required for detection and accurate quantification of enhanced GFP (EGFP) localized to these structures. In such cases, longer exposure times and/or an increased number of images per well or treatment condition may be required.

When using a laser scanning system, another means of increasing acquisition speed may be to reduce the image size. Since some systems capture several hundred cells per image, it is possible to reduce the scan area and still obtain data from enough cells to provide statistically significant results. For example, a full-sized image obtained using the IN Cell Analyzer 3000 typically captures between 300 and 500 cells, depending on the cell type and seeding density. Since many assays require only 150 cells per data point to achieve good results, reducing the image size by half in the direction of scanning can nearly double the throughput without compromising assay quality. Scanning systems that employ on-line, cell-by-cell analysis may offer a 'sufficient cell count' option that allows scanning of a sample to be terminated when data from a sufficient number of cells has been recorded. This, too, can increase acquisition rates.

2.3. Image analysis

2.3.1. Choosing and optimizing an analysis protocol

Choosing and optimizing the analysis protocol is one of the most critical steps for obtaining good assay results. Designing a successful and efficient analysis solution 'from scratch' requires considerable expertise and experience. For high-content assays, analysis routines that derive data from each individual cell in an image population are essential for examining heterogeneous cell populations, and often lead to higher signal-to-noise ratios than can be achieved using population-averaging methods that simply normalize a particular measurement to the number of cells in the image. So-called 'cell-by-cell' analysis routines typically employ a 'marker' probe, such as a nuclear dye, to identify individual cells. One or more additional fluorescent probes may also be used to identify cell features and follow responses of target molecules. Depending on the assay requirements, a range of intensity and morphological measurements can be reported for each cell. Segmenting features of interest (i.e. distinguishing them from each other and from the surrounding background) can be one of the most challenging and crucial aspects of the protocol. The ability to segment objects well will depend on overall image contrast and quality, as well as on the power of the segmentation tools available within the software. Developing a new analysis approach requires frequent and effective communication between the developer of the analysis routine and the cell biologist, and thorough validation of the solution before application to screening data.

It is far easier to work with pre-developed, well-validated analysis routines. Even from this starting point, there may still be a large number of parameters that need to be adjusted in order to tailor the protocol to a specific assay. Image pre-processing steps may need to be applied in order to correct for noise and shading artifacts. Application of object filters may be necessary to exclude from analysis any inappropriate objects – such as debris, or cells displaying little or no reporter fluorescence. A number of measures needs to be examined to determine which one(s) best characterizes the assay response. Object segmentation and measurement parameters then need to be systematically optimized against a range of assay metrics such as signal-to-noise ratio, Z′, and EC_{50}. If automated classification techniques are desired, classification protocols must be created and validated.

2.3.2. Maximizing speed of analysis

Not surprisingly, image analysis time may increase considerably with the inclusion of image processing steps, sophisticated segmentation techniques, numerous measurements, and classification algorithms. If throughput is a concern, the analysis approach should also be optimized for speed. Consider, for example, whether the need for pre-processing may be reduced or eliminated by improving image quality, whether changes in cell seeding density or preparation can reduce the need for advanced cell segmentation techniques, and whether the number of measurements required may be reduced.

3. RECOMMENDED PROTOCOLS

3.1. High-throughput, cell-based screening for GPCR antagonists

Receptor internalization can be monitored using the pH-sensitive fluorogenic dye CypHer5E (GE Healthcare), which is effectively nonfluorescent at neutral pH and maximally fluorescent in acidic environments. When cells expressing epitope-tagged receptor are incubated with a cognate antibody conjugated to CypHer dye, the receptor–antibody complex translocates to acidic endosomes, where the dye becomes fluorescent. The assay can be adapted to screen for either agonists or antagonists. The following protocol describes an antagonist-format assay, with an optional fixation step. Formaldehyde fixation preserves CypHer fluorescence, allowing the assay to be performed in either fixed- or live-cell format. Cells are challenged with agonist after briefly pre-incubating with labeled receptor–antibody complex, nuclear dye, and test compounds. In the presence of hit compounds (antagonists), endosome-associated CypHer5E fluorescence is decreased compared with agonist-only controls. Typical images and screening results are presented in *Fig. 3* (also available in color section).

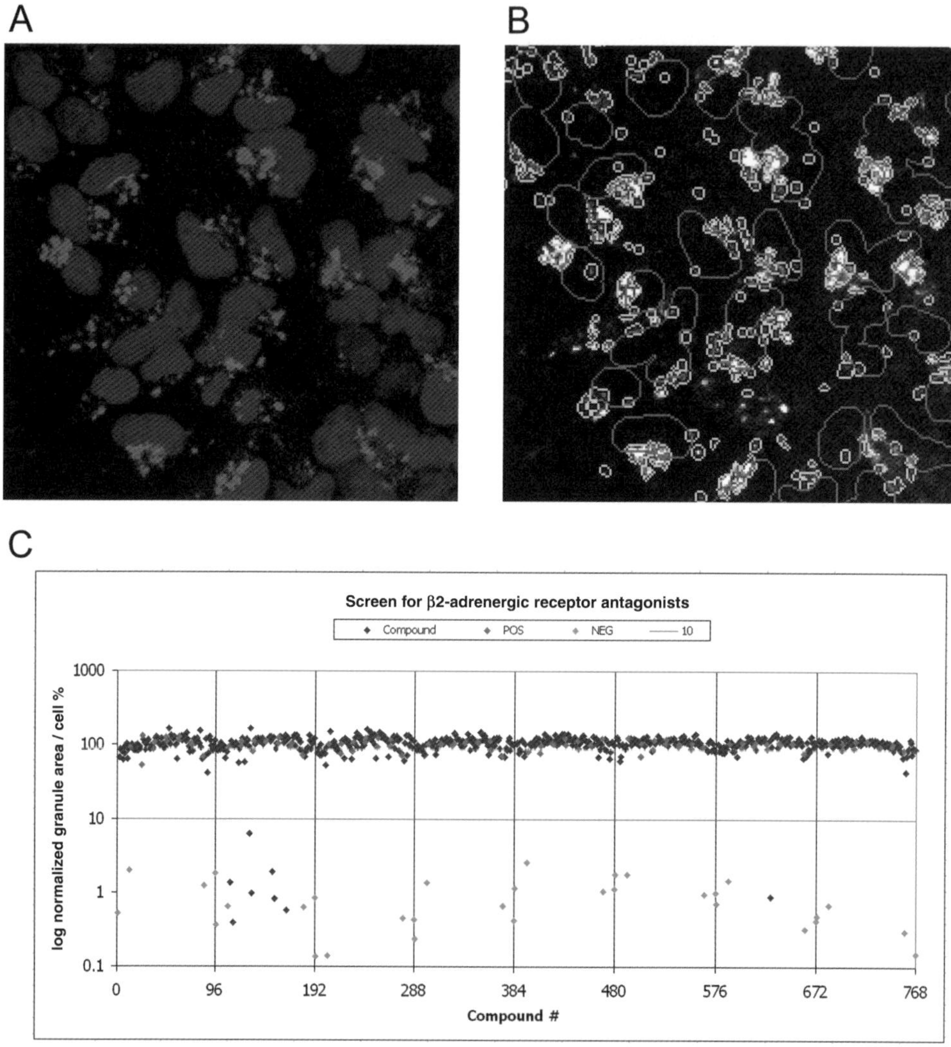

Figure 3. High-content screening for β2-adrenergic receptor antagonists (see page xxx for color version).
When cells expressing β2-adrenergic receptor are treated with agonist, pH-sensitive CypHer5E dye becomes fluorescent at the sites of receptor internalization. (*A*) Cells imaged with the IN Cell Analyzer 1000 following treatment with agonist. Nuclei are stained with Hoechst (blue) and sites of receptor internalization are identified by CypHer fluorescence (red). (*B*) Bitmap overlay indicating analysis results for the image shown in (*A*). Nuclei are outlined in magenta; identified CypHer fluorescence is outlined in yellow. (*C*) Compound-induced responses from a screen of a 640-compound library are plotted as a percentage of the agonist-only control response. Potential antagonists (hits) were defined as those with a response of less than 10% of control values. All known antagonists in the library and no false positives were identified. Each data point represents the population-averaged result from a single image, with the following color-coding: blue, test compound; pink, positive control; orange, negative control.

Protocol 1

Screening of LOPAC library compounds for β2-adrenergic receptor antagonists

Equipment and Reagents

- Poly-D-lysine (Sigma)
- 96-Well assay plates (Greiner Bio-One)
- Library of compounds organized on 96-well plates and diluted with suitable solvent[a] (e.g. dimethyl sulfoxide (DMSO)) at an appropriate stock concentration.
- HEK 293 cells expressing vesicular stomatitis virus G protein (VSV–G) epitope-tagged β2-adrenergic receptor
- Fetal bovine serum (FBS; Sigma)
- Geneticin G418 (Sigma)
- L-Glutamine (Invitrogen)
- Nonessential amino acids (Sigma)
- Culture medium: minimal essential medium (MEM) (Sigma) containing 10% (v/v) FBS, 200 μg/ml G418, with L-glutamine and nonessential amino acids
- Assay medium: phenol red-free MEM (Invitrogen)
- VSV–G antibody conjugated with CypHer5E (GE Healthcare)
- Antagonist solution (10× stock): 1 μM alprenolol (Sigma) in assay medium
- Agonist solution (10× stock): 1 μM isoproterenol (Sigma) in assay medium
- Hoechst 33342 (Invitrogen)
- Neutral buffered 4% formaldehyde solution (Sigma)
- Multimek 96 Automated 96-Channel Pipettor (Beckman)
- Cell imaging platform: IN Cell Analyzer 1000 (GE Healthcare)

Method

1. Dilute library compounds to 1 μM (i.e. 10× the desired final assay concentration[b]) in assay medium (MEM).
2. Pre-coat 96-well plates with poly-D-lysine to promote cell adhesion.
3. Seed the cells at a density of 8000 cells per well[c] in 100 μl of culture medium.
4. Incubate for 72 h in a humidified environment at 37°C and 5% CO_2 until approximately 70% confluent.
5. Equilibrate the plate to room temperature (22–25°C) for 5 min.
6. Add 60 μl of fresh assay medium containing 2–2.5 μg/ml of CypHer5E-conjugated antibody[d] and 5 μM of Hoechst nuclear stain.
7. Incubate at room temperature for 10 min.
8. Add 20 μl of test compound or control solution[e].
9. Incubate at room temperature for 10 min.
10. To stimulate receptor internalization, add 20 μl of agonist solution to all wells and incubate at 37°C for 30 min.
11. Fix cells (optional) by removing the assay solution and incubating cells with 100 μl of 4% formaldehyde for 30 min at room temperature.
12. Image on an IN Cell Analyzer 1000. Use a 620 nm excitation filter, a 700 nm emission filter, and an exposure time of 300 ms per field for CypHer5E. Use a 360 nm excitation filter, a 535 nm emission filter, and an exposure time of 500 ms per field for the Hoechst nuclear stain.

13. Analyze images using the Dual Area Object Analysis module, or an equivalent analysis routine. Set up an analysis protocol that segments individual cells using the Hoechst image channel with the top-hat segmentation method and detects endosomally localized receptor in the CypHer5E image channel using the multi-scale top-hat segmentation option (see *Fig. 3B*, also available in color section). Optionally, specify a second inclusion type in the Hoechst channel to monitor DNA condensation or fragmentation associated with compound toxicity. Select relevant output measures such as nuclear intensity, nuclear area, inclusion count, and inclusion intensity parameters. Set the protocol to report all available measures to an EXCEL spreadsheet.
14. Run the analysis.
15. When the analysis is complete, open the results in EXCEL, or export to a GRAPHPAD PRISM template. Use CypHer inclusion/cell intensity ratio or inclusion area as an indicator of receptor internalization. Normalize the response data to controls on the same assay plate and plot percentage response vs. compound ID (see *Fig. 3C*, also available in color section).
16. Assess plate quality using the Z′ value calculated from controls on each plate. Re-assay compounds from any plate with a value of $Z' < 0.3$.
17. Assess screen quality using the Z coefficient.

Notes

[a]Test the effect of compound solvents such as DMSO on cell viability and ensure that final concentrations used in the assay do not compromise cell health.

[b]Perform titration experiments on a random subset of library compounds, using the Z factor as a metric to determine the optimal final concentration for the assay. Typical final concentrations range from 100 nM to 10 μM. For screening a LOPAC library of pharmacologically active compounds, we found the optimal final assay concentration to be 100 nM.

[c]Optimize seeding density for the cell line to obtain 70–80% confluence at the start of the assay.

[d]Assess assay performance using a range of antibody concentrations to determine the optimal antibody concentration.

[e]Include replicate control wells on each plate to assess plate quality and normalize for plate-to-plate variability (typically, we position replicate positive and negative controls symmetrically in columns 1 and 12).

3.2. High-content siRNA screening

RNA interference using siRNA libraries is a powerful technology for elucidating gene function by downregulating gene expression at the post-transcriptional level. Phenotypic changes associated with siRNA knockdown can be monitored using cell lines expressing fluorescent reporter proteins. Test siRNAs are transiently transfected into the reporter cell line and behavior of the fluorescent reporter probe is monitored using a high-content imaging system. A range of cell features and additional fluorescent probes can be monitored to assess the effects of knockdown. *Protocols 2, 3* and *4* are typical methods used in optimizing and performing an siRNA study. *Protocol 2* details how to perform siRNA transfection in multi-well plates for subsequent high-content analysis. *Protocol 3* describes how to optimize siRNA knockdown in EGFP-expressing cell lines. *Protocol 4* provides a method of screening an siRNA cell-cycle array using cell lines that report cell-cycle phase. The subcellular distribution and intensity of the fluorescent phase reporter changes in a cell-cycle-dependent manner. The phase marker assay is multiplexed with an assay for bromodeoxyuridine (BrdU) incorporation, which identifies cells that have undergone DNA replication. When a G_2/M cell-cycle phase reporter cell line is multiplexed with BrdU incorporation (see *Fig. 4*, also available in color section), information from both assays enables assignment of each cell to one of five distinct phases of the cell cycle (G_1, S, G_2, prophase, or mitosis).

3.3. High-content analysis during compound profiling

High-content secondary assays can be designed to characterize the mechanism of action, specificity, potency, toxicity, and off-target effects of lead compounds. *Protocol 5* describes a method for monitoring the effects of compounds on a cell line that expresses an EGFP–FYVE domain fusion protein as a sensor for phosphatidylinositol 3-phosphate, the product of phosphatidylinositol 3-kinase (PI3K) activity (22). PI3Ks regulate key cellular processes such as signal transduction and membrane trafficking. The tandem FYVE domain of the reporter binds to phosphatidylinositol 3-phosphate, which is enriched on the outer leaflet of endosomal membranes. PI3K inhibitors deplete the binding substrate for the reporter, causing it to translocate to the cytosol. In addition to monitoring localization of the EGFP reporter, the protocol provides a means of monitoring nuclear and mitochondrial changes that may correlate with compound toxicity. Example results obtained using *Protocol 5* are presented in *Fig. 2* (also available in color section).

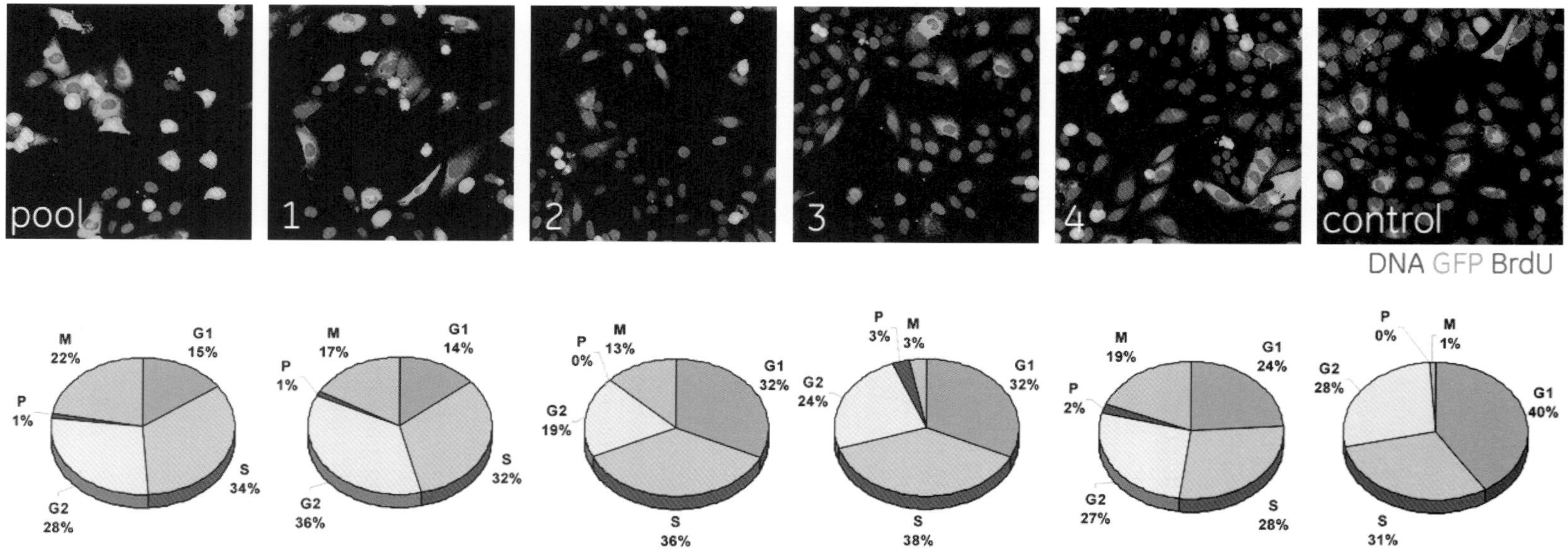

Figure 4. Analysis results from an siRNA study using *Protocol 4* with a G_2/M cell-cycle phase reporter cell line (see page xxxi for color version).
Images from wells treated singly (labelled 1–4) or in combination (labelled 'pool') with four siRNAs directed against different regions of the mRNA for polo-like kinase, which is a regulator of cell-cycle progression during mitosis. An EGFP fusion protein (green) reports cell-cycle phase; nuclei are stained with Hoechst (blue), and nuclei that have incorporated BrdU are stained with Cy5 (red). For each image, the corresponding analysis results are displayed as a pie chart that shows the proportion of cells in mitosis (M), G_1 phase, S phase, G_2 phase, and prophase (P). Changes in cell-cycle phase distribution can be assessed relative to the distribution determined from cells in the control well, which received a pool of scrambled siRNAs.

Protocol 2

Preparation of cells in 96-well plates for siRNA transfection and high-content analysis

Equipment and Reagents

- Culture medium (Sigma)
- FBS (Sigma)
- Cell line
- Cell-culture flasks
- Penicillin and streptomycin (Sigma)
- Trypsin-EDTA (Sigma)
- Sterile phosphate buffered saline (PBS; calcium- and magnesium-free; Invitrogen)
- Hemocytometer or disposable cell-counting chamber (Immune Systems)
- Tissue-culture incubator (37°C, 5% CO_2, humidified)
- Inverted tissue-culture microscope
- Imaging-grade 96-well plates (e.g. microclear black, Greiner; ViewPlate, Packard)
- DharmaFECT transfection reagent[a] (Dharmacon)
- DharmaFECT cell-culture reagent (Dharmacon)
- siARRAY RTF siRNA 96-well plates (Dharmacon)
- Multi-channel pipette

Method

1. Grow cells in flasks in an incubator in medium supplemented with 10% FBS, 100 units/ml penicillin and 100 µg/ml streptomycin. Do not allow cells to exceed 90% confluency.
2. Equilibrate the siARRAY plate to room temperature.
3. For each well to be transfected, combine 0.125 µl of DharmaFECT transfection reagent with 24.875 µl of DharmaFECT cell-culture reagent and add 25 µl to each well of the siARRAY plate. Incubate for 20–90 min at room temperature to rehydrate the siRNAs.
4. Remove the culture medium from the cells, wash cells once with PBS, and add a small volume (e.g. 5 ml) of trypsin-EDTA and incubate at 37°C to detach the cells.
5. Dilute the cell suspension with an equal volume of medium with 10% FBS but without antibiotics.
6. Count the cell suspension and dilute to 5000–10 000[b] cells/100 µl in medium with 10% FBS but without antibiotics.
7. Aliquot 100 µl of cell suspension to each well containing rehydrated siRNAs.
8. Using a multi-channel pipette, immediately transfer the cell and siRNA mixtures from the siARRAY plate into equivalent wells in the imaging plate[c].
9. Incubate the plate for the desired period before processing for imaging.

Notes

[a]Choose a transfection reagent appropriate to the cell line based on manufacturer's data or determine the optimal reagent and conditions using *Protocol 3*.

[b]The optimum cell number will depend on the cell line and the intended duration of the experiment. Seeding at 5000–10 000 cells per well should give good results for common laboratory cell lines in knockdown experiments lasting 24–48 h.

[c]Some siARRAY RTF products may be supplied in plates compatible with imaging and high-content analysis. In this case, omit step 8 of the protocol.

Protocol 3

Optimization of siRNA knockdown in EGFP-expressing cell lines using high-content analysis

Equipment and Reagents

- Cell suspension at 5000 cells/100 µl in antibiotic-free medium with 10% FBS (see *Protocol 1*)
- Sterile PBS (calcium- and magnesium-free; Invitrogen)
- Tissue-culture incubator (37°C, 5% CO_2, humidified)
- Automated fluorescence microscope and image analysis software (e.g. IN Cell Analyzer 1000 or IN Cell Analyzer 3000; GE Healthcare)
- Imaging-grade 96-well plates (e.g. microclear black, Greiner; ViewPlate, Packard)
- DharmaFECT transfection reagents 1, 2, 3, and 4 (Dharmacon)
- EGFP siRNA (Dharmacon)
- Culture medium (Sigma)
- Neutral buffered 4% formaldehyde solution (Sigma)
- Hoechst 33342 (Sigma)

Method

1. Prepare a 2 µM solution of EGFP siRNA in RNase-free buffer.
2. For each well to be transfected, dilute 0.05–0.5 µl of DharmaFECT transfection reagents with serum- and antibiotic-free medium to a total volume of 22.5 µl. Add to 2.5 µl of siRNA mix and incubate for 20 min at room temperature. Prepare transfection controls using 2.5 µl of serum- and antibiotic-free medium in place of the siRNA solution.
3. Aliquot 25 µl of the transfection mix to each well of an imaging-grade 96-well plate. Add 25 µl of serum- and antibiotic-free medium only to additional nontransfected control wells.
4. Add 100 µl of cells to each well and incubate for 24–48 h in a tissue-culture incubator.
5. Remove the medium and wash cells with PBS by gentle aspiration.
6. Fix cells with 100 µl of formaldehyde for 30 min at room temperature.
7. Remove the formaldehyde, wash cells with PBS, and stain the nuclei with 100 µl of 2 µM Hoechst 33342 in PBS.
8. Image the cells using excitation and emission filters for Hoechst and EGFP.
9. Analyze the images for cell number and EGFP expression in the nontransfected control, transfection control, and siRNA-treated wells.
10. Determine the optimum balance of siRNA concentration, transfection reagent concentration, and transfection time to yield the maximum reduction of EGFP expression with minimum cellular toxicity.

Protocol 4

Analysis of siRNA knockdown of cell-cycle control genes in G_1/S and G_2/M cell-cycle phase marker cell lines using multiplexed high-content analysis

Equipment and Reagents

- G_1/S and/or G_2/M cell-cycle phase marker cells (GE Healthcare) suspended at 5000 cells/100 μl in antibiotic-free medium with 10% FBS (see *Protocol 1*)
- DharmaFECT 3 transfection reagent (Dharmacon)
- DharmaFECT cell-culture reagent (Dharmacon)
- Cell-cycle siARRAY RTF siRNA 96-well imaging plates (GE Healthcare/Dharmacon)
- Cell Proliferation Fluorescence Assay (GE Healthcare)
- Automated fluorescence microscope and image analysis software (e.g. IN Cell Analyzer 1000 or IN Cell Analyzer 3000; GE Healthcare)
- Sterile PBS (calcium- and magnesium-free; Invitrogen)
- Tissue-culture incubator (37°C, 5% CO_2, humidified)
- Neutral buffered 4% formaldehyde solution (Sigma)
- Triton X-100 (Sigma)
- Hoechst 33342 (Sigma)

Method

1. Equilibrate the siARRAY plate to room temperature.
2. For each well to be transfected, combine 0.125 μl of DharmaFECT 3 with 24.875 μl of DharmaFECT cell-culture reagent and add 25 μl to each well of the siARRAY plate. Incubate for 20–90 min at room temperature to rehydrate the siRNAs.
3. Aliquot 100 μl of cell suspension into each well containing rehydrated siRNAs and incubate for 24–48 h in a tissue-culture incubator.
4. Dilute BrdU labeling reagent (from the Cell Proliferation Fluorescence Assay) 1:250 with tissue-culture medium, add 100 μl per well and incubate for 1 h at 37°C.
5. Remove medium and wash cells with PBS by gentle aspiration.
6. Fix cells with 100 μl of 2% formaldehyde/0.1% Triton X-100 for 30 min at room temperature or overnight at 4°C.
7. Stain cells for BrdU incorporation using the detection reagents supplied in the Cell Proliferation Fluorescence Assay.
8. Remove formaldehyde, wash cells with PBS and stain nuclei with 100 μl of 2 μM Hoechst 33342 in PBS.
9. Image the cells using excitation and emission filters for Hoechst, EGFP, and Cy5.
10. Analyze the images for cell number, cell-cycle distribution, BrdU incorporation, and cellular morphology parameters using the appropriate image analysis software.

Protocol 5

Multiplexing EGFP–FYVE domain reporter assay with toxicity indicators

Equipment and Reagents

- Culture medium: Dulbecco's MEM with GlutaMAX (Invitrogen) supplemented with 10% FBS
- Humidified incubator at 37°C, 5% CO_2
- EGFP–FYVE domain cells (GE Healthcare) in exponential-phase growth, suspended in culture medium at 5×10^4 cells/ml
- Test compound dose-series solutions[a]: 0, 0.25, 2.5, 25, and 50 μM in culture medium
- 10 mM Hoechst 33342 (Invitrogen)
- 250 μM MitoTracker (Invitrogen) in ethanol
- PBS (Invitrogen)
- Formaldehyde (Sigma) diluted to 2% in PBS
- IN Cell Analyzer 1000 imaging system with Dual Area Object Analysis module

Method

1. Transfer 100 μl of cell suspension to each well of a 96-well plate and culture at 37°C until cells reach 80–90% confluence.
2. Decant the culture medium and wash cells with 200 μl of pre-warmed culture medium.
3. Add 100 μl of each test compound dilution to each of 16 replicate wells and incubate for 24 h[a] at 37°C in a humidified incubator.
4. On the day of imaging, prepare a Hoechst/MitoTracker staining solution by mixing 100 ml of culture medium, 10 μl of Hoechst stock solution, and 100 μl of MitoTracker stock solution. Pre-warm the solution to 37°C prior to step 6.
5. Remove the culture medium containing test compounds and wash wells with 200 μl of culture medium.
6. Add 100 μl of the Hoechst/MitoTracker staining solution to each well and incubate at 37°C for 30 min.
7. Decant the staining solution and wash wells with 200 μl of PBS.
8. Add 100 μl of 2% formaldehyde to each well and incubate at room temperature for 30 min.
9. Wash wells with 200 μl of PBS.
10. Image the cells on an IN Cell Analyzer 1000 with the 10× objective. Choose dichroic 61003BS (DAPI/FITC/Cy5). Choose 360 nm excitation and 460 nm emission filters for Hoechst, 575 nm excitation and 535 nm emission filters for EGFP, and 620 nm excitation and 700 nm emission filters for MitoTracker.
11. Create an analysis protocol using the Dual Area Object Analysis module as follows: identify nuclei in the Hoechst image channel using the top-hat segmentation technique. Use the multi-scale top-hat method to identify mitochondria and endosomes in the MitoTracker and EGFP channels, respectively. Choose the EGFP channel for cell segmentation with the multi-scale top-hat method. Define a filter to exclude cells expressing little or no EGFP reporter[b]. Report all available measures to an EXCEL spreadsheet.
12. Run the analysis.
13. If appropriate, create and train a classifier to recognize subpopulations of cells and add it to the analysis protocol. Use the Feature Space Optimization tools to refine the number of

parameters. For example, in *Fig. 2* (also available in color section), an automated classifier was created to distinguish healthy from unhealthy cells using the six most influential measures.

14. When analysis is complete, open the EXCEL results file (or export results to GRAPHPAD PRISM).
15. Create plots to assess the effects of compounds on one or more relevant output measures. There are 39 available measures to choose from; inclusion count, inclusion/cell intensity ratio, nuclear intensity, and nuclear area are particularly useful for this assay.

Notes

[a]Adjust concentrations and incubation times to be suitable for the compounds being tested.
[b]For a typical assay, a filter set to exclude cells less than 250 gray levels (EGFP) is appropriate.

4. TROUBLESHOOTING

Refer to section 2 for a discussion of critical factors relevant to the protocols provided. *Protocol 1* will need to be optimized for each specific application. Cell type, antibody affinity for the epitope tag, dye-to-antibody coupling ratio, and receptor expression levels will all contribute to assay performance. Optimize each element individually during assay development. Incubation times may also need to be adjusted for the cell system. Detailed suggestions for optimization are provided with the CypHer product. Assay noise is frequently introduced during automated liquid addition; adjust the dispense speed and height to avoid damaging cells. The use of controls on each plate helps to control plate-to-plate variability and results in better screening data.

When working with the cell-cycle phase marker cell lines used in *Protocols 2–4*, take care to minimize cell disruption during wash, permeabilization, and antibody incubation steps. Cells that are dead, dying, dividing, or blocked in mitosis may be weakly adherent and therefore particularly susceptible to disturbance. Selective loss of these cell subpopulations will skew the analysis results. Avoid creating vortices when adding reagents to wells containing live cells, and keep the formaldehyde incubation time between 20 and 30 min to avoid under- or overfixing the cells. To minimize background fluorescence, use the plates, serum, and culture medium from the suppliers recommended in *Protocol 2*.

For all of the protocols involving image analysis, high image quality is essential. Inspect images to ensure that field illumination is even and that features to be quantified are bright and in focus. If not, adjust the instrument and acquisition set-up and re-image the samples. If re-acquisition is not possible, try applying noise reduction or shading correction techniques before analysis. Overexposure of images can be as detrimental as underexposure. Check pixel gray-level values of representative cell features to ensure that they fall within the dynamic range of the camera. If not, re-image the sample using a shorter exposure time, decrease the laser power, or add a neutral density filter.

Good segmentation of nuclei, cell bodies, and features of interest is also critical for success. When creating a new analysis protocol, assess the analysis results and ensure that what you want to measure is what is actually being identified by the software. Always cast a critical eye on numerical outputs. For example, if cell counts look unusually high, check that the analysis routine is not mistaking cell debris for legitimate cells. Most analysis software provides visualization tools such as bitmap overlays that allow you to assess the results. When using IN Cell Analyzer 1000 software, if too few or too many objects are detected, try systematically varying the sensitivity settings during object segmentation. For cell identification in particular, try varying the segmentation method as well. When creating a classification protocol, ensure that the image quality of the training data set is excellent and that all objects of interest are well segmented before proceeding to annotation.

Acknowledgements

The author gratefully acknowledges Lynne Smith, Mike Looker, and Ian Davies for the CypHer screening protocol and data; Nick Thomas and Mike Kenrick for the cell-cycle phase marker methods and data; and Angela Williams and Ray Ismail for *Protocol 5* and associated data.

5. REFERENCES

★★ 1. **Clemons PA** (2004) *Curr. Opin. Chem. Biol.* **8**, 334–338. – *Review of complex phenotypic assays in high-throughput screening.*

★ 2. **Liebel U, Starkuviene V, Holger E, *et al.*** (2003) *FEBS Lett.* **554**, 394–398. – *Description of a microscope-based screening platform for large-scale functional protein analysis in intact cells.*

3. **Yarrow JC, Feng Y, Perlman ZE, Kirchhausen T & Mitchison TJ** (2003) *Comb. Chem. High Throughput Screen.* **6**, 279–286

★★★ 4. **Mitchison TJ** (2005) *ChemBioChem*, **6**, 33–39. – *Mini-review of small-molecule screening and profiling using automated microscopy.*

★★★ 5. **Erfle H, Simpson JC, Bastiaens PIH & Pepperkok R** (2004) *BioTechniques*, **37**, 454–462. – *siRNA cell arrays for high-content screening microscopy.*

★★★ 6. **Guiliano KA, Chen Y-T & Taylor DL** (2004) *J. Biomol. Screen.* **9**, 557–568. – *A study of the role of p53 activation using high-content screening with siRNA.*

7. **Perlman ZE, Mitchison TJ & Mayer TU** (2005) *ChemBioChem*, **6**, 145–151. Erratum **6**, 218.

★ 8. **Kittler R, Putz G, Pelletier L, *et al.*** (2004) *Nature*, **432**, 1036–1040. – *siRNA screening to identify genes essential for cell division using a primary high-throughput cell viability screen, followed by a secondary high-content assay.*

★★★ 9. **Almholt DLC, Loechel R, Nielsen SJ, *et al.*** (2004). *Assay Drug Dev. Technol.* **2**, 7–20. – *Demonstration of high-content primary screening using a nuclear translocation assay to screen 183 375 compounds at a rate of 19 000 compounds per day. Compound availability and toxicity were assessed during the primary screen. A range of secondary assays, some high-content, were used to deconvolute the mode of action of hit compounds.*

★ 10. **Lundholt BK, Linde V, Loechel F, *et al.*** (2005). *J. Biomol. Screen.* **10**, 20–29. – *Use of a low-content, high-throughput, cell-based primary screen combined with several high-content secondary assays to identify Akt pathway inhibitors.*

★ 11. **Gasparri F, Mariani M, Sola F & Galvani A** (2004) *J. Biomol. Screen.* **9**, 232–243. – *Quantitation of the proliferation index of human dermal fibroblast cultures.*

★ 12. **Beske O, Guo J, Li J, *et al.*** (2004). *J. Biomol. Screen.* **9**, 173–185. – *Demonstration of the use of encoded particle technology for simultaneous analysis of multiple cell types.*

13. **Castelain P, van Hummelen P, Deleener A & Kirsch-Volders M** (1993) *Mutagenesis,* **8**, 285–293.
14. **Varga D, Johannes T, Jainta S, *et al.*** (2004) *Mutagenesis,* **19**, 391–397.
15. **Edwards BS, Oprea T, Prossnitz ER & Sklar LA** (2004) *Curr. Opin. Chem. Biol.* **8**, 392–398.
★★ 16. **Yarrow JC, Perlman ZE, Westwood NJ & Mitchison TJ** (2004) *BMC Biotechnol.* **4**, 21–29. *– Comparison of image-based readout methods for a high-throughput cell migration assay.*
17. **Lindqvist A, Kallstrom H & Karlsson Rosenthal C** (2004) *J. Cell Sci.* **117**, 4979–4990.
18. **Goodyer ID, Brophy G & Roquemore EP** (2002) *J. Clin. Ligand Assay,* **25**, 294–304.
★ 19. **Oakley RH, Hudson CC, Cruickshank RD, *et al.*** (2002) *Assay Drug Dev. Technol.* **1**, 21–30.
20. **Lundholt BK, Scudder KM & Pagliaro L** (2003) *J. Biomol. Screen.* **8**, 566–570.
★★★ 21. **Zhang J-H, Chung TDY & Oldenburg KR** (1999) *J. Biomol. Screen.* **4**, 67–73. *– Seminal work describing Z and Z′ metrics that have been widely adopted within the pharmaceutical industry to assess screen and assay quality, and are now being applied to assess cell-based, high-content assays and screens.*
22. **Gillooly DJ, Morrow IC, Lindsay M, *et al.*** (2000) *EMBO J.* **19**, 4577–4588.
23. **Adie EJ, Francis MJ, Davies J, *et al.*** (2003) *Assay Drug Dev. Technol.* **1**, 251–259.
★★ 24. **Milligan G** (2003) *Drug Discovery Today,* **8**, 579–585. *– Mini-review of high-content assays for ligand regulation of GPCRs.*
25. **Ghosh RN, Chen Y-T, DeBiasio R, *et al.* (2000)** *BioTechniques,* **29**, 170–175.
★ 26. **Mackem S, Baumann CT & Hager GL** (2001) *J. Biol. Chem.* **276**, 45501–45504. *– A glucocorticoid/retinoic acid receptor chimera that displays cytoplasmic/nuclear translocation in response to retinoic acid.*
★★ 27. **Starkuviene V, Liebel U, Simpson JC, *et al.*** (2004) *Genome Res.* **14**, 1948–1956. *– High-content microscopy to identify novel proteins with roles in secretory membrane trafficking.*
28. **Ding G, Fischer PA, Boltz RC, *et al.*** (1998) *J. Biol. Chem.* **273**, 28897–28905.
★ 29. **Ghosh RN, Grove L & Lapets O** (2004) *Assay Drug Dev. Technol.* **2**, 473–481. *– Quantitative cell-based, high-content screening assay for epidermal growth factor receptor-specific activation of mitogen-activated protein kinase.*
★★ 30. **Yu H, West M, Keon BH, *et al.*** (2003) *Assay Drug Dev. Technol.* **1**, 811–822. *– Measurement of drug action in a cellular context using protein-fragment complementation assays.*
★★ 31. **Pagliaro L, Felding J, Audouze K, *et al.*** (2004) *Curr. Opin. Chem. Biol.* **8**, 442–449. *– Review of emerging classes of protein–protein interaction inhibitors and tools for their discovery and profiling.*
32. **Vogt A, Cooley KA, Brisson M, *et al.*** (2003) *Chem. Biol.* **10**, 733–742.
33. **Milligan G, Feng G, Ward RJ, *et al.*** (2004) *Curr. Pharm. Des.* **10**, 1989–2001
34. **Lovborg H, Nygren P & Larsson R** (2004*) Mol. Cancer Ther.* **3**, 521–526.
35. **Ramm P, Alexandrov Y, Cholewinski A, Cybuch Y, Nadon R & Soltys BJ** (2003) *J. Biomol. Screen.* **8**, 7–18.
36. **Conway BR, Minor LK, Xu JZ, *et al.*** (1999) *J. Biomol. Screen.* **4**, 75–86.
37. **Mastyugin V, McWhinnie E, Labow M & Buxton F** (2004*) J. Biomol. Screen.* **9**, 712–718.
★ 38. **Li Z, Yan Y, Powers EA, *et al.*** (2003) *J. Biomol. Screen.* **8**, 489–499. *– High-throughput image-based screening of 486 000 compounds to identify gap junction blockers using calcein acetoxymethyl ester.*
39. **Niemisto A, Dunmire V, Yli-Harja O, Zhang W & Shmulevich I** (2005) *IEEE Trans. Med. Imaging,* **24**, 549–553.
40. **Odeyale CO & Hook GR** (1990) *J. Leukoc. Biol.* **48**, 403–411.

APPENDIX 1
List of suppliers

ABgene – www.abgene.com
Agar Scientific – www.agarscientific.com
Alexis Corporation – www.alexis-corp.com
Amersham Biosciences – www.amershambiosciences.com
Anachem Ltd – www.anachem.co.uk
Appleton Woods Ltd – www.appletonwoods.co.uk
Applied Biosystems – www.appliedbiosystems.com
Applied Precision – www.api.com
AutoGen, Inc. – www.autogen.com
Axon Instruments – www.axon.com

Barloworld Scientific – www.barloworld-scientific.com
Becker and Hickl – www.becker-hickl.de
Beckman Coulter, Inc. – www.beckman.com
Becton, Dickinson and Company – www.bd.com
Bio-Rad Laboratories, Inc. – www.bio-rad.com
BOC Group – www.boc.com
British Biocell International – www.british-biocell.co.uk
Brosch direct Ltd – www.broschdirect.com

Calbiochem – www.calbiochem.com
Cambridge Research & Instrumentation, Inc. – www.cri-inc.com
Cambridge Scientific Products – www.cambridgescientific.com
Carl Zeiss – www.zeiss.com
Cellomics, Inc. – www.cellomics.com (now part of Fisher Scientific)
Chemicon International, Inc. – www.chemicon.com
Chroma – www.chroma.com
Corning, Inc. – www.corning.com

DakoCytomation – www.dakocytomation.com
Dharmacon – www.dharmacon.com
Difco Laboratories – www.difco.com
Dionex Corporation – www.dionex.com
DuPont – www.dupont.com

Elliot Scientific Ltd – www.elliotscientific.com
Edinburgh Instruments – www.edinst.com
European Collection of Animal Cell Culture – www.ecacc.org.uk
Evotec Technologies GmbH – www.evotec-technologies.com
Evrogen – www.evrogen.com

Findel Education Ltd – www.fipd.co.uk
Fisher Scientific International – www.fishersci.com
Fluka – www.sigma-aldrich.com
Fluorochem – www.fluorochem.co.uk

GE Healthcare – www.gehealthcare.com
Gibco BRL – www.invitrogen.com
Goodfellow Cambridge Ltd – www.goodfellow.com
GraphPad Prism – www.graphpad.com
Greiner Bio-One – www.gbo.com

Hamamatsu – www.hamamatsu.co.uk
Harlan – www.harlan.com
Hybaid – www.hybaid.com
HyClone Laboratories – www.hyclone.com

ICN Biomedicals, Inc. – www.icnbiomed.com
Improvision – www.improvision.com
Insight Biotechnology – www.insightbio.com
Integrated BioDiagnostics – www.ibidi.com
Invitrogen Corporation – www.invitrogen.com

Jencons-PLS – www.jencons.co.uk

Kendro Laboratory Products – www.kendro.com
Kentech Instruments – www.kentech.co.uk
Kodak: Eastman Fine Chemicals – www.eastman.com

Lab-Plant Ltd – www.labplant.com
Lambert Instruments – www.lambert-instruments.com
Lancaster – www.lancastersynthesis.com
LaVision Biotech – www.lavisionbiotec.de
Leica – www.leica.com
Leica Microsystems – www.leica-microsystems.com
Life Technologies, Inc. – www.lifetech.com
LOT-Oriel – www.lot-oriel.com

Mattek – www.glass-bottom-dishes.com
MBL International – www.mblintl.com
Media Cybernetics, Inc. – www.mediacy.com

Merck, Sharp and Dohme – www.msd.com
MetaChem – www.metachem.com
Millipore Corporation – www.millipore.com
Miltenyi Biotec – www.miltenyibiotec.com
Molecular Devices Corporation – www.moleculardevices.com
Molecular Probes – www.invitrogen.com
MWG Biotech – www.mwg-biotech.com

National Diagnostics – www.nationaldiagnostics.com
Nevtek – www.nevtek.com
New England BioLabs, Inc. – www.neb.com
Nikon Corporation – www.nikon.com
Nunc – www.nuncbrand.com

Olympus Corporation – www.olympus-global.com
Optical Insights – www.optical-insights.com
Optivision Ltd – www.optivision.co.uk

Packard – www.perkin-elmer.com
Perbio Science – www.perbio.com
PerkinElmer, Inc. – www.perkinelmer.com
Pharmacia Biotech Europe – www.biochrom.co.uk
Photek – www.photek.com
Photonic Research Systems – www.prsbio.com
Photonic Solutions plc – www.psplc.com
Picoquant – www.picoquant.com
Polysciences, Inc. – www.polysciences.com
Prior Scientific – www.prior.com
Promega Corporation – www.promega.com
Pyser-SGI – www.pyser-sgi.com

QBiogene – www.qbiogene.com
Qiagen N.V. – www.qiagen.com

R&D Systems – www.rndsystems.com
Roche Diagnostics Corporation – www.roche-applied-science.com
Roper Scientific – www.roperscientific.com

Sanyo Gallenkamp – www.sanyogallenkamp.com
Sarstedt – www.sarstedt.com
Schleicher and Schuell Bioscience, Inc. – www.schleicher-schuell.com
Scientifica – www.scientifica.uk.com
Serotec – www.serotec.com
Serva Electrophoresis – www.serva.de
Shandon Scientific Ltd – www.shandon.com
Sigma-Aldrich Company Ltd – www.sigma-aldrich.com

Solent Scientific – www.solentsci.com
Sorvall – www.sorvall.com
Stratagene Corporation – www.stratagene.com
Synergy Software – www.synergy.com
Systat Software, Inc. – www.systat.com

TAAB – www.taab.co.uk
Thales Optem – www.thales-optem.com
Thames Restek – www.thamesrestek.co.uk
Thermo Electron Corporation – www.thermo.com
Thistle Scientific – www.thistlescientific.co.uk

Universal Imaging Corporation – www.moleculardevices.com

Vector Laboratories – www.vectorlabs.com
VWR International Ltd – www.bdh.com

Willco Glass – www.willcowells.com
Wolf Laboratories – www.wolflabs.co.uk

Xsira Pharmaceuticals – www.xsira.com

York Glassware Services Ltd – www.ygs.net

Zeiss – www.zeiss.com

Index